Botanical Pesticides for Pest Management

Botanical Pesticides for Pest Management

Dr. Prashant Kumar Sirohi

RANDOM

RANDOM PUBLICATIONS

NEW DELHI (INDIA)

Botanical Pesticides for Pest Management

ISBN 978-93-5111-588-5

Published in 2015 in India by

RANDOM PUBLICATIONS

4376-A/4B, Gali Murari Lal, Ansari Road
New Delhi-110 002
Phone : +9111-43580356, 011-23289044, 011-43142548
e-mail: sales@randompublications.com,
info@randompublications.com, randomexports@gmail.com

Reprinted 2021

Type Setting by : Friends Media, Delhi-110089
Digitally Printed at : Replika Press Pvt. Ltd.

Preface

Botanicals are pesticides derived from plants. They degrade rapidly and therefore are considered safer to the environment than the common synthetic chemicals. However, as with any pesticide, botanicals must be used properly. Many plants and minerals have insecticidal properties; that is, they are toxic to insects. Botanical insecticides are naturally occurring chemicals extracted or derived from plants or minerals. They are also called natural insecticides. Organic gardeners will choose these insecticides, in some cases, over synthetic organic materials.

Botanical insecticides are naturally occurring chemicals extracted from plants. Natural pesticidal products are available as an alternative to synthetic chemical formulations but they are not necessarily less toxic to humans. Some of the most deadly, fast acting toxins and potent carcinogens occur naturally. Some of the botanical pesticides are very toxic to fish and other cold-blooded creatures and should be treated with care. Protective clothing should be worn when spraying, even though their toxicity is normally low to warm-blooded animals. Botanical insecticides break down readily in soil and are not stored in plant or animal tissue. Often their effects are not as long lasting as those of synthetic pesticides and some of these products may be very difficult to find.

I would like to thank my team for standing beside me throughout my career and writing this book. My special thanks go to "Random Publications" who have published the book.

– Dr. Prashant Kumar Sirohi

Contents

1

Pest Management: Monitoring Systems

Field monitoring is the key to an integrated pest management programme. When performed systematically on a quantitative basis, it provides the information needed to detect important changes in pest activity and disease incidence; evaluate pesticide performance; and establish seasonal records for evaluating and upgrading managerial decisions. When weather data and cultural practices that influence pest behaviour are considered, pest behaviour and populations can better be anticipated.

Monitoring systems may be limited to a single primary pest Such as the citrus rust mite, and to environmental conditions that influence rust mite behaviour or they may include all major pests and beneficials. Monitoring techniques vary depending on individual preferences and management intensity, but should provide a consistent, repeatable means to numerically measure pest activity. Follow monitoring procedures upon which the control action thresholds were designed.

Monitoring frequency depends on the pests' reproductive capacities and time needed under optimum conditions for populations to reach economic damage thresholds. Because of the reproductive potential of citrus rust mite, biweekly monitoring of susceptible fruit is required from petal fall until harvest or arrival of cool weather. Other pests generally require less frequent monitoring. Outside of the petal-fall to harvest interval, monthly monitoring is adequate for the citrus mite complex and other potential pest or disease situations.

Sampling Methods: Sample size, location and selection are important to any sample programme. Each citrus orchard is unique and requires individual monitoring; the first step in developing a monitoring system is to decide how many trees need to be examined and, within this parameter, how many fruit, leaf or soil samples need to be observed or collected for adequate representation and sampling reliability. Sampling may involve trees randomly selected for each monitoring period or permanently designated station trees. Trees should be dispersed sufficiently to achieve representative sampling of the entire orchard. This can be accomplished in most rectangular orchards by sampling in a diagonal direction. For square orchards, a Z-shaped pattern is appropriate. The station tree concept involves selection of representative trees at the time of orchard establishment or at the beginning of each season. Selected trees are flagged, numbered and revisited each monitoring period throughout the season.

Random samples provide information on a larger proportion of trees within an orchard over time. Station trees more closely follow pest or disease dynamics on a few individual trees. Both systems have merit and, depending on the desired result and amount of time to be invested, a combination of both methods may prove most useful.

Systematic recording of data in the field is time consuming but necessary for clarity in review and interpretation of results. If field counts are considered impractical and some form of rating scale is employed (*e.g.*, trace, low, medium, high, very high), the assignment of numerical values from 1-5 is acceptable. Organization of the data in summary form following field sampling allows for id review and assessment of individual orchard situations. Information gathered during the monitoring process aids in planning future pest management strategies.

CONTROL ACTION CRITERIA

Control action criteria can be defined as specific pest population levels or threshold levels that signal impending economic damage to the crop or trees. Ideally, each threshold value considers the pest's life cycle, its potential for rapid population increases under existing or anticipated environmental conditions, orchard history, the type of damage likely to occur in proportion to control costs, other stresses the trees are experiencing and other relevant factors. Progress has been made in establishing threshold values for citrus rust mite, the major economic pest of Valley citrus, but standardized criteria for other common pests are lacking. Suggestions in this text for key pests are based on guidelines developed in other citrus production states and observational data recently accumulated by local IPM programmes. Action criteria in any pest management programme are flexible and should be altered to fit individual management conditions.

Monitoring Guidelines

These guidelines provide information on how to monitor specific pests and offer insight on what constitutes reason for control action. Pre-season monitoring suggestions are provided to encourage early season inventory of pest populations and disease potential in advance of post-bloom control effects. Early monitoring provides information helpful in selecting the most appropriate and cost-effective pesticides. This early information may be useful in making volume purchases of pesticide materials and for determining which orchards merit immediate attention.

Citrus Mite Complex

Field studies of rust mite behaviour have indicated that populations in excess of 50 mites per square centimetre (cm^2) on fruit for 2 weeks or more are needed for rind damage to become visible. Also, rust mite populations on infested fruit may more than quadruple in a 2-week period under favourable conditions. These data support the need for a 2 week monitoring frequency for rust mite and are useful in establishing action thresholds.

As a general reference, a threshold value of 3 to 5 mites/cm^2, depending on optimal or suboptimal mite reproductive conditions, has been used successfully for several years in the local TAEX-IPM programme. Structured to allow the grower 2 weeks' lead time before significant damage Could occur, and provide a cushion for any sampling weakness, this threshold value applies to the following sampling regime:

- Sampling is from selected station trees, one tree per two acres.
- Six fruits sampled per station tree; two fruits on each of east, north and west quandrants, selected at random from partial shade areas.
- Mite counts taken from shaded side of fruit using a 14X hand lens, counting all live unites present in four separate lens fields per fruit.
- Each lens field is roughly 1/4 cm^2 in area, for a total of 1 cm^2 per fruit or 6 cm^2 sample per station tree.

To avoid rust mite management decisions being biased by two or three heavily infested fruits, consider the proportion of infested fruit. Also, consider mite distribution within trees or within orchards. Under this particular sampling regime, less than 20 per cent incidence of infested fruit or of station trees would ameliorate average rust mite density values, and an incidence of more than 20 per cent would substantiate or override them. The following criteria are utilized:

- Less than three mites/cm^2 and less than 20 per cent of sample fruit with mites-no action necessary.
- Three to five mites/cm^2 and 20 per cent of sample fruit with mites-borderline situation.
- More than five mites/cm^2 and more than 20 per cent of sample fruit with mites-control should be initiated.

A sampling technique involving different action criteria has been used successfully by growers for several years. It involves random sampling of a single fruit per tree on 50 randomly selected trees within an orchard. Rust mite counts are expressed in numbers per cm^2. Action criteria vary according to irrigation schedules and weather outlook, but generally action is indicated when average mite density exceeds 20 mites/cm^2 and/or 25 per cent of sample fruit are observed to be infested. Factors to consider when making control decisions regarding rust mites include:

- Variety-given similar rust mite densities, oranges are downgraded less by mite damage than grapefruit.

- Lead time needed by producer to spray.
- Weather outlook and time of year-heaviest rainfall periods are May/June and August/September. Evaluate borderline mite densities during these periods in light of possible re-entry difficulties.
- Irrigation scheduling-orchards expected to need water within 10 to 14 days of an anticipated action threshold should be sprayed prior to irrigation.
- Miticide used in previous spray treatment-mite populations may rebound quickly following use of low residual materials.
- Orchard history-certain orchards, because of location, cultural practices, tree spacing or lack of uniformity in tree sizes and other reasons, consistently experience rapid re-infestation following control efforts.

Citrus red mites, first observed in 1980, are distributed throughout the Valley and have become a major citrus pest, particularly in the mid-Valley area between US 77 and US 281. Red mites feed on tender new leaves and populations increase with new growth flushes. Population levels decline during hot, dry July/August weather and resurge with the fall flush. Information about natural control agents in the Valley is not sufficient to determine when control action is most appropriate.

Control action guidelines for red mite on Texas citrus are incomplete pending further research on natural control agents and tree tolerance levels. Factors to consider in evaluating need for red mite control include:

- General tree vigore.
- Mite reproductive potential and environmental conditions.

Texas citrus mites do not require specific control action in most seasons if trees and fruit are protected from other mite and insect pests. Threshold levels vary according to tree vigour and should be determined by period during which heavy feeding is sustained rather than by specific mite numbers. Numbers exceeding 10 adult mites per leaf over a period of 30 days have been associated with winter defoliation of crown leaves.

Major considerations in Texas citrus mite management include:

- Tree vigour.
- Current and projected climatic conditions.
- Deleterious effect of fungicides applied for melanose or rust mite control on beneficial fungi.
- Effect of certain fall scalicide applications on over-wintering mite populations (azinphos-methyl and methidathion have been implicated with increased mite populations).
- Time of year-predisposition of mite-damaged leaves to defoliation by strong, dry winds greatest during fall/winter months.

False spider mites' increase in the past was associated with the use of organophosphate pesticides to control other pests and situations involving use of zineb alone for control of rust mite. With the present trend to use specific miticides such as dicofol, chlorobenzilate and fenbutatin-oxide, primarily for rust mite control, recent false spider mite problems have been associated with orchards where aldicarb soil treatment replaced post-bloom and late spring sprays. In aldicarb-treated orchards where the need for rust mite control may not arise until late summer, start monitoring for false spider mite in July and continue through September.

Armored Scales: California red scale is a major pest of citrus in the Rio Grande Valley. It attacks all parts of the tree aboveground and is a common cause of limb dieback and tree death in new trees. Populations usually are kept in check by natural enemies, but tend to reach damaging levels during abnormally dry seasons on trees adjacent to dusty roads. A small parasitic wasp, Aphytis lingnanensis, provides effective control in most seasons if not restricted by dust or by application of broadspectrum insecticides. Begin monitoring for red scale in late spring to early summer to advantageously schedule a spray application during periods of high crawler activity or the early growth stages when scales are most susceptible to control. Orchards having a history of heavy red scale infestations the previous summer should be monitored early following warm winters. Monitoring always is required from July through October when colonization on fruit is easily seen and infestation levels are more easily determined.

To limit scale damage below tolerances set for federal grade standards, begin control action when 5 per cent of the fruit examined have 10 or more live scales. Examine at least 10 outside fruits on both east and west quandrants of sample trees to calculate infestation levels. A hand lens will help distinguish live scale insects from dead or parasitized scales, which appear partially loosened from the peel or exhibit holes cut by adult parasites as they exited the scale body. Estimates of scale parasites from leaf samples are less reliable than those based on fruit samples because of the difficulty in determining leaf age and when parasitism occurred.

Consider the following for monitoring and control of California red scale:

- Monitor new tree plantings frequently as they are subject to relatively rapid infestation.
- Inspect every young tree in the planting at least twice during summer and fall months.
- Natural control by scale parasites often is reduced during the heat of July and August.
- Increased incidence of spherical yellow crawlers encountered when searching fruit for rust mites signals impending infestations in bearing orchards.
- Oil sprays are least injurious of all scalicides to scale parasite populations.

Chaff scale is the most common widespread armored scale pest of Valley citrus. It attacks all aboveground parts of trees and can result in serious downgrading of fruit, but infestations rarely cause twig and limb dieback. Several species of parasites are known natural control agents.

Monitoring for chaff scale is suggested during pre-season and biweekly scouting for rust mite. Grade standards for fruit are important in deciding if summer or early fall scalicide application is warranted. Control action criteria are the same as described for California red scale.

Secondary Pests: Various minor citrus pests capable of causing economic damage under favourable circumstances require consideration during monitoring of major pests. Secondary pest eruptions usually result from loss of natural enemies through improper pesticide use, exposure to off-target pesticide drift or weather extremes. The two most common secondary pests are brown soft scale and citrus mealybug. Outbreaks of these pests occur most frequently and persistently near fields traditionally planted close to cotton and vegetable crops. Brown soft scale infestations generally develop during the summer and fall months. Early signs of infestation are easily overlooked during monitoring activities until blackening of leaves by sooty mold fungi becomes noticeable. Tree vigour is diminished by feeding activities and inhibition of photosynthesis by sooty mold growing on the leaves. When localized infestations are found, spot treatment of trees on an "as needed" basis may be appropriate to reduce costs and prevent possible deleterious effects on natural enemies.

Citrus mealybug is controlled most effectively with pesticides during early stages before protective wax secretions and sooty mold form. Monitoring activities must be concerned with identification and quantitative assessment of first and second stage larvae. Mature mealybugs are highly resistant to contact pesticides. Clusters of fruit-usually the higher quality, inner canopy fruit-are preferred by mealybugs. Heavy infestations interfere with fruit development and may cause fruit drop. Sooty mold is difficult to remove at the packing shed, often resulting in downgrading of fruit.

Mealybugs have several predatory and parasitic natural enemies. Severe outbreaks in orchards not normally affected by mealybugs usually are the result of pesticide-related reductions or other unfavourable circumstances affecting natural enemies.

Whitefly infestations (citrus whitefly, citrus blackfly, cloudy-winged whitefly or woolly whitefly) occurring in isolated situations also can be associated with disruption of natural control agents.

Texas leafcutting ants can cause serious damage to citrus, particularly In young plantings, so monitoring notes on location and severity of infestations will facilitate control efforts. The leafcutting ant poses a special problem in well-drained sandy loam soils because it may nest some distance from the orchards Repeated defoliation of trees reduces tree vigour, eventually killing young trees. Locate the nest by tracking at dawn or dusk or during periods of cloudy weather because aboveground activity ceases during direct sunlight.

The information given herein is for educational purposes only. Reference to commercial products or trade names is made with the understanding that no discrimination is intended and no endorsement by the Cooperative Extension Service is implied.

Texas Citrus Mite

Tetranychidae: Eutetranychus banksi (Banks)

Texas citrus mites (TCM) are 0.3 to 0.4 mm long and the eight-legged adults vary in colour from yellow to dark green, with characteristic dark internal food spots on each side down the back. Female mites are broadly oval and somewhat flattened. Males are more slender with longer legs and are triangular in shape. Larvae have six legs and are light yellow. Eggs are disc shaped and usually are laid near the midrib or branching veins. The life cycle can be very short under high temperatures, leading to very rapid population explosions. Unlike the citrus rust mite, the Texas citrus mite prefers low humidity conditions. Most of the time, greater numbers are found on leaves on the south side of the tree, and in the top of the tree. They usually prefer to attack upper surfaces of leaves. Leaves damaged by TCM appear gray, show mesophyll collapse and may drop during periods of drying winds or when trees are stressed by lack of water or other factors.

Mites also appear on fruit but to a lesser degree than on leaves, and rarely cause noticeable feeding damage on fruit. Texas citrus mites usually are more prevalent in the May to July period, although they may increase rapidly at other times when conditions are favourable.

Insect and Mites Pests

Mites: Description and Biology

Mites represent the single most economically important group of Texas citrus pests. The citrus rust mite together with the spider mite complex, including citrus red mite and Texas citrus mite, are responsible for a major portion of pest losses and require the bulk of pesticides used on citrus in Texas. Mites feed on fruit and foliage by piercing individual plant cells and sucking out the contents, thus, destroying the cell. Their pest potential lies in their ability to develop rapidly and their tremendously great reproductive potential, allowing for enormous populations to develop in a short period of time.

Citrus Rust Mite

Eriophyidae: Phyllocoptruta oleivora (Ashmead)

The citrus rust mite (CRM) is considered the most important pest of Texas citrus. Rust mites are long, wedge-shaped and light yellow, measuring about

0.1 to 0.2 mm long, generally not visible to the unaided eye. The citrus rust mite feeds on the exposed surfaces of fruit, destroying the rind cells leading to the characteristic russeting of fruit.

This damage may result in reduced grade and size, thereby drastically reducing the marketability and value of the crop. Rust mites feed and develop year-round on citrus foliage and, when available, will develop on fruit. Rust mites may complete a generation in 7 to 10 days during the summer while development is drastically slowed or may halt during cool periods.

Early rust mite damage to fruit in late spring or early summer is evidenced by bronzing of the rind, a condition commonly called "buckskin" fruit. Severely affected fruits do not expand normally and frequently exhibit rind cracking. Damage occurring later in fruit development is evidenced by a darkening of the peel where heavy feeding took place, generally dark brown to black on green fruit and rust-brown on mature fruit. Similar symptoms of melanose disease on fruit, particularly grapefruit, may be confused with mite-induced russeting, but usually can be differentiated by noting peel texture and checking for presence of melanose pustules. Grapefruit generally are more susceptible than oranges to fruit damage by citrus rust mites.

High populations of CRM may reduce tree vigour. Affected leaves develop a dull green colour and show reduced turgidity. Rust mites pose a threat to the fruit crop from April to October, particularly during periods of warm, humid weather with frequent, light showers. If the latter conditions extend into early winter, fruit russeting may occur as late as December.

A reliable sign of impending rust mite damage to fruit is the dull, cloudy appearance of heavily infested fruit caused by thousands of mite skins cast off during the molting process. The distribution of citrus rust mites within trees varies during the year and is generally determined by environmental conditions.

Rust mites tend to seek out high humidity areas away from direct sunlight but also avoid areas where dew forms. Rust mites overwinter on foliage and in bark crevasses. On foliage, mites are most likely to be found on undersides of dry, inner canopy leaves. Mites appear to prefer fruit on the west or northwest quadrant of trees during May to June when cloudy conditions prevail. As summer temperatures increase, mites move inward and towards the north quadrant, away from the direct sunlight. After harvest during winter months, highest mite populations have been observed on east or northeast quadrants.

CONCEPT OF NON-TARGET INVERTEBRATES

The concept of a non-target organism is a relative one and therefore requires some clarification. The term non-target organism generally refers to organisms outside the main target group. For example, with most organophosphate, carbamate and pyrethroid insecticides, because they often are capable of killing many different types of insects as well as other invertebrates such as spiders and crustacea, non-target organism usually refers to non-insect species.

With Bt insecticides, owing to their high selectivity, the definition of a non-target organism typically is much broader and includes all insects outside the taxonomic order to which the primary target insects belong. Bt insecticides are so specific, even against insects, that their spectrum of activity is typically identified in a very narrow manner, such as 'lepidopteran-active', 'dipteran-active' or 'coleopteran-active'. Even then, Bt insecticides are so specific that a Bt subspecies generally characterized as 'lepidopteran-active' may be highly toxic to some lepidopteran species, but have only low or no toxicity to others.

This point can be illustrated with the HD1 isolate of B. thuringiensis subsp. kurstaki (Btk), the isolate used widely in commercial formulations to control lepidopteran pests. Btk is highly toxic and very effective against larvae of the cabbage looper, Trichoplusia ni, a common pest of vegetable crops, but typically exhibits poor activity against the beet armyworm, Spodoptera exigua, another important caterpillar pest that belongs to the same taxonomic family. This is because none of the toxins produced by Btk (Cry1Aa, Cry1Ab, Cry1Ac and Cry2A) are very toxic to Spodoptera species.

For this reason, the product XenTari (Valent BioSciences) based on B. thuringiensis susbp. aizawai, which produces a protein (Cry1Ca) of moderate toxicity to Spodoptera species, was developed for control of Spodoptera species. With respect to the 'dipteran-active' B. thuringiensis subsp. israelensis, this subspecies is only highly toxic to species of the suborder Nematocera ('long-horned flies' meaning flies with long antennae), a subdivision of the fly group, which, in addition to mosquitoes and blackflies, contains the mushroom flies, craneflies and chironomid midges. Flies such as houseflies, horseflies, stable flies, and the many types of fruitflies are not sensitive to toxins or spores of this subspecies. A high degree of specificity and hence safety is attributed to each Bt insecticide, meaning that a Bt subspecies that serves as the active ingredient is limited to being toxic primarily to the insect species of only one taxonomic order.

Nevertheless, this would still mean that many non-target species of this order would be sensitive to the Bt endotoxins by the normal route of entry (*i.e.* ingestion). Thus what we consider a pest is an arbitrary concept as opposed to one based on taxonomy. This has led to considerable misunderstanding about the effects of 'lepidopteran-active' Bt subspecies used as insecticides, or the proteins derived from these that are used in Bt crops. An isolate like HD1 of B. thuringiensis subsp. kurstaki has a broad host range against lepidopteran species, due primarily to the four insecticidal proteins it produces.

Therefore, when used in the field it will be capable of killing larvae of target as well as certain non-target lepidopterans. Among the targets are larvae of many moth species, especially those of the family Noctuidae (*e.g.* the corn earworm, the cotton budworm and bollworm, and the cabbage looper). Among the non-targets in certain geographical areas are the larvae of non-pest lepidopterans including larvae of the monarch butterfly, and many other species of moths and butterflies, some of which are endangered species.

This can pose a dilemma for farmers as well as the governmental agencies, both regulatory agencies and local governments, in making decisions about the effects of Bt insecticides, and now Bt crops, on non-target organisms. The choices in the case of endangered species are particularly difficult, and can be illustrated with control of the gypsy moth, Lymantria dispar.

Most communities in the eastern and mid-western regions of the United States have banned the use of synthetic chemical insecticides to control this pest in forests and residential areas. Bt is used despite the protests of many environmentalists. The reason is that Bt represents the most environmentally compatible choice, even for endangered lepidopteran species.

If synthetic chemical insecticides were used, their impact on all non-target invertebrates would be much more devastating than the use of Bt. If Bt is used, populations of the gypsy moth are controlled, at least temporarily, but there are also losses of sensitive non-target lepidopterans in the treated habitats.

However, the option of not treating with Bt is worse in that gypsy moth larvae periodically defoliate most plants, thereby eliminating the food sources of all leaf-eating insects. Alternatively, at least some of the larvae of the non-target species will survive the Bt treatment, whereas few, if any would survive the use of a chemical insecticide. With respect to specific evaluations of Bt insecticides against non-target invertebrates, there have been numerous studies in the labouratory as well as in field situations under operational pest and vector control conditions.

Literally thousands of tons of Bt insecticides have been applied in the environment over the past four decades, and the overall record, especially considering the amounts applied, is one of remarkable safety. For more specific details and references to the extensive primary literature on non-target studies, the interested reader is referred to the recent comprehensive text by Glare and O'Callaghan, as well as review articles by Couch and Foss, Dejoux and Elouard, Lacey and Mulla, Meadows, Melin and Cozzi, Mulla and Vinson.

Bacterial insecticides based on different subspecies of Bt have been tested extensively in the labouratory against non-target invertebrates to meet registration requirements, and have also been evaluated in field situations to assess effects of formulated products under operational conditions. Both short-term (*i.e.* from a few days to several weeks) as well as long-term studies of more than a year have been conducted. In the labouratory studies, doses used to evaluate the effects on non-targets are typically as much as 1000-fold the amount that these invertebrates would encounter in the field, and in many cases the doses are much higher.

Representative non-target invertebrates that have been studied include earthworms and microcrustacea such as daphnids and copepods that make up much of the zooplankton in treated areas. In addition, insects tested have included non-target Coleoptera (beetles), Diptera (flies), Neuroptera (lacewings), Odonata (dragonflies and damselflies), Trichoptera (caddisflies) and Hymenoptera (parasitic wasps), especially species that constitute the major predator and parasite groups that attack the insect pests or disease vectors that are the targets of the Bt applications.

Larvae and adults of beneficial insects such as the honeybee, Apis mellifera, are also tested. In testing Bt products used against caterpillar pests, more emphasis has been placed on evaluating the effects on terrestrial non-target invertebrates. However, because these products can drift or be washed into streams and ponds, many aquatic invertebrates have been tested in labouratory studies and in natural habitats. In the case of B. thuringiensis subsp. israelensis, used to control mosquito and blackfly larvae, greater emphasis has been placed on evaluating the effects on aquatic non-target insects and other arthropods. Summaries of these results and those of other studies carried out over the past 30 years show virtually no adverse direct or indirect effects, especially long-term effects, of Bt or formulated products of Bt on non-target populations. The obvious exceptions are non-target species that are closely related to the target pests or vectors, or insects such as endoparasitic hymenopteran species that require the target lepidopteran pests as hosts. But even these are not affected in some cases.

Even in 'forced' feeding studies, the Bt subspecies did not have an effect on insects or non-target invertebrates, such as shrimp, that were outside the order of insects designated as the target group. In some of the earliest studies, effects were seen on earthworms and flies. But these early studies were conducted with strains that may have had the β-exotoxin, which has not been permitted in commercial formulations for decades.

In cases where the effects of Bt on non-target populations have been monitored under field conditions, the effects on non-target organisms were much less than those resulting from the use of chemical insecticides. The effects of Bt therefore must be viewed from the perspective of the consequences of using alternative control technologies. An appropriate example is the use of B. thuringiensis subsp. israelensis (Bti) in the Volta River Basin to control the larvae of Simulium damnosum, the blackfly vector of onchocerciasis, a blinding eye disease of humans.

The onchocerciasis control programme is sponsored by the World Health Organization and United Nations Development Programme. After more than a decade of intensive use, it was concluded that Bti was of 'only the slightest of hazards' to any of the non-target organisms tested. More specifically, when Bti formulations were applied to rivers, the 'drift' of invertebrates (*i.e.* the target and non-target invertebrates found floating in the rivers and presumably killed or disturbed by the application) increased two- to three-fold in comparison with untreated rivers.

However, when chemical insecticides were applied under similar ecological conditions, the drift increased 20- to 40-fold. In other words, the application of chemical insecticides was approximately ten times more detrimental to the non-target invertebrate populations than the use of Bti. In addition to the much greater impact of the chemical insecticides on non-target invertebrates in the rivers, the blackfly population began to develop resistance to these chemicals. Replacement of the latter with Bti-based insecticides, to which no resistance has developed, during the drier periods of the year ensured the success of this programme, and allowed large fertile areas of the river valleys in West Africa to be returned to productive agriculture.

SAFETY TESTING

The genetic engineering of crops to make them resistant to insects through the production of Bt Cry proteins has raised concerns about the safety of these crops to non-target invertebrates and vertebrates, especially their safety to humans.

Proponents of Bt crops argue that the safety of Bt crops has been established by direct feeding studies as well as by the 40-year safety record established by Bt insecticides. Moreover, they argue that Bt crops are even better because they typically contain only one Cry toxin and thus lack the other components of Bt insecticides such as spores, fermentation residues and formulation additives.

Alternatively, opponents of Bt crops counter that their safety concerns are valid because Bt crops produce Cry proteins in a form that is modified from what the bacteria produce. For example, Cry proteins do not form crystals in plants, but remain in solution, thereby eliminating the dissolution step from the different levels of specificity. In addition, they argue that Bt proteins are synthesized in a truncated form making them already active, and that it is possible that Cry proteins undergo processing in plants, such as glycosylation (which has not been demonstrated), that could make them toxic or allergenic to mammals. The opponents of Bt crops state therefore that the safety record of Bt insecticides cannot be relied upon to validate the safety of Bt crops. The concerns raised by the opponents of Bt crops have been widely reported in the popular press, creating a public backlash against Bt crops, especially in several European countries. The backlash has been minimal in the United States and countries such as Canada, China and Argentina. However, the finding that sufficient quantities of Bt maize pollen can kill larvae of the popular monarch butterfly under certain labouratory conditions was reported in most newspapers 1999.

This resulted in a higher level of concern by the public about Bt crops, at least about their potential effects on non-target organisms. These concerns have resulted in additional scrutiny of existing tests and data used by scientists and governmental agencies to determine the safety of Bt crops, as well as additional investigations under typical field conditions. Based on a recent series of studies conducted under field conditions, it was concluded that pollen from Bt maize would have only a negligible effect on monarch populations.

Before reviewing the types of studies and scientific evidence used to access Bt crop safety, some additional background is provided here about the forms of Cry proteins produced by crops. Even if in solution, the toxins still must undergo activation and meet the other criteria for toxicity summarized earlier. Most crops currently on the market produce full-length Cry or partially truncated proteins, typically Cry1Ab or Cry1Ac. In crops that produce truncated Cry proteins, the truncated version lacks a major portion of the C-terminal half of the molecule.

It has been assumed in some studies that this truncation results in the production of active toxin in plants. Activation of the toxin by midgut proteases also cleaves 26-29 amino acids from the N-terminus of the molecule, so even the truncated Cry proteins produced by plants are not fully activated, unless this fragment has been removed by plant proteases. It should also be realised that even in bacterial insecticides, some activated toxin is present due to the action of bacterial proteases produced during sporulation.

Most of the toxin in crystalline form is protected from activation, but toxin in solution at the time of lysis is subject to activation. Thus the safety tests carried out with the various Bt subspecies and formulated products based on these would have included activated Cry proteins. If these were toxic to non-target organisms, this toxicity should have been detected in the safety tests conducted over the past 3-4 decades, especially considering the high doses used in these tests.

The possibility remains that some plants may modify Cry proteins in a way that the bacteria from which they were derived do not. The developers of Bt crops have examined Bt proteins produced by various Bt crops for post-translational modification, including glycosylation, phosphorylation and acetylation. No indication has been found that these proteins have been modified post-translationally.

Even if post-translational modification of Bt proteins is eventually found to occur in some Bt crops, there is no reason to assume that these modifications would make a Bt crop less safe to mammals or non-target invertebrates. Many plant proteins, for example, seed storage proteins that are widely consumed by humans, are glycosylated, yet cause no harmful effects.

Furthermore, if post-translational modifications were to occur, owing to the complex mode of action of Cry proteins, they are just as likely to interfere with the toxicity of Cry proteins to target insects, possibly making the crops ineffective. Present Bt crops represent an early phase of a new technology, and it is easy to exaggerate their potential benefits and shortcomings. Given the 40-year safety record of Bt insecticides along with the well-accepted empirical methods for testing the safety of chemical insecticides, bacterial insecticides and drugs, it is appropriate that a combination of prior studies and empirical methods be used to establish the safety or lack thereof of Bt crops. Over the past few years, studies of Bt crop safety using empirical methods have begun to appear in the scientific literature.

These studies have examined the effects on non-target invertebrates and vertebrates including mammals in the labouratory and field. Under operational growing conditions, without exception, these studies show that Bt crops have no significant adverse consequences for non-target invertebrate populations, and if anything their use is beneficial because the amount of broad-spectrum chemical insecticides used is reduced.

Replacement of chemical pesticides with Bt crops provides better protection of beneficial insect populations due to the much greater specificity of Bt proteins. The labouratory studies also provide a variety of evidence that Bt crops are safe for human consumption.

SAFETY OF BT TO MAMMALS

Commercial formulations of Bt insecticides are used to control caterpillar pests on many vegetable crops and in forests near residential areas. They are also used to control mosquito and blackfly larvae in rivers, reservoirs and other bodies of water used for human consumption. As a result, and because other bacilli such as B. cereus and B. anthracis can infect humans, the Bt subspecies that serve as the active ingredients as well as the formulations made from these have been tested extensively for their safety to vertebrates, especially mammals.

These tests are designed primarily to assess whether the bacteria have the potential to infect mammals, and whether the endotoxins are toxic when administered through different routes of entry. In addition, eye and dermal irritability tests have also been conducted. In some cases, preparations have even been tested on human volunteers. Standard tests required by governmental agencies include feeding studies, pulmonary studies, and intraocular and intraperitoneal inoculation studies.

Standard test animals include mice, rats and guinea-pigs, but non-mammals including chickens and other birds as well as fish have also undergone similar tests. A typical test involves a replicated series of trials in which mice and/or rats are injected intravenously or intraperitoneally with the candidate preparation (one million Bt cells per mouse, 10 million cells per rat). The test animals are then observed for several weeks for infection or toxic reactions.

Attempts are made during and at the end of the study to isolate bacteria as a measure of the rate at which inoculated cells grow in, or are cleared from, the test animals. Successful isolation, however, is not an indication of infection because this can be due to the persistence of spores within tissues, as noted by Siegel and Shadduck.

Therefore, criteria for infection are isolation of vegetative cells and an increase in the Bt population. Tests have also been done where large numbers of cells have been injected into the mouse brain. The results of numerous feeding, ocular, pulmonary and interperitoneal injection tests conducted with B. thuringiensis subsp. kurstaki, B. thuringiensis subsp. israelensis and B. thuringiensis subsp. morrisoni (strain tenebrionis) against mice and rats have shown no evidence of infection or toxicity to the test animals. Even in immunodeficient mice, no infection was detected after interperitoneal injection of as many as 10 million colony forming units. In pulmonary tests, B. thuringiensis subsp. israelensis cells were cleared from the lungs within four days. However, in the intraperitoneal tests, viable Bt cells could be isolated from mice up to 10 weeks after inoculation. Nevertheless, there was no evidence of infection,or toxicity after 10 weeks. There are limits to such aggressive testing procedures where large doses of microorganisms are administered.

For example, Siegel and Shadduck found that intracerebral injection of B. thuringiensis subsp. kurstaki and B. thuringiensis subsp. israelensis caused high mortality when 10 million cells were injected into the brains of weanling rats. But even in these tests, most rats injected with one million or fewer cells survived and showed no signs of infection. Intraperitoneal injection of mice with 100 million colony forming units did cause high mortality with some isolates of Btk and Bti, but the results were not consistent, and lower amounts of 10 million cells caused no illness or mortality.

Similar tests with B. subtilis, which does not produce Cry toxins, produced comparable results. In contrast to these results, there is one study in which an unidentified Bt Cry protein putatively caused abnormalities in the ileum of mice fed potato treated with Bt.

The micrographs presented in this paper showed clear evidence of cell damage. But the cause of this damage remains unclear. The strain was not characterized with respect to endotoxin content, and it is possible that the damage was due to β-exotoxin, which is present in many Bt isolates and known to damage vertebrate intestinal mucosa.

This study is also suspect because the authors erroneously attribute cytolytic proteins and properties to B. thuringiensis subsp. kurstaki that are not characteristic of this subspecies, but rather are properties of B. thuringiensis subsp. israelensis. In addition, in other studies in which the ileum of mice was examined using immunocytochemical methods, no receptors for the Cry1Ab protein were found.

Only a few tests have been conducted on human volunteers, and these were feeding studies in which each individual was fed 1 g of Bt (10 billion spores along with endotoxin crystals) per day for three consecutive days. This is at least 1000 times the dose a human would consume by eating fresh vegetables immediately after treatment with a Bt formulation at the recommended rate. Yet no infections or illness occurred. In addition, there have been no confirmed human infections or deaths during the 40 years that commercial preparations have been used for pest and vector control around the world. This statement cannot be made for chemical insecticides, where hundreds of cases of illness and 5-10 deaths occur yearly as a result of the misuse of methyl parathion alone. An overwhelming amount of evidence indicates that Bt products, owing to their highly specific mode of action, cannot be 'misused' in the sense that they could accidentally cause illness or death.

There have been a few reports of humans 'infected' with Bt, or Bt causing infections in domesticated animals. However, aside from being extremely rare, these reports have not been substantiated, and the presence of Bt may have been due to persistence rather than infection. McClintock *et al.* concluded that Bt proteins were not involved in any of these rare reports. Considering the widespread use of Bt, and its simple nutritional requirements for growth, if it had even marginal potential as a pathogen for humans and domesticated animals it should have been isolated as the cause of infection much more frequently.

It is appropriate to note that the consumption of fresh vegetables, especially organically grown lettuce, cabbage, broccoli and tomatoes, even if washed, results in routine human consumption of viable Bt spores and endotoxin crystals. In addition to being allowable as a spray on crops just prior to harvesting, Bt is common in the environment.

THE GENERAL BIOLOGY OF BACILLUS THURINGIENSIS

Bacillus thuringiensis (Bt) is a common Gram-positive, spore-forming bacterium that can be readily isolated on simple media such as nutrient agar from a variety of habitats including soil, water, plants, grain dust, dead insects and insect feces. Its life cycle is simple. When nutrients are sufficient for growth, the spore germinates, producing a vegetative cell that grows and reproduces by binary fission.

The bacterium continues to multiply until one or more nutrients, such as sugars, certain amino acids and oxygen, become insufficient for continued vegetative growth. Under these conditions, the bacterium sporulates to produce a spore and parasporal body, the latter composed primarily of insecticidal proteins. These are commonly referred to in the literature as protein toxins or endotoxins.

Although Bt can be isolated from many environmental sources, and is often referred to as a 'soil bacterium', several features indicate that its principal ecological niche is insects. The original isolations of Bt, for example, were made from diseased caterpillars, and these remain a good source of isolates. More importantly, Bt produces a range of protein toxins and toxin synergists that are very effective at killing certain species of insects, especially larvae of lepidopterous insects, providing a rich substrate for Bt's reproduction.

REPRESENTATIVE ENDOTOXIN-CONTAINING PARASPORAL BODIES

The principal toxins are protein -endotoxins, the " designating a particular class of toxins, and endotoxin referring to their localization within the bacterial cell after production. Many isolates of Bt also produce the β-exotoxin (a competitive inhibitor of messenger RNA polymerase), insecticidal proteins produced during vegetative growth (Vips), the synergist zwittermicin A, and enzymes such as phospholipases that enhance the activity of the -endotoxins.

In addition, the spore itself can synergize the activity of Bt proteins in some insects, especially those not very sensitive to -endotoxins. In many insect species, especially grain-feeding lepidopterans, Bt reproduces to very high levels after insect death, with millions of spores being produced per insect. Thus, Bt's ecology and reproductive biology suggest that its toxins and toxin synergists evolved to debilitate or kill directly a range of insect species, thereby providing a substrate for reproduction of this bacterium.

Insecticidal Proteins of B. Thuringiensis

Because the β-exotoxin is not permitted in commercial bacterial insecticides in the United States, the two principal active components in commercial Bt preparations are the spore and parasporal body. The endotoxins in the parasporal body account for most of the formulation's insecticidal activity, including initial paralysis followed by death. The lethal effects of these proteins have been known since their discovery in the 1950s. In the early 1980s, shortly after the development of recombinant DNA techniques, it was discovered that Bt -endotoxins were encoded by genes carried on plasmids. This discovery led to a major research effort in many labouratories around the world aimed at understanding the genetic and molecular biology of these toxins. This effort resulted in the cloning and sequencing of numerous Bt genes, over 100 to date, and characterization of the toxicity of the proteins each encodes.

At the time Hofte and Whiteley wrote their review, a wide variety of confusing names and acronyms were being used to refer to Bt's endotoxin genes and proteins. Computer analyses showed that the nucleotide sequences of most of these genes were quite similar. To standardize the terminology, Hofte and Whiteley proposed a simplified nomenclature for naming all insecticidal Bt genes and proteins. In this nomenclature, the proteins are referred to as Cry (for crystal) and Cyt (for cytolytic) proteins. Though modified recently, this system is still in use today, and is described briefly here along with its modifications and additions. The Cry and Cyt nomenclature developed by Hofte and Whiteley was originally based on the spectrum of activity of the proteins as well as their size and apparent relatedness as deduced from nucleotide and amino acid sequence data, and protein gel analyses.

At that time, with the exception of a 27-kDa cytolytic protein from B. thuringiensis subsp. israelensis, all proteins appeared to be related, and probably derived from the same ancestral protein. Therefore Hofte and Whiteley termed the encoding genes 'cry' genes and the proteins they encode 'Cry' proteins. This designation was followed by a Roman numeral which indicates pathotype (I and II for toxicity to lepidopterans, III for coleopterans and IV for dipterans), followed by an upper-case letter indicating the chronological order in which genes with significant differences in nucleotide sequences were described.

The I and II for lepidopterantoxic proteins also indicate size differences, with the I referring to proteins with a mass of about 130 kDa and the II designating those of 65-70 kDa. Some epithets also include a lower-case letter in parentheses which indicates minor differences in the nucleotide sequence within a gene/protein type. Thus, CryIA referred to a 130 kDa protein toxic to lepidopterous insects for which the first gene (cryIA) was sequenced, whereas CryIVD referred to a 72-kDa protein with mosquitocidal activity for which the encoding gene was the fourth from this pathotype sequenced.

The 27-kDa Cyt1A protein first isolated from B. thuringiensis subsp. israelensis differs from other Bt proteins not only in its smaller size, but also in that it is highly cytolytic to a wide range of cell types in vitro, including those of vertebrates.

In addition, it shares no apparent relatedness to Cry proteins. Owing to these differences and its broad cytolytic activity, Hofte and Whiteley referred to this as the CytA protein encoded by the cytA gene. In their revision of Bt gene nomenclature, Hofte and Whiteley listed 38 published gene sequences that encoded 13 different Cry proteins and the single CytA protein. Since their publication, the number of Bt Cry endotoxins has more than tripled, and several new cyt genes have also been described.

As more and more cry genes were sequenced and analysed, it was decided to name genes based on their relatedness as determined primarily from the degree of their deduced amino acid identity. As a result, the nomenclature developed by Hofte and Whiteley was modified in the following manner. The cry and cyt descriptors have been maintained, but the Roman numerals have been replaced with Arabic numbers to indicate major relationships (90 per cent identity), with higher degrees of identity being indicated by following upper-case letters (95 per cent identity), and minor variations of these alleles being designated by lower-case letters, with the parentheses around the latter eliminated.

For example, what was CryIA(c) is now Cry1Ac (and the corresponding gene, cry1Ac), a relatively minor change. However, for some genes/proteins the changes are greater. For example, CryIVD does not cluster with the earlier CryIV (now Cry4) proteins, and is now the taxon Cry11Aa. Although the new designations carry no specific information regarding insecticidal spectrum, primary activity is still inferred because the numbers have been maintained for many of the genes, and there remains a high degree of correlation between relatedness and the spectrum of activity. For example, Cry1 still refers largely to toxicity to lepidopteran insects; Cry2 to lepidopteran toxicity and in some, dipteran activity; Cry3 to coleopteran toxicity, and Cry4 to dipteran toxicity.

Bacillus Thuringiensis Subspecies

While commonly referred to in the singular as 'Bt', B. thuringiensis is actually a large group of subspecies, all characterized by the production of one or a few parasporal bodies per cell during sporulation. Each parasporal body contains one or more protein endotoxins, typically as crystalline inclusions, and each of these is typically toxic to a limited range of insects. The endotoxins occur in the parasporal body as protoxins, which after ingestion, dissolve and are converted to active toxins through cleavage by proteolytic enzymes in the insect stomach (midgut).

The activated toxins bind to receptors on the midgut microvillar membrane in sensitive insects, lysing the cells and destroying much of the midgut epithelium, causing insect death.

In essence, Bt endotoxins are stomach poisons selective for insects and certain other invertebrates. At present, there are more than 70 subspecies of Bt, distinguished from one another by immunological differences in flagellar (H antigen) serotype. Each subspecific name corresponds with a specific H antigen number.

For example, B. thuringiensis subspecies kurstaki is H3a3b, whereas B. thuringiensis subspecies morrisoni is H8a8b. In the literature, the term 'variety' is also used in place of subspecies, as is occasionally the term 'strain'. Because the H antigen serotype-subspecific name often does not correlate with insecticidal properties, acronyms and numbers are often used to designate specific isolates, especially those with important insecticidal properties.

For example, HD1 (isolate number 1 from Howard Dulmage), is used to designate a specific isolate of B. thuringiensis subsp. kurstaki (H3a3b) that produces four major endotoxin proteins and has a broad spectrum of activity against lepidopteran pests. Another isolate of B. thuringiensis subsp. kurstaki (H3a3b) is HD73. This produces only a single endotoxin protein (Cry1Ac) and as a result has a much narrower spectrum of activity against insects than HD1.

Historically, HD1 was the first Bt isolate developed commercially for the control of lepidopterous pests, and remains the most widely used in commercial products today. This isolate has also been the source of much of our knowledge of Bt genetics and molecular biology as well as endotoxin genes used in transgenic bacteria and plants.

Despite the large number of Bt subspecies that have been described, the taxonomic validity of these as well as maintaining B. thuringiensis as a species separate from B. cereus has been in question for many years. At the species level, the primary problem is that the only phenotypic character that clearly differentiates B. cereus from B. thuringiensis is the parasporal body synthesized by the latter species during sporulation. In most Bt subspecies, the information for endotoxin production is encoded on large transmissible plasmids.

When these plasmids are lost naturally or cured from Bt subspecies by growing cells at 42°C, no endotoxins are produced. Cured Bt strains that hosted these plasmids cannot be reliably distinguished from B. cereus. At the subspecies level, phenotypic biochemical differences among many are minor, indicating that many subspecies may not be valid using accepted standards of differentiating bacterial subspecies. In other cases, the differences in biochemical properties and insecticidal activity are so significant that some subspecies could be viewed as distinct species.

From the standpoint of Bt's use as an insecticide, taxonomic studies, especially flagellar serotyping, have aided isolate classification but have proven unreliable as accurate predictors of insecticidal activity. Perhaps the best example of this is found in B. thuringiensis subsp. morrisoni (H8a8b), which includes isolates active against lepidopterous (isolate HD12), coleopterous or dipterous (isolate PG14) insects. Another example of this is found in the occurrence of the 128-kb plasmid that encodes mosquitocidal Cry and Cyt proteins among numerous subspecies of Bt, including israelensis, morrisoni, entomocidus, kenyae and thompsoni.

To summarize the current status of Bt systematics, most molecular evidence is in agreement with more classical biochemical and physiological studies which indicate that B. thuringiensis and B. cereus are the same species, and that the latter becomes the former when it acquires one or more plasmids that express genes for insecticidal proteins. Nevertheless, maintaining Bt as a separate species has practical value because of its insecticidal properties, as does dividing Bt isolates into subspecies based on flagellar antigens.

The latter is useful in cataloguing the more than 20 000 isolates of Bt collected to date. However, to understand the insecticidal properties of an isolate, regardless of its subspecies/serotype or other designation, knowledge of the insecticidal protein genes encoded and expressed is required.

Mode of Action of Cry Proteins

Aside from allergenicity, knowledge of the mode of action and structure of Bt's insecticidal proteins are the two interrelated topics that have most bearing on understanding the safety and risks of insecticides or transgenic crops based on these proteins. Although the spore can play an important role in the pathogenicity of B. thuringiensis to certain insect species, the Cry and Cyt proteins are responsible for the paralysis and death of most target insects.

The three-dimentional structure of three of these is known from X-ray crystallographic studies. Although the mode of action remains to be resolved at a detailed molecular level, these structures along with studies on the effects of the toxins in vivo and in vitro show how Bt proteins kill insects while being safe for most non-target organisms, especially vertebrates. In the typical Bt, B. thuringiensis subsp. kurstaki, for example, the endotoxin crystals dissolve after ingestion upon encountering the alkaline (pH 8-10) juices of the midgut. Dissolution requires the reduction of disulfide bridges that stabilize the Cry molecules in the parasporal crystal. Most Cry toxins are actually protoxins of about 130-140 kDa (*e.g.* Cry1 and Cry4 proteins) from which an active toxin 'core' in the range of 60-70 kDa is released into the midgut by proteolytic cleavage.

Proper activation results in cleavage of 26-29 amino acids from the N-terminus and about 600 amino acids from the C-terminus. Activated toxin molecules pass through the peritrophic membrane and bind to specific receptors on the apical microvillar brush border membrane of midgut epithelial cells, which lies just beneath the peritrophic membrane.

Binding is an essential step in the intoxication process, and in susceptible insects the toxicity of a particular Bt protein is correlated with the number of specific binding sites (*i.e.* receptors) on microvilli as well as the affinity of the Bt molecules for these sites. However, binding by itself, even high-affinity binding, does not always lead to toxicity, indicating that insertion and probably post-insertional processing in the midgut membrane is required to obtain toxicity.

For example, in three different studies it has been shown that the Cry1Ac toxin can bind with high affinity to microvillar membrane vesicles from Lymantria dispar, Spodoptera frugiperda and Heliothis virescens, but with little or no subsequent toxicity. This indicates that insertion and likely post-insertional processing of Cry proteins is essential to intoxication. In highly sensitive insect species, the microvilli lose their characteristic structure within minutes of toxin insertion, and the cells become vacuolated and begin to swell.

This swelling continues until the cells lyse and slough from the midgut basement membrane. As more and more cells slough, the alkaline gut juices leak into the hemocoel where, as a result, the hemolymph pH rises by a half unit or more. This causes the paralysis and eventual death of the insect. Though this general picture of the mode of action has been known for some time, the details of the insecticidal process at the molecular level remain unresolved, especially the series of events that occur after the toxin binds to a receptor.

There is good evidence for an immediate influx of potassium and calcium, in response to which the cell takes in water in an effort to balance these cations. To explain this cationic influx, it has been proposed that Bt molecules insert into the microvillar membrane forming transmembrane cation pores. For pores to form, it is postulated that during membrane insertion, the molecules undergo conforma-tional changes that permit insertion.

After insertion the toxin molecules associate to form a pore composed of six toxin molecules. As more and more Cry molecules enter the membrane additional pores form, leading to the influx of cations and water followed by cell hypertrophy and lysis. For a more in depth discussion of the studies dealing with the mode of action of Bt, the interested reader is referred to the reviews by Gill *et al.*, Knowles and Dow and Schnepf *et al.*

Structure of Cry Proteins

Analysis of cry gene sequences showed that the active portion of the Cry toxin molecule (*i.e.* essentially amino acids 30-630) contains five blocks of conserved amino acids distributed along the molecule, and a highly variable region within the C-terminal half. The variable region was thought to be responsible for the insect spectrum of activity and experimental evidence from recombinant DNA studies was obtained in support of this hypothesis.

For example, by swapping the highly variable regions of Cry1Aa and Cry1Ac, the insect spectrum of these molecules could be reversed. In addition, binding studies showed that in most cases the degree of sensitivity of an insect to a particular Cry molecule was directly correlated with the number of high-affinity binding sites on the midgut microvillar membrane. Studies over the past decade have improved our knowledge of Cry protein molecular biology through determination of the structure of Cry3A, Cry1Aa and Cry2A, identification of regions on the Cry molecule involved in midgut binding and specificity, and identification of several glycoprotein receptors on the insect microvillar membrane.

The structure of the Cry3A molecule was the first to be solved, and similarities in conserved amino acid blocks among Cry toxins and conservation in hydrophobicity indicate that this structure is a good general model for Cry toxins. The solution of the crystal structure of the Cry3A molecule showed that this protein is basically wedge-shaped and consists of three domains.

Domain I is composed of amino acids 1-290 and contains a hydrophobic seven-helix amphipathic bundle, with six helices surrounding a central helix. This domain contains all of the first conserved block and a major portion of the second conserved block of amino acids. Theoretical computer models of the helix bundle of this domain show that after insertion and rearrangement, aggregations of six of these domains could form a pore through the microvillar membrane. Domain II extends through amino acids 291-500 and contains three antiparallel β-sheets around a hydrophobic core. This domain contains most of the hypervariable region and most of conserved blocks 3 and 4. The crystal structure of the molecule together with recombinant DNA experiments and binding studies indicate that the three extended loop structures in the β-sheets are responsible for initial recognition and binding of the toxin to binding sites on the microvillar membrane. Domain III comprises amino acids 501-644 and consists of two antiparallel β-sheets, within which are found the remainder of conserved block number 3 along with blocks 4 and 5.

The structure resolved by Li *et al.* indicates that this domain provides structural integrity to the molecule. Later site-directed mutagenesis studies of conserved amino acid block 5 in the Cry1Aa molecule suggest that this domain may also play a role in pore formation. Though the structure of the molecule, experiments and computer modeling indicate that binding is attributed primarily to the hypervariable region of this domain, studies by Wu and Aronson also show that positively charged amino acids in domain I are important to binding and toxicity.

Moreover, domain III has also been shown to play a role in receptor binding. Domain I is the pore-forming domain, whereas domain II is responsible for the recognition and binding of the molecule to receptors on the microvilli of the insect midgut epithelium. Domain II therefore accounts for much of the high selectivity and safety of Bt-Cry proteins because mammals and other vertebrates lack receptors for this domain.

Domain III assists in stabilizing the structure of the molecule, and also can play a role in binding. An important aspect of understanding Bt's molecular biology and determining its mode of action at the molecular level is identification of the proteins to which the toxin molecules bind on the microvillar membrane. Sangadala *et al.* contributed to this area with the identification of the first two proteins in the midgut of Manduca sexta to which the Cry1Ac toxin binds.

They identified an amino peptidase of 120 kDa as a major binding protein. More recently, cadherin-like proteins and glycoconjugates have also been identified as receptors for Bt endotoxins. These findings indicate that the initial recognition signals on the midgut of a sensitive insect are not ion channels, but one or more proteins or glycoproteins within or which extend from the microvillar membrane. Thus, if these observations hold for other Cry proteins, the midgut receptors are then essentially docking proteins for the toxins, indicating that the toxins do not directly affect ion permeability by binding to ion channels. These studies of the structure of Bt toxins and their binding properties have identified key regions of the molecule responsible for binding and, putatively, the first binding proteins on the insect midgut epithelium. While further studies are needed to more clearly define the specificity of binding and binding sites, the most enigmatic area of Bt's molecular biology with respect to its mode of action is how the toxin causes toxicity after binding to the membrane. An understanding of this could provide information that might enable the development of insecticidal Cry proteins for use against insects against which we currently have none, such as cockroaches and grasshoppers.

MONITORING AND SAMPLING ON DAMAGE RELATIONSHIPS

In order to establish damage relationships, several issues should be considered. A decision has to be made on what and how to monitor and sample. Accurate assessment of disease severity or pest density is essential. Decisions have to be made on the size, method and frequency of sampling. Besides monitoring the pest or disease, it is also important to monitor the host plants and the environmental parameters. Plant growth may vary from season to season and thus potential yield in a crop free of pests and diseases will vary. Often, disease or pest infestation are assessed as number of lesions, pustules or insects, for example, without monitoring the size of the host plant. From a physiological viewpoint, the amount of healthy plant tissue is more important than the amount of affected plant tissue, since the healthy tissue produces the yield.

When yield loss is expressed as a percentage of the yield in the uninfested control, a problem may arise when comparing different growing seasons or predicting future losses.

When the yield in the uninfested control varies, the slopes of the regression lines describing percentage yield loss will be different from each other, since both lines are forced through the origin. When the damage relationship is described as yield compared to the control, the level of the lines will vary from season to season, but the slopes of the regression lines will be similar and give a more accurate description of the actual situation. Modern computer programmes enable fairly easy stepwise multiple regression analyses. However, the choice of parameters should be restricted to those that can logically be expected to play a role in the damage relationship in order to provide a more predictive relationship.

PEST DENSITIES FOR DETERMINING THRESHOLDS

Various methods can be used to create different epidemics or pest densities. The time of inoculation/infestation can be modified, which will result in different levels of disease and pest attack at any given time point. However, climatic conditions will also vary and may interfere with an adequate analysis of the impact of severity on yield.

Furthermore, it is possible to use different levels of inoculum/initial pest density or create a gradient of disease/pest attack by putting infested plants on one side of the greenhouse. Disease or pest level can also be modified by a variation in environmental parameters, but this method is not preferred because the environmental parameters may influence yield regardless of disease or pest.

The most frequently used method is the utilization of selective pesticides. A disease or pest is allowed to reach a certain predetermined level at which time it is stopped with a chemical pesticide.

DESIGN OF THE EXPERIMENTS

The design of the experiments largely depends on the type of model to be developed. For descriptive models based on regression analysis, the experiment should resemble commercial practices as much as possible. To assess yield, the plots should be large enough to rule out significant edge-effects. Sometimes, damage relationships are derived from a comparison of different greenhouses. However, this is not recommended, since factors other than the level of disease or pest may also vary. For the development of simulation models, the experiments are usually of smaller scale and can partly be done under controlled conditions. The effect of one or two factors on plant physiology and on pests or pathogens can thus be determined and this effect is then quantified and incorporated into the model. Thus, prediction of yield is done by using information from an immediately lower level.

2

Greenhouse Environment and the Pathogen or Pest Organism

Humans can affect both the host plants, the greenhouse environment and the pathogen or pest organism by cultural practices and by application of chemical, biological and other types of control. Cultural practices may be aimed at rendering the plant less susceptible or acceptable to diseases and pests, or eliminating (or at least reducing) infection sources. Furthermore, any cultural practice changing plant growth, such as leaf area, will influence the microclimatic conditions for the pathogens and pest organisms in the crop. Greenhouse crops tend to be labour-intensive and this can lead to a more frequent spread of pathogens and pests over the crop by workers than in field crops.

DISEASE EPIDEMICS AND PEST POPULATION DYNAMICS

The amount of disease or the pest density resulting from the interactions between environment, plant and pathogen/pest, and the influence of humans on these factors, is the subject of plant disease epidemiology and pest population dynamics research.

Plant disease epidemiologists and agricultural entomologists have approached these studies in seemingly different ways. However, given that both types of scientist deal with populations of living organisms that are subjected to common phenomena such as birth, death, development and migration, approaches of plant disease epidemiology and pest population dynamics are basically similar.

Whereas the plant pathologist deals mainly with the effects of the harmful agent (the pathogen), that is the disease, the entomologist is more concerned with the agent itself (the pest). Consequently, plant pathologists usually measure the disease incidence or severity, whereas entomologists generally estimate the number of pest individuals.

AMOUNT OF DISEASE CHANGES

The amount of disease changes over time. The curve of the amount of disease against time is called the disease progress curve (DPC). It is typically of a sigmoid shape, with a slow increase in the beginning, followed by a logistic increase and a levelling off at the maximum level of disease.

Vanderplank showed that DPCs can be described by logit:

$$(y_1) = \text{logit}(y_0) + r^* t$$

where logit(y) = ln [$y/(1 - y)$], y_t = disease on time t, y_0 = initial disease on time 0, r = rate of increase, and t = time. The logit transformation will turn a sigmoid curve into a straight line, which enables an easier comparison of DPCs than the original sigmoid curves.

For some diseases, transformation by gompits is better than logits:

$$\text{Gompit}(y) = -1n\ [-1n(y)]$$

Disease can be reduced by reducing initial inoculums (y_0) lowering the rate of increase of the disease (r) or limiting the duration of the epidemic (t) by delaying its start through preventive measures.

Reduction of initial inoculum is the purpose of sanitation measures before planting the new crop, but inoculum can also be reduced during the epidemic by removing diseased plants or plant parts. Reduction of inoculum will delay the epidemic.

It depends on the kind of pathogen whether this delay provides sufficient control. In the case of a pathogen whose spores are abundantly present in the environment, for example powdery mildew fungi, the reduction of initial inoculum will only give a small delay in disease progress. In the case of rare inocula, sanitation may provide almost complete control.

Sanitation takes place both in the nurseries and in the greenhouses. Most manipulations of either the pathogen, the environment or host-plant susceptibility are aimed at a reduction of the rate of disease progress. This is achieved by slowing down any of the processes in the infection cycle.

HUMAN POPULATION GROWTH

Malthus' equation, initially developed to describe human population growth, was soon adopted by entomologists to study insect demography.

The equation predicts that a population will grow exponentially according to:

$$N_t = n_0 e^r$$

where N_t is the number of pest organisms at a specified time, N_0 is that number at an initial time (0), e is the base of Naperian logarithms, r is the rate of population increase, and t is the elapsed time. If r is assumed to be constant and independent of conditions that affect pest development, survival and reproduction, population growth is unlimited.

This rate of increase r is also called intrinsic or maximal rate of increase and as such is referred to as r_m and depicts the rate at which the population would increase under permanently favourable conditions. In nature, however, favourable conditions are never indefinitely maintained and several – usually many – factors limit or retard population growth.

To reflect this, the so-called Verhulst' model predicts that populations will grow until reaching a maximum following a logistic or sigmoid curve that can be mathematically expressed by:

$$N_1 = K/\{1 + [(K\text{-}N_0)/N_0]e^{-r}\}$$

where *K*, called *carrying capacity* (the maximum population size that the environment's resources can sustain), is the asymptote of the sigmoid curve and the rest of elements as in the formula of exponential growth.

The parameter *K* is a measure of the global effect of all environmental factors that limit the growth of a population, the so-called environmental resistance.

Note that it can also represent the logistic model of Vanderplank and, although biologically unrealistic, it allows to show how pest control procedures may prevent pest populations to reach damaging densities.

Like in disease control, pest population growth may be reduced by decreasing or delaying immigration of first pest individuals into crop plants (lowering N_0 or *t*), or by decreasing the rate of population increase via integrated enhancement of environmental resistance, for example by release of natural enemies (lowering *K*). In the chapter on plant resistance, the reader will find a discussion on how different effects on pest development and reproduction influences $r_{m.}$

EPIDEMIOLOGY OF PLANT PATHOGENS AND POPULATION BIOLOGY OF PEST ORGANISMS

Knowledge of the epidemiology of plant pathogens and population biology of pest organisms in greenhouse crops enables the development of integrated control measures. More than in field crops, cultural practices and the environment can be manipulated to prevent epidemics. Until now, growers were mostly interested in high yields and therefore, cultural practices and greenhouse climate regimes would not primarily be chosen for disease/pest damage prevention. However, the increasing awareness of the need to limit the input of energy and chemical pesticides, as well as the increasing problems with pesticide resistance, have made growers more willing to consider adaptations to limit diseases and pests.

It is important that this knowledge is available. Integrated control can consist of any combination of control measures, including chemical control. Usually, chemical control is limited to an absolute minimum in IPM systems and it is considered as the last defensive barrier. Integration of cultural practices, such as cultivar choice, the composition of the nutrient solution and climate control, together with biological control measures offer good perspectives for the future. It depends on the crop and on the greenhouse facilities to determine which measures can be incorporated into an IPM programme. In general, all components of the tetrahedron may be modified. As long as the control measures have no negative influence on each other, generally speaking, the amount of control will be greater when more than one component of the tetrahedron is modified. Biological control will generally be enhanced by cultural practices that prevent a too explosive disease epidemic or pest outbreak, or by practices that favour the activity of natural enemies. For example, biological control of powdery mildew fungi is more suitable in partially resistant cucumber cultivars than in very susceptible cultivars, and the biological control of greenhouse whitefly works better in cultivars that are less prone to pest development.

In heated greenhouses, biological control can be combined with a climate regime that promotes the development of the biocontrol agents. This combination of climate control and biological control is also a form of IPM. As mentioned before, integrated control is more complicated than chemical control because more components of the tetrahedron are usually involved and more detailed knowledge on interactions is needed. However, several successful integrated control programmes have been developed.

In this text, several aspects of plant disease epidemiology and population biology of pest organisms have been discussed, mainly to show how many factors play a role in the occurrence and management of an epidemic or pest. The factors which should be studied largely depend on the disease or pest concerned.

In general, a combination of small-scale experiments under controlled conditions and larger-scale experiments under semi-commercial conditions gives a good insight into the relationships between the host plant, the environment, the pathogen or pest and human activity. The knowledge of such relationships is fundamental to understand why and when a pathogen or pest population may grow and cause yield loss.

The identification of the factors and relationships that permit a species to achieve pest status can help the researcher to design and evaluate methods to manipulate such factors in an integrated way to prevent diseases and pests from reaching the damage threshold.

Damage thresholds are a basic tool for decision-making in integrated pest management. Reliable damage thresholds are derived from a full understanding of damage relationships. However, the complexity of damage relationships means that relatively few damage thresholds are nowadays available, even in greenhouse crops in which several IPM programmes have been successfully implemented. Further knowledge on damage relationships would permit pest control decisions to be based on a cost/benefit analysis. This is particularly relevant for diseases and pests that leave visible injuries on the plant and force growers to spray chemicals – or to adopt any other control measure – unnecessarily. Furthermore, awareness of the plant tolerance to certain levels of diseases and pests would help to apply control methods – like plant resistance or biological control – that do not seek to eradicate the disease or the pest, but to optimize their control in an economical, social and environmental context.

MODE OF ACTION AND STRUCTURE OF CYT PROTEINS

Several Cyt proteins are known, including Cyt1A, Cyt2A, Cyt1B and Cyt2B, though most of our knowledge of these proteins is based on studies of Cyt1A and Cyt2A, and the crystal structure of the latter toxin. These proteins are highly hydrophobic and all have a mass in the range of 24-28 kDa. They share no significant amino acid sequence to identify with Cry proteins, and are thus unrelated. The first Cyt protein, Cyt1A, was identified as a component of the parasporal body of B. thuringiensis subsp. israelensis.

Initially, it was thought that Cyt1A played little role in the toxicity of this subspecies to mosquito and blackfly larvae. However, there is now general agreement that it is not only toxic to these and related flies belonging to the dipteran suborder Nematocera, but that it is important as a synergist of Cry4 and Cry11 toxins, and in avoiding the development of insect resistance.

While rare in comparison to Cry proteins, since the early work on Cyt1A in B. thuringiensis subsp. israelensis, Cyt proteins have been reported from mosquitocidal isolates of B. thuringiensis subsp. kyushuensis (H11a11c), morrisoni (H8a8b), medellin (H30) and jegatheson (H28a28c).

As in the case of Cry proteins, Cyt proteins are synthesized during sporulation as protoxins and assembled into crystalline inclusions that make up a portion of the parasporal body. They are always associated with Cry proteins, but do not cocrystallize with these, instead forming a separate inclusion.

The shape of inclusion formed by Cyt proteins varies from hemispherical to angular because they assemble in a spherical parasporal body simultaneously along with inclusions formed by the other Cry proteins with which they occur. However, when produced alone in a recombinant B. thuringiensis, the Cyt1A protein forms a bipyramidal crystal with 12 faces. Based on studies of Cyt1A, after ingestion the 27.3-kDa protoxin molecules dissolve from the crystal under the alkaline conditions present in the midgut of susceptible insects such as mosquito larvae.

Proteolytic enzymes then activate the protoxin by cleaving amino acids at both the C- and N-terminus, releasing an active toxin of 24 kDa. This molecule passes through the peritrophic membrane where it then inserts into the microvillar brush border membrane of midgut epithelial cells. Unlike Cry molecules, it appears that Cyt proteins do not require a glycoprotein receptor for binding. Instead, Cyt proteins have a high affinity for the lipid portion of the membrane, specifically for unsaturated phospholipids such as cholesterol, phospholipidcholine and sphigomyelin.

After binding to the microvillar membrane, the cells hypertrophy and subsequently lyse. The specific mode of action of Cyt toxins at the molecular level is not known. The leading hypothesis is that Cyt molecules insert into the microvillar membrane and then assemble into clusters of as many as 18 molecules to form pores that act much like those proposed for Cry toxins. Lysis then results from the influx of cations and water. However, evidence has also been provided that Cyt molecules do not form pores, but rather act like detergents, binding to and perturbing the lipid bilayer and disrupting the structural arrangement of membrane proteins, which would also cause lysis.

Unlike Cry proteins, Cyt proteins exhibit cytolytic activity in vitro against a wide range of invertebrate and vertebrate cell types, though at higher concentrations. However, toxicity is not observed when ingested in vivo by most insects tested, including larvae of lepidopterous species, non-nematoceran flies such as houseflies and fruitflies, and vertebrates such as mice.

The mechanism of this in vivo specificity is not known, but may involve specific combinations of unsaturated phospholipids. Though Cyt proteins have been thought to be only toxic to nematocerous dipterans, it has recently been shown that at least one beetle species, Chrysomela scripta, is sensitive to the Cyt1A protein. An important property of Cyt proteins, based primarily on studies of Cyt1A, is their ability to synergize the toxicity of the Cry proteins with which they occur.

For example, combination of Cyt1A with the Cry4A, Cry4B or Cry11A proteins with which it occurs in the parasporal body of B. thuringiensis subsp. israelensis results in toxicities three- to fivefold higher than the specific toxicity of Cyt1A or any of the Cry proteins alone. The mechanism of this synergism is not known but probably involves cooperativity of the Cyt and Cry proteins in binding to and/or inserting into the microvillar membrane.

SAFETY OF BACILLUS THURINGIENSIS: GENERAL CONSIDERATIONS

Knowledge of Bt's biology, especially the information on the mode of action and structure of its -endotoxin proteins, forms a sound foundation for understanding and evaluating Bt safety. From these studies, it can be seen that these proteins have a series of 'built-in' levels of specificity that should provide Bt bacterial insecticides and Bt crops with a high degree of safety for vertebrates and most non-target invertebrates. For Bt insecticides based on Cry proteins, these specificities include the following:

- Endotoxin crystals and spores must be ingested to have an effect; there is no 'contact' activity, as occurs with chemical insecticides. This is the reason sucking insects and other invertebrates such as spiders and mites are not sensitive to Bt.
- After ingestion, the endotoxin crystals of Bt subspecies used to control

lepidopterous pests and mosquitoes and blackflies require alkaline conditions, typically a pH in the range of 8-11 in the midgut lumen, to be solubilized in a form conducive to activation by midgut proteases. Under the highly acidic conditions that exist in the stomachs of many vertebrates, including humans, the endotoxin crystals may dissolve, but most of the solubilized protein is rapidly degraded, often within minutes, by gastric juices to non-toxic peptides.

- Once in solution in the insect midgut lumen, the toxin must be properly activated through cleavage by midgut proteases. Improper cleavage results in inactive toxin.
- Once activated, the toxin must bind to 'receptors' on midgut microvilli. Most chewing insects that ingest toxin crystals, even if they have alkaline midguts, including many lepidopterans, do not have the appropriate receptors, and thus are not sensitive to the toxins. Even insects sensitive to one class of Bt proteins, such as lepidopterans sensitive to Cry1 proteins, are not sensitive to the Cry3A proteins active against coleopterans, one reason being they lack the receptors for these.
- After binding to a midgut receptor, the toxin must enter the membrane, change conformation in the process, and oligomerize to form pores that will be toxic.

With respect to level E, at present, the specific conformational changes that must take place to exert toxicity are not known. It is known, however, that high-affinity irreversible binding can occur in some insects, yet not lead to toxicity. This implies that a specific type of processing (*i.e.* another level of specificity) is required for toxicity that occurs as or after the toxin inserts in the membrane.

In Bt crops, only a portion of the second level (*i.e.* B) of the first five levels of specificity has been circumvented. When synthesized in plants, full-length and truncated Cry proteins do not form crystals, and even if quasi-crystalline inclusions do form, toxin remains in solution within the plant cells.

Nevertheless, whether full-length or truncated protoxins, Cry proteins produced by Bt crops must still be properly activated after ingestion, and must successfully meet the criteria for binding and membrane insertion by levels D and E to be toxic. Furthermore, with the one exception of the Cry9C protein, which was engineered to resist rapid proteolytic cleavage, Bt proteins produced in Bt crops are rapidly degraded under conditions that mimic the mammalian digestive system. Therefore, most of the inherent levels of specificity that account for the safety of Cry proteins used in commercial bacterial insecticides apply to these same proteins when used to make Bt crops resistant to insects.

For Cyt proteins, which are not used in any Bt crops at present, the specificity in vivo has the same levels or stages in A-C. However, membrane insertion does not appear to require a protein receptor, but rather the presence of specific unsaturated fatty acids in the lipid portion of the microvillar membrane. Lastly, an important concept of evaluating safety is to consider the route by which an organism is likely to encounter a toxin.

Even though pulmonary (inhalation) and intraperitoneal injection studies are done with microbial Bt insecticides and proteins, their normal route of entry by target and non-target organisms is by ingestion. This is even more so for Bt proteins produced in Bt crops.

Bacterial Insecticides and Transgenic Crops

More than 30 Bt formulations are on the market in the United States, Europe, Asia and South America. Most of these are complex mixtures of sporulated cells produced by large-scale fermentation and formulation ingredients (stabilizing agents, spreaders, stickers, fillers).

The fermentation products contain spores and an array of toxins found in either B. thuringiensis subsp. kurstaki, which is used to control a wide range of lepidopteran pests, or B. thuringiensis subsp. israelensis, used to control the larvae of numerous mosquito and blackfly species. Both owe their success to their broad spectrum of activity and the absence of any widespread insect resistance after many years of use.

These characteristics are due to the complex composition of the parasporal body, which in each case contains four major proteins, and to the moderate frequency of usage. With respect to Bt transgenic crops, those on the market are based on full-length or truncated Cry proteins. These crops were constructed by transforming the plants with Bt cry genes modified to optimize their expression in plants by changing the codons to reflect plant codon usage.

All Bt crops currently on the market produce only a single Cry protein, Cry1Ab, Cry1Ac or Cry3A. However, crops which produce two Cry proteins, such as Bt cotton producing Cry1Ac and Cry2Ab, are being tested in the field. In addition, a variety of insect-resistant crops which produce Bt non-Cry or non-Bt insecticidal proteins are under development, and submissions for regulatory approval have been made in the United States and Australia.

Safety Testing for Bacterial Insecticides

In addition to their insecticidal efficacy, a major impetus for using Cry proteins in Bt crops was their long history of safety to non-target organisms, especially to vertebrates. The most important levels of Bt endotoxin specificity, *i.e.* activation, binding and membrane insertion, apply equally to evaluating the safety of Cry proteins whether used in Bt crops or bacterial insecticides. Therefore, the tests and data that support a very high degree of safety for bacterial insecticides containing Cry proteins are relevant to assessing the safety of Bt crops.

Extensive testing has been and remains required to meet rigorous safety requirements established by governmental agencies such as the US Environmental Protection Agency. Use of data from these tests is valid as a major approach to evaluating Bt crop safety, especially considering that many hundreds of safety tests have been conducted over several decades to register numerous bacterial insecticides based on different subspecies of Bt.

The principal Bt subspecies evaluated in these tests have been B. thuringiensis subsp. kurstaki, B. thuringiensis subsp. israelensis, B. thuringiensis subsp. aizawai and B. thuringiensis subsp. morrisoni (strain tenebrionis). The materials evaluated have been the active ingredients, *i.e.* sporulated cultures containing spores and crystals of Cry and Cyt proteins, as well as formulated products. Among the materials tested are all of the Cry proteins used in commercial Bt crops currently on the market. In determining what types of tests should be done to evaluate the safety of bacterial insecticides, early tests were based primarily on those used to evaluate chemical insecticides. However, the tests have evolved over the decades and are now designed to evaluate the risks of Bt, specifically the infectivity of the bacteria and toxicological properties of proteins used as active ingredients.

The tests are grouped into three tiers, I-III. Tier I consists of a series of tests aimed primarily at determining whether an isolate of a Bt subspecies, as the unformulated material, poses a risk if used at high levels, typically at least 100 times the amount recommended for field use, to different classes of non-target organisms. The principal tests include acute oral, acute pulmonary (inhalation) and acute intraperitoneal evaluations of the material against different vertebrate species, with durations from a week to more than a month, the length depending on the organism. In the most critical tests, the mammals are fed, injected with, and forced to inhale millions of Bt cells in a vegetative or sporulated form.

Against invertebrates, the tests are primarily feeding and contact studies. Representative non-target vertebrates and invertebrates include mice, rats, rabbits, guinea-pigs, various bird species, fish, predatory and parasitic insects, beneficial insects such as the honeybee, aquatic and marine invertebrates, and plants. If infectivity or toxicity clearly results in any of these tests, then the candidate bacterium would be rejected. If uncertainty exists, then tier II tests must be conducted. These tests are similar to those of tier I, but require multiple consecutive exposures, especially to organisms where there was evidence of toxicity or infectivity in the tier I tests, as well as tests to determine if and when the bacterium was cleared from non-target tissues.

If infectivity, toxicity, mutagenicity or teratogenicity is detected, then tier III tests must be undertaken. These consist of tests such as two-year feeding studies and additional testing of teratogenicity and mutagenicity. The tests can be tailored to further evaluate the hazard based on the organisms in which hazards were detected in the tier I and II tests. To date, none of the registered bacterial insecticides based on Bt have had to undergo tier II testing. In other words, no moderate or significant hazards or risks have been detected with any Bt subspecies against any of the non-target organisms studied so far.

As a result, all Bt insecticides are exempted from a tolerance requirement, *i.e.* a specific level of insecticide residue allowed on a crop just prior to harvest. Moreover, no washing or other requirements to reduce levels consumed by humans are required. In fact, Bt insecticides can be applied to crops such as lettuce, cabbage and tomatoes just prior to harvest. It is important to realise that such a statement cannot be made for just about any synthetic chemical insecticide. This does not mean that registered bacterial insecticides do not have any negative impacts on non-target populations or do not harm non-target organisms, but rather that these materials do not pose a significant risk to populations of these organisms or the environment.

MODE OF INSECT CONTROL

The bacterium Bacillus thuringiensis (Bt) is distinguished from other bacilli by the production of large crystals during sporulation composed of one or more insecticidal proteins. These proteins have served as the principal active ingredient of numerous commercial bacterial insecticides used worldwide for more than 40 years to protect crops against caterpillar pests, and more recently to control coleopterous insects and mosquito and blackfly vectors of human and animal diseases.

The first genes encoding insecticidal Bt proteins were cloned early in the 1980s. This led quickly to their use to construct recombinant bacterial insecticides containing novel combinations of these proteins and insect-resistant Bt crops. The first Bt crops became commercially available in the United States in the mid-1990s, and have been widely adopted by farmers despite higher seed costs in comparison with conventional crops.

During the 1999 growing season, farmers in the United States planted over 20 million hectares of insect-resistant Bt transgenic crops, including over 10 million hectares of Bt maize (corn), 1.5 million hectares of Bt cotton, and about 30 000 hectares of Bt potatoes. This acreage is expected to grow to about 15 million hectares of corn and 3 million hectares of cotton within five years, representing, respectively, about a third of the corn and half of the cotton acreage in the United States.

Bt crops are also being grown in China and Argentina, and their potential deployment is being assessed in several other countries. In addition, more than 30 other Bt crops, including rice, sorghum, most major vegetable crops as well as many tree crops used for fruit, nut and fibre production are under development. For farmers, Bt crops offer advantages over conventional crops in that the insecticidal proteins are produced directly by the plants, and continuously during most of the growing season, thereby reducing the material and application costs of using synthetic chemical insecticides.

Growers have reported improved profit margins over conventional varieties averaging $60 to $100 per hectare during the first few years of Bt cotton plantings in the southeastern United States. Consumers and the environment benefit from reductions in the use of synthetic chemical insecticides. For example, it is estimated that the use of Bt cotton in the United States reduced chemical insecticide applications in cotton for lepidopterous pests by 2.7 million pounds in 1999.

This represented a reduction of 15 million insecticide applications (22 per cent) in comparison to the number used in 1995 prior to the use of Bt cotton. These reductions preserve populations of insect natural enemies such as spiders and predatory and parasitic insects, and eliminate exposure of non-target vertebrates, including humans, to these chemicals. The agronomic and environmental advantages of Bt crops are widely recognized in the agricultural community. In fact, the United States Environmental Protection Agency has declared that Bt crops are an "environmental asset".

However, concerns about the long-term impact of their use have been raised by environmentalists and scientists in regard to the development of resistance in insect populations, and the safety of insecticidal proteins produced by Bt crops to humans and other non-target organisms. Important questions have been raised about the safety of Bt crops, but most of the specific test protocols and resultant data relevant to Bt proteins produced by these crops are scarce and are only now being published in refereed scientific journals.

This is because these data were developed primarily to fulfill governmental registration processes required in most countries to sell products, processes which do not require the data to be published in journals. Nevertheless, an extensive database exists in the literature on Bt strains, proteins and products and this information is reviewed here along with data available on Bt crop safety. To address the issue of safety, it is important to understand the role that Bt's insecticidal proteins play in its general biology and the mechanisms by which these proteins exert their toxic effects on insects.

Bt and its insecticidal proteins have a high degree of specificity that accounts for their safety to most organisms, especially in comparison with synthetic chemical insecticides. An understanding of Bt's mode of action, specificity and toxicology, only rarely treated in discussions of safety, is important in assessing any risks. Thus, to place the safety of Bt insecticidal proteins and Bt crops to non-target vertebrates and invertebrates in perspective.

3

The Potential Benefits of GIS Techniques in Disease and Pest Management

Spatial distribution of diseased units in a pathosystem is the most important factor affecting the field estimation of disease intensity. Spatial distribution includes the way in which disease lesions are distributed among healthy units and the way in which diseased host units are distributed among healthy units. Distribution of diseased units may be random, aggregated or regular. With randomly distributed disease, the variance is theoretically equal to the mean. In aggregated patterns, the variance in the number of lesions per leaf is greater than the mean number of lesions per leaf, but when there is a regular pattern the variance is smaller than the mean. When a large number of host units are sampled for disease, a frequency distribution showing the number of diseased units in each severity category may be determined. The sample frequency distribution can be compared with theoretical distributions using the goodness-of-fit test, and the parameters of the empirical distribution may be defined. Theoretical distributions applied to biological systems include the normal, log normal, Poisson, Weibull, Gamma and negative binomial ones. Knowledge of the frequency distribution is essential for the design of sound sampling procedures. When estimating disease intensity per field, the sampling unit, sample size, sampling point, sampling fraction and sampling method must be considered. In most disease assessments, the sampling unit is a plant. Often, only selected parts (such as individual leaves) may be assessed for disease intensity.

For each field, a predetermined number of sampling units is selected to give a mean value representative of that field; this is the sample size. Sample size is determined by the cost of sampling, the precision required and the available time, as well as by the spatial distribution of the disease; that is, the sample size should be empirically defined. Many sampling methods have been reported for plant disease assessments. Samples may be taken at intervals along predetermined lines in the field or greenhouse and these may be either one diagonal, both diagonals (forming a big letter X), or (if a more representative sample is required) a large W or Z pattern. With a disease that is randomly distributed, all the above methods will give comparable results and reduced variance in the sample mean may be better achieved by increasing sample size. If the diseased units are aggregated, the sampling method will be more important than sample size, and the large X or W sampling pattern is preferable to the single diagonal.

MONITORING PATHOGEN POPULATIONS

Monitoring the pathogen, primarily by trapping air-borne spores, has been used as a measurement of disease intensity and development and could serve either as an alternative, or a complement, to disease assessment. Given the current technology, the use of spore counts of pathogen populations for field measurement of disease is unlikely to replace the main conventional methods of measuring disease severity (disease symptoms), unless its accuracy can be shown to override the ease and low cost of symptom assessment. The monitoring of pathogen populations may serve another purpose.

As fungicides still remain an important tool for control of plant pathogens in the greenhouse, it is important to monitor populations of the pathogen for their resistance to potential fungicides. The term "monitoring resistance" is used to denote testing for sensitivity of target organisms in field populations. This can range in scope from continuous surveillance programmes over several years and involving many locations to short-term investigations into individual cases of suspected resistance. Good monitoring is the cornerstone of fungicide resistance research.

Without such work, we would know virtually nothing about the occurrence of resistance in crop pathogens. Moreover, resistance monitoring, together with monitoring for changes in practical performance, is a vital component of integrated resistance management. Several tools have been developed for such a purpose. For example, a tool for estimating the resistance of populations of *Botrytis cinerea* Pers.:Fr. to common fungicides has recently been developed. Tested fungicides are added to a selective medium in Petri dishes. The plates are exposed in the greenhouse at approximately midday, when *B. cinerea* conidia are released into the air. Plates are exposed for 30–60 min, according to the intensity of the disease in the greenhouse and then incubated for 4–7 days. Counts of typical *B. cinerea* colonies in the media supplemented with the fungicides are compared with those from fungicidefree plates. The data may then be used to make a recommendation on fungicide use.

BENEFICIAL PLANTS (ALLELOPATHY)

Some higher plants exhibit biological activity against other plants, a phenomenon known as allelopathy. For example, tomato will not grow under walnut trees or on land that has recently been cleared of walnut because of the phytotoxin juglone, which is secreted by walnut roots. Some higher plants exert a similar influence over soil microorganisms that cause disease. In addition, residues of allelopathic plants act, as do many green manures, to enhance the populations of soil microorganisms, many of which compete with or are antagonistic to pathogens. The use of allelopathic plants for disease control does not require registration. Several examples of allelopathy are associated with lettuce, crucifers, marigold and onion.

LETTUCE AND ALLIED PLANTS

Fusarium crown and root rot of tomato can be controlled by growing lettuce between two successive tomato crops in the greenhouse or by planting dandelion beside tomato. Chemicals in lettuce, dandelion and allied plants, such as endive and chicory, interfere with the growth of the fungus *Fusarium oxysporum* f. sp. *radicis-lycopersici* in the soil. The active chemicals in this case are phenolic materials, which retard the growth of *F. oxysporum* f. sp. *radicis-lycopersici*. Some of these compounds, for example cichoric acid, also bind strongly to iron, which is required by the fungus. Cabbage and other brassicas-Residues of cabbage and other brassicas decompose in the soil, releasing volatile, sulfur-containing isothiocyanates and ammonia. These chemicals are toxic to several pathogens, including: *Aphanomyces euteiches*, the root rot fungus of pea and bean; *Rhizoctonia solani,* a widespread root pathogen; and *Fusarium oxysporum* f. sp. *conglutinans*, the causal agent of cabbage yellows. The toxic effects are enhanced when these metabolites are confined by a plastic tarp, as during soil solarization.

The isothiocyanates released by *Brassica* residues are chemically related to methylisothiocyanate, the active ingredient of some commercial soil fumigants. A recent study suggested that the release of isothiocyanates from chopped rapeseed tissue incorporated into root-knot nematode-infested soil also suppressed nematode populations; however, other researchers failed to confirm this and suggested that success may vary with cultivar, nematode species, geographic location and climate.

The African marigold *(Tagetes erecta* L.) and the French marigold *(Tagetes patula* L.) are inhibitory to the nematodes *Pratylenchus, Tylenchorhynchus* and *Rotylenchus* spp., and their use with certain vegetable crops may reduce root lesioning and virus transmission. The toxic compounds secreted by marigolds are sulfur-containing polythienyls.

White rot caused by the fungus *Sclerotium cepivorum* is one of the most difficult diseases of onion to control. Its sclerotia are very long lived in soil and germinate only in the presence of the host, stimulated by various oils (disulfides and polysulfides) secreted by onion. Synthetic onion oils have been found to stimulate the germination of sclerotia of *S. cepivorum* in the absence of the host. The resulting mycelium is very susceptible to a variety of natural biocontrol mechanisms operated by soil organisms, and the population of *S. cepivorum* is considerably reduced. Unfortunately, synthetic onion oils are too expensive to be used on a field scale.

BENEFICIAL INSECTS, MITES AND PATHOGENS

Besides allelopathic plants, beneficial organisms include parasitic or predaceous insects, mites, spiders, daddy longlegs, centipedes, nematodes, protozoa, bacteria, viruses, and fungi. Indigenous populations of these organisms are considered to have a major role in the natural regulation of many species of pest insects, mites, slugs and snails, and some nematodes and weeds. Unfortunately, their role in natural regulation of pests in commercial production of vegetables has not been assessed or exploited to a significant extent, and the impact of chemical and other control practices on their effectiveness is seldom considered in present management systems.

Although there has been extensive research on the introduction of beneficial organisms for control of insects on vegetables, their use in commercial production has been limited to the control of insects and mites on greenhouse crops, and to the application of formulations of the bacterium *Bacillus thuringiensis* Berliner to control insects in various field-grown crops. Some indigenous and introduced species also are beneficial as pollinators. The major groups of predatory insects and mites, as well as parasites of pest insects, are discussed here, starting with the predators.

The adult and larval stages of many beetles are predatory. For instance, ground beetles (family Carabidae) are found in or on the surface of the soil and on low plants and under stones. The larvae of certain ground beetles are beneficial, but the adults may be injurious if they feed upon small fruits. Rove beetles (family Staphylinidae) attack all types of insects and some prey on slugs and snails. Their larvae live in the soil or in damp places.

The Vedalia lady beetle, introduced from Australia to California in the late 1800s to control an introduced scale insect on citrus trees, is a classic example of pest management by introduction of a biological control agent. In less than two years, the scale was under control. Most lady beetles or ladybird beetles, are entirely beneficial because both the adult and larva are predaceous. The two-spotted, seven-spotted, and convergent lady beetles are common in Canada. Aphids are their main food source but the eggs and nymphs of other insects also are accepted.

Lady beetles overwinter as adults in any sheltered location, becoming active in spring. The females deposit yellow, cigar-shaped eggs in small groups in the midst of aphid colonies. Lady beetle larvae start to feed on the prey as soon as they hatch. Usually one generation and sometimes a partial second generation occur per year. During summer, all stages from egg to adult may be present simultaneously. Imported lady beetles show promise for use in greenhouses and may be available commercially. However, when released outdoors they usually disperse beyond the target crop and become ineffective.

Lacewings (Neuroptera; several families): Lacewings are fragile-looking insects with four lace-patterned wings and a green or brown body. Both the adult and larva of many species feed chiefly upon aphids, though they also attack other insects and mites. Lacewing eggs are deposited singly or in groups of three or four on leaves and stems, often near a colony of aphids.

The green lacewings (family Chrysopidae) may suspend their eggs on long, thread-like stalks. Lacewing larvae are active and voracious. Their jaws are designed for probing crevices, piercing prey, and sucking body fluids. Some lacewings hibernate as adults. Others overwinter as mature larvae in small cocoons of dense, white silk. Lacewings tend to be habitat-specific, and certain species are more common than others in horticultural crops.

Hover flies (family Syrphidae): Hover flies, sometimes called flower or syrphid flies, have the ability to remain stationary while in flight. The eggs are pale yellow and cylindrical, with small, spine-like projections; they are laid on leaves and shoots where there are colonies of aphids. The legless, slug-like maggot is pale green or red-green and creeps over the plant. Aphids are its primary food source. Hover flies overwinter among dead leaves either as a maggot or pupa. They can be very abundant in some seasons.

The midge *Aphidoletes aphidimyza* (Rondani) (family Cecidomyiidae) shows promise as a biocontrol agent for use against the melon aphid in greenhouses.

Predatory mites: Predaceous mites occur naturally, often in large numbers, and some species are available commercially for biological control of small insects and phytophagous mites in greenhouses.

One such mite, *Phytoseiulus persimilis* Athias-Henriot (family Phytoseiidae), has been used effectively in greenhouses against the two-spotted spider mite. Other predatory mites mentioned in this book are *Amblyseius* (syn. *Neo-seiulus*) *cucumeris* Oudemans, *Amblyseius* (syn. *Neo-seiulus*) *barkeri* Schuster and Pritchard, and *Metaseiulus* (syn. *Typhlodromus*) *occidentalis* (Nesbitt). These predatory, foliagedwelling mites are being used against thrips in greenhouses. A soil-dwelling mite, *Hypoaspis* (syn. *Geolaelaps*) sp. also is being used in greenhouses.

Parasitic wasps (Hymenoptera; several families): More than three quarters of all Hymenoptera are parasitic. They are important in the natural regulation of many insect populations, and species of several wasp families have been introduced into Canada to control crop pests. The ichneumonids, braconids, chalcids, eulophids, pteromalids and scelionids include the major, naturally occurring biocontrol agents.

Their bodies are long and slender, and in females often end in an ovipositor. Parasitic wasps attack the immature stages of other insects; some are host specific, whereas others attack a wide range of insects.

The female parasite lays an egg in or near the host's body and the parasite larva feeds on the body fluids of the host, slowly killing it. Only one parasite larva is associated with a host in the Ichneumonidae. Braconids lay many eggs on a single host. The parasitic wasps that are available commercially for use in greenhouses are *Encarsia formosa* Gahan against the greenhouse whitefly.

Parasitic flies (notably the family Tachinidae): Tachinid flies parasitize moths, butterflies, sawflies, leaf beetles, and some land slugs and snails. Their host range includes many pests of horticultural crops. The larval stages of tachinid flies are parasitic, usually on other insects. Some tachinid females lay their eggs on their hosts, but the majority of them retain their eggs and deposit them fully developed with an active maggot ready to hatch. The females of some tachinids pierce the host and inject their eggs. Others broadcast their eggs, or lay them separately on a plant for the host to swallow. The larval stage passes inside a single host, which is killed in the process. When mature, the larva usually abandons the carcass of its host, pupating and overwintering in the soil. Adult tachinid flies are active, fast-flying insects that feed on nectar or honeydew and search erratically for suitable hosts.

True bugs (Heteroptera; several families): Many true bugs are predaceous, although some, such as the tarnished plant bug are pests. The nymphs and adults of true bugs are very active and have a varied diet. The immature and adult stages occur together throughout the season. Some species hibernate as adults under loose bark, among dead leaves or in other sheltered situations, becoming active again in the spring. Predaceous true bugs eat young caterpillars, aphids and spider mites. The minute pirate bug *Drius tristicolor* (White) is an example of a predatory bug (family Anthocoridae) that shows promise for use against the western flower thrips in greenhouses.

Mantids (Dictyoptera): The introduced European praying mantis *Mantis religiosa* L. is a non-discriminating predator that now may be found in parts of eastern Canada and in the southern interior of British Columbia, particularly around home gardens where its egg masses are left undisturbed.

Ants (family Formicidae) and social wasps (families Sphecidae and Vespidae): Most ants are opportunistic predators and scavengers on other arthropods. Ants and social wasps may greatly reduce populations of caterpillars and other insects, and they can be found almost everywhere. Sphecid wasps include species that are solitary and others that occur in nesting aggregations. Adult sphecids consume plant nectar and fluids from their prey, using the bodies of their prey to provision a nest for their larvae. In contrast, all hornets and yellow jacket wasps (Vespidae) feed their larvae the masticated remains of their prey, which may be insects or other animals.

Pathogens and other parasites of insects: Naturally occurring pathogenic bacteria, viruses, fungi, protozoa and nematodes contribute substantially to regulation of many species of insects on vegetables. For example, several species of naturally occurring bacteria kill insect pests of vegetable crops; however, factors that influence their pathogenicity and procedures required to increase their effectiveness in reducing populations of insect pests are not well understood. The bacterium *Bacillus thuringiensis* Berliner is the only pathogen that is used commercially in Canada to protect vegetable crops from insect pests. Crystals formed in the sporangium of the bacterium contain an endotoxin which, upon ingestion by a susceptible insect, causes paralysis and death. Strains of *B. thuringiensis* are used quite extensively to control the larvae of several foliage-eating butterfly and moth pests of vegetables, particularly on cruciferous crops and, more recently, to control the Colorado potato beetle on potato and tomato.

Insect pests of vegetables are killed by various species of indigenous fungi, notably species of *Beauveria, Entomophthora* and *Metarhizium*. Under favourable conditions, these and other fungi may decimate insect populations. *Beauveria bassiana* (Bals.) Vuill., for example, kills substantial numbers of adult Colorado potato beetles in some years, and species of *Entomophthora* kill various species of bugs and flies, such as root maggot adults. The green muscardine fungus *Metarhizium anisopliae* (Metsch.) Sorokin shows promise for control of white grubs in the soil. Other entomopathogenic fungi include *Aschersonia aleyrodis* Web. and *Verticillium lecanii* (A. Zimmerm.) Viegas; both infect pupae of the greenhouse whitefly.

Viruses kill several species of insects on vegetable crops, in some cases causing high mortality. The most common are the Baculoviruses, that is, the polyhedrosis viruses and the granulosis viruses, which kill larvae of several species of butterflies and moths (order Lepidoptera).

For example, a naturally occurring granulosis virus is considered to be instrumental in regulation of the imported cabbageworm, and polyhedrosis viruses contribute substantially to suppression of the cabbage looper and to the oscillations in populations of the fall armyworm. Applications of these viruses are very effective in controlling the cabbage looper and the imported cabbageworm on cruciferous crops. However, the viruses are not registered for use at this time. Several species of protozoa (order Microsporidia) are parasitic on insect pests of vegetables, particularly Lepidoptera. Parasitism by microsporidia may suppress populations of susceptible insects by reducing vigour and fecundity, often without killing the host. Several species have potential for use in pest management; *Nosema locustae* Canning, a parasite of grasshoppers, is of particular interest in Canada.

Many species of pest insects are susceptible to parasitism by nematodes, principally of the families Steinernematidae and Heterorhabditidae. There is particular interest in these insectparasitic nematodes for control of soil-inhabiting pest insects. One possible disadvantage in using these nematodes is that they are readily killed by desiccation, thus limiting their usefulness.

PEST AND DISEASE CONTROL

Using a pesticide wisely requires forethought and analysis of a particular pest problem or situation to determine the best choice under those particular conditions. Often, more than one choice exists for a particular pest. Factors that should be considered in selecting one pesticide over another include:

- *Product efficacy* or effectiveness;
- *Application frequency* needed or amount of residual control provided;
- *Cost* per acre, factoring in the number of applications required to achieve control;
- *Potential hazards* to applicator, mixer and grove workers; signal words on the product label are a good indication of this hazard;
- *Hazard to beneficials* and other non-target organisms and
- *Use restriction*, including re-entry intervals, preharvest intervals, number of allowable applications, time required between applications and restriction on mixing with other products, particularly oils.

Use of the information contained in the guide. This publication is designed to provide, to the best extent possible, a listing of the insecticides and miticides available for use against selected citrus crop pests. It is written for comparison and educational purposes only and is not intended to provide complete information for applications of a product.

The product label attached to the container always must be consulted before use. The pesticide listing for control of citrus pests was compiled from the most recent product labels available at the time this guide was prepared. No guarantee of effectiveness is intended by listing a particular product for use against a particular pest; only that the pest is listed on the label for the product.

For selected pests, where a sufficient research base exists, an efficacy rating (ER) is provided to give a performance index of various prove products for controlling a particular pest. Ratings are based on field tests conducted in Texas and other citrus growing areas. Local conditions and pest pressure may cause actual performance to vary.

EFFECTS OF PH ON SPRAY SOLUTIONS

Water sources used for pesticide applications in the Lower Rio Grande Valley are alkaline with pH readings typically ranging from 7.8 to 8.2. Use of high pH water, particularly above 8.0, may affect the performance of many pesticides, particularly organophosphates and carbamates. Degradation of certain pesticides is more rapid at higher pH, especially at higher temperatures (>95°F). Little information is available on the degradation rate of specific pesticides in high pH water; some pesticides are relatively unaffected. Generally, it is a good idea to check pH levels with a pH meter on a regular basis and buffer spray water with suitable acid buffer down to a range of 6-7 pH. When tank mixes of more than one pesticide or when fertilizer is combined with a pesticide, the final sprav solution should be checked to determine pH.

Use of Spray Adjuvants

Adjuvants or spray additives are intended to enhance product performance or coverage. Surfactants or spreading agents lower the spray particle surface tension allowing droplets to spread out on plant surfaces instead of combining to form large droplets subject to run-off. They are especially useful when spraying hard-to-wet surfaces such as waxy citrus leaves, armored scale insects, mite colonies with webbing or hairy-surfaced weeds. Some pesticide products are formulated with surfactant included. Stickers and extenders function to lengthen pesticide performance life by increasing product tenacity or weathering qualities of materials sprayed on plant surfaces. These frequently are combined with surfactants and are available as "spreader-stickers." Other adjuvants available for special applications include anti-foam agents and suspending agents that improve performance and enhance application of pesticide suspensions. In general, adjuvants properly used increase product performance, reduce waste and help minimize non-target contamination.

Pesticide Safety

The key to using pesticides safely is to be aware of potential hazards and take adequate precautions before and during use to prevent any unwanted effects. Before using any pesticide, read the label. Besides application instructions, the label provides information on re-entry intervals, hazards to the environment and non-target organisms, a statement of practical treatment or first-aid in case of exposure, storage and disposal procedures for containers, pre-harvest intervals to avoid excessive residues and the symptoms of poisoning. Users also should consult *Texas Pesticide Laws and Regulations* for state provisions that affect the application of pesticides to citrus.

Protecting Bees from Pesticides

With the exception of some tangerines and tangelos, most citrus cultivars do not require honeybee pollination. However, blooming citrus is highly attractive to honeybees as a food Source and may provide a good nectar source for a high quality honey sought by beekeepers. Although insecticides are not routinely applied during bloom, it is important to recognize that many pesticides applied to citrus are toxic to honeybees. The insecticides may cause significant bee death while they are foraging in sprayed orchards or through contamination of water or pollen brought back to the hive. The grower and beekeeper should cooperate closely to minimize bee losses. The following guidelines will reduce bee losses:

- If at all possible, avoid spraying during bloom. If applications are needed, apply when the fewest bees are foraging. Applications in the evening are least hazardous, early morning sprays are second best and mid-day sprays are the worst relative to bee hazards.

- To prevent heavy losses of bees, don't allow pesticides to drift onto colonies. Bees often cluster on the front of their hives on hot evenings. Pesticide drift or direct spray at this time generally results in high levels of mortality.
- Where insecticides are needed, consider their toxicity. "Highly toxic" pesticides present an extreme hazard to bees and should be used during bloom only if no other alternative exists. Insecticides categorized as "Moderately toxic" or "Relatively non-toxic" pose a much reduced hazard to bees and may be applied with relative safety in the evening or morning hours. For specific hazard ratings of pesticides to honeybees.

Precautions for Certain Miticides and Insecticides

Many pesticides have special precautions listed on the label that should be considered prior to their use. Failure to completely follow label guidelines may result in excessive residues at harvest, plant injury, poor pest control or undue hazards to non-target organisms including the applicator. The following are some of the major precautions for some of the most commonly used citrus pesticides:

Petroleum Oil Sprays: When applied properly, oils provide a very useful tool for controlling some citrus pests without damage to beneficial organisms but the sprays require careful use to avoid plant Injury. Soil moisture, humidity, temperatures, growth stage of leaves and fruit and the type of oil used can influence whether or not tree injury occurs.

Spray oils should not be applied when soil moisture is low or when relative humidity is below 30 per cent and daytime temperatures are likely to exceed 95^{0} F. Effective use of oils for pest control requires that the oil persist on the tree long enough to kill the pest but not long enough to cause tree injury. Persistence of oil film on foliage beyond the tree's tolerance can interfere with photosynthesis and transpiration. Such damage can result in twig dieback, leaf drop, reduced yield and the reduction in amount of fruit sugars (soluble solids).

Tangerines and tangelos tend to be more sensitive to oil damage than grapefruit and oranges. Late summer or fall applications of oil sprays i-nay delay maturity, which will interfere with proper colouring of early harvested fruit and may cause increased cold Susceptibility. Persistence of oils on leaf surfaces increases with higher oil dosages or through the use of oil with higher distillation temperatures such as a narrow range 44O-type oil. When choosing the type or rate of oils, keep in mind that oil dissipates more quickly under higher temperatures.

In situations where damage is likely from a regular oil spray because of hot, dry conditions, oil can be added at a low percentage to certain insecticide sprays to improve coverage. Before combining oil with any pesticide, read the label carefully to make sure the risk of phytotoxicity is not excessive. Do not combine oil sprays with sulfur or follow an application of sulfur within at least 30 days or plant injury may occur. When combining pesticides with oil, use of emulsifiable concentrates or other liquid formulations is preferred over wettable or soluble powders.

Comite: Complete coverage is essential. Leaf distortion and/or fruit spotting may occur when used in the spring or it may occur if Comite is tank mixed with oil or applied within 2 weeks of an oil application. Do not use in a spray solution above pH 10 or mix with alkaline materials (such as lime).

Ethion: Do not apply within 90 days of a previous application.

Guthion: With one application per season, a 7-day preharvest interval should be observed; if two applications are required, PHI = 28 days.

Lorsban 4E: Do not apply when temperatures exceed 95° F. Do not apply when trees are stressed by drought or high temperature. Lorsban 4E should not be used in combination with Difolatan 80. For orchard floor application, do not tank mix with Roundup or Paraquat herbicides.

Morestan: One application per season.

Sulfur: Do not make sulfur applications during hot weather. Applications are restricted to November through May. Do not apply within 30 days of an oil application.

Supracide: Make no more than two applications per growing season, with at least 45 days between applications. Do not apply within 14 days of harvest. If applied with oil, the preharvest interval is 60 days.

Vendex: Do not apply more than four times in 12 months or within 7 days of harvest.

Making Pest Management Suggestions: The information and suggestions included in this publication reflect the opinions of Extension entomologists and pathologists based on field tests or use experience. The management suggestions are a product of research and are believed to be reliable. However, it is impossible to eliminate all risks. Conditions or circumstances that are unforeseen or unexpected may result in less than satisfactory results even when these suggestions are used. Texas Cooperative Extension will not assume responsibility for such risks. Such responsibility shall be assumed by the user of this publication.

Suggested pesticides must be registered and labelled for use by the Environmental Protection and the Texas Department of Agriculture. The status of pesticide label clearances is subject to and may have changed since this publication was printed. County Extension agents and appropriate specialists are advised of changes as they occur.

The user is always responsible for the effects of pesticide residues on his livestock and crops, as well as problems that could arise from drift or movement of the pesticide from his property to that of others. Always read and follow carefully the instructions on the container label.

CHEMICAL CONTROL OF PESTICIDES

Pesticides are regarded by many as controls of last resort because they add considerably to the cost of production and their misuse creates high-profile environmental problems. Nevertheless, in well-planned, integrated pest management programmes, pesticides have a valuable role if used judiciously.

A major problem with many pesticides is the rapid development of pathogen races or insect and mite populations that are resistant to them. When that happens, their continued use is pointless; not only do they have little or no efficacy, but they can aggravate the situation by killing other organisms that contribute to the natural biological control of pathogens and other pests. Blanket, season-long pesticide programmes used as insurance in anticipation of pest outbreaks are seldom justified if other precautions have been taken.

The application of crop protection chemicals to vegetable crops or to the soil is a widely used method of controlling diseases, insects, weeds and many other pests. Because of the environmental impact of the use of chemical pesticides and because of their toxicity to humans, management procedures to minimize their use and their effects on non-target species are being developed and implemented. Accordingly, applications of effective pesticidal chemicals should be made in response to indications that populations of the pest have developed or will develop to economically significant densities. The threshold at which action is taken is dependent on the type of damage inflicted on the crop, especially the marketed portion of the crop. Populations of pests on crop plants are monitored to determine the timing and necessity for applications. This procedure of pest management minimizes the amount of chemical applied throughout the season.

Unfortunately, many growers prefer to apply pesticides indiscriminately to crops according to a predetermined schedule as insurance against damage by pests. This practice increases the detrimental environmental impact of adding pesticide chemicals to the field habitat and increases the cost of crop production. Also, the application of some chemicals causes stress to crop plants and may make them more susceptible to disease and other pests. On the other hand, scheduled applications of pesticides may be necessary to prevent development of economically damaging populations of pests, such as root maggots and other pests within roots or stems, that cannot readily be killed when crop damage is occurring, or pests for which populations cannot be predicted with acceptable accuracy.

Environmental impact: The unnecessary use of pesticides is discouraged because of their direct and indirect effects on the field habitat and environment, on the consumer and the grower, and on the cost and efficiency of production. Most insecticidal and acaricidal chemicals kill or are harmful to several species of non-target arthropods, including parasitic and predaceous species that contribute to regulation of primary or potential pests. There are cases where additional applications of pesticides were required to control pests whose populations increased to a damaging level following application of a pesticide to control a different species.

Applications of chemicals may affect populations of birds and other animals that feed on pest species, and residues of some chemicals in soil and on plant debris may affect microbial activity in soil, thus influencing the availability of plant nutrients. In addition, there is concern for the possible harmful effects of residues of chemical pesticides in food and for the effects on growers of excessive exposure. To chemicals. Also, the tendency for pest species to become resistant or tolerant to a pesticide is enhanced by repeated use of the chemical, necessitating the use of higher concentrations or the use of other chemicals, often resulting in greater crop loss and increased cost of production.

MANAGEMENT BY EXCLUSION AND REGULATION

Federal legislation administered by Agriculture and Agri-Food Canada and complemented by provincial legislation is the means by which certain diseases and pests are regulated in Canada. Such regulations are aimed at preventing the introduction of new diseases and pests and their movement within Canada. The Canada Seeds Act and the Plant Protection Act provide for field surveys; inspection of produce, premises, containers and packages; destruction or treatment; restriction on movement; and other safeguards and requirements as specified in regulations under these Acts.

Treatment of infested commodities, short of destruction, consists of fumigation, cleaning in a manner that is satisfactory to an inspector, or diversion to an approved industrial process. There are substantial reasons for concern about pests that may be associated with vegetable commodities at the time of harvest.

For instance, pests may be disseminated to other production regions within the country, a possibility that is greatly enhanced by modern, rapid transportation facilities. Pests also have resting stages that may survive cleaning and packaging operations and storage. The harvested, unprocessed commodity itself may serve as a pathway for the spread of pests. Apart from cold storage, which may affect the survival of some pests, Canada does not yet have the necessary facilities for bulk treatment and disinfestation of commercially produced quantities of vegetable commodities. Research into such alternative pest control technologies has only just begun.

Further information about the current status of federally regulated pests can be obtained from Agriculture and Agri-Food Canada.

Potato Viruses

- Andean potato latent virus
- Andean potato mottle virus
- Arracacha virus B
- Tobacco ringspot virus, potato calico strain
- Potato deforming mosaic virus
- Potato mop top virus
- Potato virus T
- Potato virus V
- Potato virus X, resistance breaking strain.

These and other viruses that occur in other countries are of concern to Canada because of their potential effects on crop yield and impact on trade. Their importation into Canada poses a threat to the potato industry in Canada, and thus they are cause for regulatory concern. To protect Canadian producers from potentially harmful viruses, the importation of potatoes from all countries other than the USA is prohibited. However, germplasm for research or commercial evaluation can be imported through Agriculture Canada's post-entry quarantine programme. The Columbia root-knot nematode has not been reported in Canada, but it is widely distributed in the western United States. It was described as a distinct species in 1980 after years of confusion with the northern root-knot nematode, which is widespread in Canada. The Columbia root-knot nematode is a pathogen of bean, carrot, corn, pea and potato. It also affects canola, sugar beet and cereals, such as wheat, barley and oat. On potato, it induces galls on the surface of the tubers, greatly downgrading them. When tubers are sliced, the body of the female nematode can be seen as a tiny, water-soaked dot, and masses of eggs appear as small brown spots.

Potato-rot Nematode Ditylenchus Destructor Thorne

The potato-rot nematode is a widespread and serious pest of potato in Europe; it also occurs in many Asian countries and has been reported from New Zealand and Peru. In South Africa it is an important pest of peanut. In the United States, it has been found in Arizona, California, Hawaii, Idaho, Indiana, New Jersey, Oregon, Washington and Wisconsin, and it is subject to quarantine regulations in some of those states. In Canada, the nematode was detected in potato in 1945 in three small areas of Prince Edward Island, and in 1952 in a small potato planting on Vancouver Island, British Columbia. Following eradication procedures and provincial legislation, the infested fields in P.E.I. have been kept out of potato production, and the nematode has not been detected in surveys conducted since the early 1960s. The potato-rot nematode also attacks bulbous iris and over 80 other hosts, including such vegetables as beet, carrot, celery, crucifers, cucurbits, eggplant, onion, pepper, rhubarb and tomato. Information on host range, diagnostic techniques, feeding habits, ecology and persistence of *D. destructor* in the field is incomplete; most infestations are detected only when found in mature potato tubers.

The potato-rot nematode feeds chiefly on fungi and can survive in weed hosts and in fallow soil for several years, but it seldom survives in rotted tubers in storage or in the field. Infestations are spread mainly by seed tubers. The nematode feeds first on the roots and later invades the tubers, in which it feeds just under the skin. Initial symptoms include small (0.3 mm diameter) pits surrounded by snow-white rings that are visible when the tuber is peeled. Continued feeding results in a depressed area in the skin, which becomes necrotic and dries out, resulting in characteristic irregular, triangular cracks. Severely affected tubers have large, sunken areas covered by dried-out skin (16.37). The cracks in the skin provide ready access to bacteria and fungi, and affected tubers usually rot in storage or in the soil.

Tomato Pinworm Keiferia Lycopersicella (Walsingham)

The tomato pinworm (family Gelechiidae) occurs in Mexico, the southern United States, the Caribbean, and Hawaii. Larvae of this moth feed in developing tomato fruits and may be encountered on post-harvest tomatoes as an imported pest. The most recent importation into Canada occurred during 1988–89, when larvae were found in or on tomatoes originating from

California. This moth was unknown in Canada prior to 1946, when larvae were confirmed in field tomato crops and in greenhouses in southwestern Ontario; however, these infestations did not survive. There since have been two isolated infestations in Canada: in 1970 in a greenhouse on Vancouver Island, and in 1975 in a greenhouse and surrounding home gardens at Kamloops,

British Columbia. Eradication at both locations was accomplished by exposure to the elements and by planting cucumber, which is a non-host crop. These occurrences probably had their origin in post-harvest tomatoes that then were being imported during the off-season from Mexico and the southern and western United States. These experiences show that the tomato pinworm cannot survive the winter outdoors.

Canada. Temporary greenhouse infestations can be avoided by destroying used shipping containers and by screening greenhouse openings, and they can be managed by including cucumber or other non-solanaceous plants in the greenhouse cropping cycle.

INTEGRATED PEST MANAGEMENT

For many years, Canadian vegetable producers relied heavily on protective applications of pesticides to reduce the impact of diseases and pests on their crops. In some cases, excessive amounts of chemicals were applied with little or no regard to effects on non-target organisms or contamination of the produce, soil and water.

In the late 1950s, however, scientists, particularly those dealing with control of arthropod pests, began to develop alternative procedures for crop protection, using a minimum of broad-spectrum chemical insecticides in combination with selective chemicals, biological agents, and modified cultural practices. This work stimulated interest in regulation of plant pathogens by biocontrol agents. In addition, by monitoring population levels of pests and pathogens in relation to crop damage, action thresholds were established for use in crop protection programmes as a guide to the timing of control measures. Thus, the present holistic approach to disease and pest control, termed integrated pest management, is evolving and is being widely implemented. Predictive programmes, such as Botcast for onion botrytis leaf blight, Tomcast for tomato early blight, and Blitecast for potato late blight, are resulting in a more precise approach to disease and pest control and are within the spirit of sustainable agriculture. Similarly, action thresholds for application of chemical insecticides to regulate insect pests on such crops as carrot, celery, crucifers, onion, potato, sweet corn and tomato have been or are being developed.

The effective integration of cultural practices and other methods of disease control is illustrated by the development of a system for the control of pea root rot. This system takes into account four pathogenic fungi and their disease cycles, tillage practices, drainage, soil compaction, cultivar susceptibility, green manure crops, and the predisposition of pea to these fungal pathogens by herbicides. A more limited programme for the control of potato late blight has been developed, and an integrated system for the control of cucumber powdery mildew in the greenhouse has been described, along with several ways of manipulating the greenhouse crop and its environment to escape diseases. Thus far, none of these systems considers the simultaneous control of insect and other pests, so the term 'integrated' is only partially valid, but it is encouraging to note that many greenhouse tomato growers have not used chemical pesticides for several years.

These integrated programmes can be improved by the development and use of selective pesticides to minimize the environmental impact of crop protection practices, particularly on indigenous and introduced agents for biological control. Undoubtedly, the most successful selective pesticides developed to date are formulations of the bacterium *Bacillus thuringiensis*

Applications of these formulations will control most foliage-feeding lepidopterous larvae, including those on cruciferous crops, lettuce and celery, and some coleopterous larvae, notably the Colorado potato beetle on potato and tomato. Such applications have no direct effects on non-target, parasitic or predaceous arthropods and do not contribute to pollution of the soil, water or crop. Furthermore, residues on the crop are not hazardous to humans and other animals. The potential for using viruses, protozoa, fungi and nematodes that kill insects and mites also is promising. Introduced parasitic or predaceous insects and mites are especially successful for control of pests of greenhouse-grown vegetables. Introductions of beneficial insects and mites, in combination with environmental manipulation and hydroponic systems to minimize the occurrence of diseases, have permitted pesticide-free production of greenhouse vegetables, particularly tomato and lettuce, by many growers. Unfortunately, introductions of beneficial insects and mites, and manipulation of those that occur naturally have been exploited very little for control of insect pests of vegetable crops under commercial field conditions.

METHODS TO CONTROL PEST INSECTS AND MITES

Other methods to control pest insects and mites, such as introduction of attractant and repellant materials, application of growth-regulating chemicals, or mass release of pheromones to cause mating disruption, which have been tested or applied to other crop systems, have not appeared promising or have not been assessed for controlling pests on vegetable crops.

Similarly, the release of sterilized males to reduce development of pest species has not proven economically practical for control of insects on vegetables.

In greenhouse vegetable crops, the high degree of environmental control available to growers enables fully integrated programmes to be devised that give good productivity as well as control of insects, mites and diseases. Biological control, using natural and introduced parasites and predators of pathogens and pests, plays a major role in greenhouse crop protection, but it is more difficult to realize in field vegetable crops because of the lack of control over the environment.

Management of viral diseases depends largely on vector control, crop hygiene, resistant cultivars, and regulatory systems of quarantine, inspection and certification.

Monitoring Certain Pests

Monitoring involves some form of surveillance to detect the presence of a disease or pest before economic damage has occurred and in time to economically apply eradicative or protective measures. Two such methods involve visual inspection and trapping. Visual inspection is the simplest method of monitoring and is used for insect and mite pests and for certain diseases, especially in combination with weather data and epidemiological information concerning sporulation, spore dispersal, infection periods and host plant growth stages. Surveillance is done on a regular basis, beginning soon after seeding or planting out, and includes watching for signs of feeding or irregularities of plant growth. Trapping is necessary for monitoring certain pests; traps may be purchased for certain pests or some may be constructed at home. The type of trap and its location must be suitable for the target species. Sticky strips or glass slides smeared with petrolatum or other adhesive are used for trapping fungal spores, aphids and thrips; coloured pans with water in the bottom are used for aphids; coloured sticky strips or ribbon-like tapes are available commercially for whiteflies in greenhouses; a variety of lights and light traps are used for moths of many types; and a pheromone in a rubber septum is available for such insects as the European corn borer. Determining the density in soil of nematode pests before planting is important for some crops.

Threshold levels have been established for a number of vegetable crops with regard to the rootlesion and root-knot nematodes.

Cultural Practices

After genetic resistance, the most important strategy for managing plant diseases and pests is best thought of as escape or prevention rather than cure. Such measures are aimed at avoiding the conditions that predispose the crop to infection or infestation; for example, selection of healthy seed of high germinability from a reputable source; treating seed if necessary to eradicate pathogens; checking the health of transplants at their source; practicing adequate crop rotation; sowing and transplanting only when soil conditions (temperature, moisture, nutrients, tilth) are adequate; maintaining adequate spacing and row orientation parallel to prevailing winds to permit good crop ventilation; avoiding overhead irrigation when the crop is in a vulnerable stage (*e.g.* at flowering in bean and pea); avoiding working in fields when the foliage is wet with dew or rain; roguing infected plants from seed crops; picking fruit before it is overripe; avoiding mechanical damage and removing field heat as quickly as possible at harvest; controlling weeds that may harbour pathogens and arthropod pests and contribute to humid microclimates in the crop; and controlling insects that may cause infectible wounds and transmit viruses.

Crop hygiene is a major factor in disease escape because it removes many sources of pathogens. Trash piles, contaminated seed trays, pots and stakes, and infected weeds are important sources of wind- and water-borne fungi and bacteria. Personal hygiene and equipment sanitation are also very important. Bacteria and viruses are easily transmitted on hands, clothing, tools and machinery. Tomato mosaic virus, for example, can persist for several months or years on sapdrenched overalls if they are hung dry in a dark closet. In bean fields, bacterial diseases often follow the tracks of workers walking among wet plants.

In nature, populations of insects and mites are regulated primarily by climate-related factors; by predation and parasitism by other arthropods and by other animals, such as birds; by diseases; by preference for food source; by competition with other plant feeders; and by numerous other factors. Because of the significant mortality of insects and mites by predation, parasitism and diseases, the best control practice for many insects and mites is to do nothing that could reduce the impact of these naturally occurring beneficial organisms and other beneficial agents in the environment. Cultural practices usually influence the impact of pests indirectly. Tillage, crop nutrition, crop rotation, timing of planting and other management practices affect the vigour of the crop plant. A vigorous, healthy plant is more likely to withstand damage from feeding by sap-sucking insects and mites. Attack by a pest sometimes can be circumvented or damage reduced by tillage, by timing of planting and by selection of field conditions. For example, planting cruciferous crops in well-drained soils and delaying planting to reduce exposure of young plants to cold, damp soil conditions will reduce damage by root maggots. The removal of crop residue from the field and other sanitation procedures may reduce infestations of some pests, such as the European corn borer, in the following crop.

INTEGRATED PEST MANAGEMENT STRATEGIES

Integrated Pest Management strategies require detailed studies which can be broken down into three steps:

1. A precise description of pest population dynamics in space and time in order to assess damage thresholds, to determine key points for control (possibly by modelling), and to evaluate control efficiency;
2. A general survey to estimate the variability in the first step between seasons or across a region;
3. The control strategy, including a survey by the grower of population dynamics.

Each of these steps requires particular sampling methods that differ in accuracy: precise measurements for detailed studies, less precise measurements but which can be used on a larger scale for variability evaluation in the second step, and quick and simple methods for final use by the growers.

Pest and disease intensity may be quantified using two different measurements:

1. Estimation of the population size, *e.g.* number of aphids per leaf, number of fungal spores in a cubic meter of air, etc.;
2. Quantification of the injury caused to the host plant, *e.g.* the proportion of leaf tissue infested by larvae, the relative leaf area covered with disease symptoms, etc. The methods should be easy to use, allow rapid estimation, be applicable over a wide range of conditions, and most of all, be accurate and reproducible. This chapter presents some of these methods which may be used for greenhouse crops.

INSECT PESTS

Estimating Insect Numbers in Samples

We will examine different ways of reducing pest assessment time. Methods based on visual abundance indices will be developed in particular, and examples of their application to insect pests will be given. Since many authors have developed methods which may be used with species other than those that they have studied, the references do not always concern greenhouse species. At each step of a study, the spatial distribution of most pests will be very patchy. Sampling plans will thus require numerous data to reach the required level of accuracy. Particular attention should be given to the evaluation of pest densities at each sampling point. This is a bottleneck which will define the "cost" of the sampling.

Methods are often available to reduce the cost of counts, but, except when automatic counting (*i.e.* picture analysis) is possible, these methods lead to a drastic decrease in accuracy. This loss of accuracy is cumulated with the error induced by the sampling itself, to define the final value of the density estimates. Moreover, several methods that are used to accelerate pest counts (*e.g.* field visual observation) underestimate pest densities per area or volume. Such systematic bias, as well as the accuracy of estimates, must be evaluated before using this kind of method.

The most accurate way of obtaining quantitative data on a pest population is to collect the substrate (*e.g.* host plants) and to take these samples to the laboratory where individuals may be isolated and counted under a stereoscopic microscope. As this method is time-consuming, numerous authors have attempted to reduce the time required.

The first step is the mechanical extraction of the individuals from the substrate. They can be extracted by washing, by flotation in high-density medium such as saccharose, or by the use of Berlesse-Tullgren funnels. Once the insects are separated from mud, sand or plant fragments, the clean extract can be fractionated into sub-samples. Both steps give results of varying precision, depending on the medium surrounding the insects and the species involved. The time required for counting the insects is reduced two to five times, but is often still too great (*e.g.* up to two hours per leaf, including washing, sub-sampling and counting, for aphids on cucumber).

Collecting insect substrate is not efficient for species such as thrips, which are very mobile. In this case, the extraction has to be made directly in the field, by using sweepnets, direct picking, mouth vacuum devices or vacuum nets like the Dietrick vacuum (D-vac). The numbers are then calculated from the part of the population which can be recovered. This part is often highly variable and the precision of the method is difficult to evaluate. However, successive sampling at the same sites may enable insect density to be estimated with accuracy. The method of Suber and Le Cren, frequently used to evaluate fish densities in rivers, was adapted to benthic insect counts by Lapchin and Ingouf-Le Thiec, and then by Lapchin to estimate larval and adult coccinellid densities in wheat fields. It has recently been used in cucumber greenhouses by Boll to estimate the number of thrips on leaves after several successive strikes.

Estimating Insect Population Densities

Insect densities may be estimated *in situ,* without collecting samples. The characteristics of the species distribution on the host plants should be taken into account and the most representative leaves or stems should be observed. This method eliminates the laboratory stage of density estimation, but does not significantly reduce counting time. Observation time may sometimes be drastically decreased if rough counting is performed.

This method is useful if the strong systematic under-estimation of the numbers it produces is constant or depends on known parameters. In the previous example of *Coccinella septempunctata* L. in wheat fields, Lapchin developed a "quick visual method", *i.e.* the observer walked within the 25 m^2 sub-plot for 2 min and counted all the adult coccinellids he saw. The numbers obtained by this method correlated well with the density estimated with the Seber and Le Cren method, and allowed the development of a sequential sampling plan for adult coccinellids in wheat fields. However, such a quick method is not automatically appropriate, even for closely related species, as Frazer and Raworth did not find that "walking counts" of adult coccinellids in strawberry fields were reliable. Variance of insect numbers is closely related to the mean.

This property implies that variations in numbers may be more easily perceived by observers who follow a geometric rather than arithmetic scale of density. In the field, such orders of magnitude can be easily translated into abundance classes. This was probably the reason why the use of categorical data was soon considered to be a good way of drastically reducing sampling costs.

In the case of aphid populations, precise density estimates of *Myzus persicae* (Sulzer) were related to the proportion of infested leaves in different parts of potato plants. This method was used on other host plants and statistically developed by several authors. As far back as 1954, presence-absence methods were improved by use of a set of abundance classes. Such classes can be purely arbitrary and, for instance, the sampling units distributed into poor, medium or heavy infestation classes. Several authors used more precisely defined classes, according to the number of colonies, their size and their localization. Different kinds of abundance class systems may be developed, according to the insect studied and its environment.

Building a System of Abundance Classes

Three main types of class can be considered. Firstly, there are classes whose limits are defined by the number of individuals that are seen during one sample unit of observation. A logarithmic scale of these limits was first used by Leclant and Remaudière to estimate *M. persicae* densities on peach trees. Another scale which is based on the approximate powers of has successive classes such as: no insect seen, 1 to 3, 4 to 10, 11 to 30, etc. This scale was used by Ferran to evaluate the density of the rose aphid [*Macrosiphum rosae*

(L.)] on rose bushes, by Boll and Lapchin for *Macrosiphum euphorbiae* (Thomas) in tomato greenhouses, and by Lapchin to estimate mummified *Aphis gossypii* Glover on cucumber plants. Secondly, purely qualitative classes, based on size and number of insect patches or on the percentage of contaminated shoots, may be used. Such classes are generally used for large sampling units such as trees.

Finally, there are intermediate systems which are based on the number of sub-units (*e.g.* leaves of a plant) in each class of a set of qualitative classes.

This system was used by Lapchin to evaluate non-mummified *A. gossypii* on cucumber plants and by Boll for *A. gossypii* in open-field melon crops. A visual class system must be both simple and complete. Ease of use depends on the number of classes and therefore there should be a sufficient number to describe accurately the trends of variability in insect density, but not so high as to be difficult to remember. An optimal class number is generally between five and eight. Another condition required for the use of visual classes is that there must be a biological basis for their definition.

To be representative, a qualitative class set must cover the different kinds of patchiness of the species which may be encountered in the field. For example, on cucumber plants isolated colonies of *A. gossypii* a few centimetres in diameter will first develop around winged immigrants (slightly infested leaves). After several days of development, the colonies suddenly spread all over the leaf area (heavily infested leaves). These simple characteristics, which are associated with the size of the leaves that are heavily infested, define the classes of abundance. Simplicity of the class system determines both the robustness of the results and the time required for field observations. In the example cited above, the observation of one sampling unit takes approximately 30 sec for each cucumber plant. The class system must cover the whole range of insect densities per sampling unit which may be encountered within the observation period and under different conditions. Thus, this range must be evaluated either from previous knowledge or from trials prior to defining the classes.

Calibration of Visual Abundance Classes

The results of visual observations may often be used without any reference to the number of individuals that they represent and, as such, these ranked qualitative data may be analysed using a large set of non-parametric statistic tools. This method has been used, for example, to evaluate the efficiency of biological control of the rose aphid on rose bushes in public gardens.

When more precise data are needed, each visual class must be calibrated by computing the mean and variability of the number of individuals actually present in the sampling units. This step is very timeconsuming because a large set of precise counts must be gathered so as to represent accurately the variability of the situations in which a given class may be chosen.

Classes Using Environmental Descriptive Variables

Calibration of the classes may be viewed as a statistical model having as a response variable the density of aphids, and as a categorical explanatory variable the visual abundance classes. A complex multivariate regression method, "projection pursuit regression", was adapted to this statistical data. Predictions of these models may be further improved by complementary explanatory variables.

This work, including calibration with complementary variables, was performed with four different class systems which were used to evaluate the density of the aphid *A. gossypii* and its parasitoid *Lysiphlebus testaceipes* (Cresson) on cucumber plants in greenhouses.

Two visual methods were used to estimate densities: "the detailed visual method" (DVM) for a leaf, and the "quick visual method" (QVM) for the whole plant. The class sets of DVM and QVM were built according to the apparent numbers of individuals in the observed sampling units. When the QVM was applied to healthy aphids, the four classes were based on the proportion of the area of leaf infested and on the size of the leaves. Precise counts were also made on the same sampling units and used as the response variable, and the data were divided into reference and validation sets. The reference sets were used to develop the regression models, and the validation sets to test their robustness. The choice of complementary explanatory variables was crucial to the development of these regression models. These variables were selected for their influence on the goodness of fit of the models as well as for the time required for collecting.

For example, when using the detailed visual method and when the target population of the model was either healthy or mummified aphids, seven explanatory variables were used:

- The visual class of the leaf;
- The visual class of the non-target population on the leaf;
- The visual classes of the target population on the upper and lower neighbouring leaves on the same plant;
- The vertical rank of the leaf on the plant;
- The number of leaves on the plant;
- The number of leaves infested by the target aphid on the plant.

Such data can be easily gathered during sampling without significant additional cost. QVM sampling of whole plants yielded a mean error of approximately one class per plant (the limits of each class are in the ratio of). The DVM had a mean error of less than one class. The range of the residuals was generally the same for both the reference and the validation data sets, confirming the robustness of these models.

The same method has now been used to calibrate visual class systems for different pest species on vegetable crops in greenhouses (thrips on cucumber plants, aphids on tomato, melon, eggplant and sprout). Each time that a new regression model is tested, particular attention must be given to the development and sampling of the reference data sets, *i.e.* they must include the same combinations of variables which will be used in further field sampling.

TIME SPENT ON EVALUATING INSECT DENSITIES

Reducing the time spent on evaluating insect densities in sampling units has a cost, which is a decrease in the precision of the estimation. However, the time saved allows the observer to increase the number of units taken into account in a sampling plan, and thus to increase the precision of the mean and variance estimates of the density. The gain in time is greatly increased when visual methods are used (a ratio of 1:10, when compared with precise counts). Mechanical methods, such as washing (ratio of 1:2), are much slower. However, the evaluation of the precision of visual methods requires a time-consuming calibration of the classes, which must be repeated for each study that deals with a different species, a different environment of the insects or a different scale of observation (*i.e.* plant or leaf). The decision to undertake such work will depend on the chance of building a "good" visual system.

We can summarize the four criteria of this evaluation as follows:

1. Insect density must be highly variable from one sampling unit (often a host plant) to another, and from one sampling date to another (this condition is easily met for numerous phytophagous insects whose densities vary on logarithmic scales from one host plant to another);
2. Most insects must be visible (for instance, such methods cannot be used for certain aphid species mainly located inside rolled leaves);
3. The visual classes must be simple and distinct, *i.e.* their boundaries have to be easily recognizable in the field;
4. These boundaries must be stable in time and space (for example, independent of the host plant growth stage).

The benefit of building such calibrated visual scales depends on which species is being observed and its environment. The scales are particularly useful for most aphid species, as they can reach very high densities and have strongly aggregated distribution patterns. Since the sampling units are not destroyed, crops can be easily monitored and the population dynamics studied separately in different fields. Such a method permits large-scale surveys.

PEST MONITORING

Most phytophagous insects are highly aggregative. Thus, the number of elementary units that are needed in a sampling plan to reach a given reliability of density estimates is drastically increased. This problem is particularly serious at the beginning of the crop season when insects are clumped around early immigrants and for species with a very high rate of increase. This is the case for most insect pests that exist under greenhouse conditions. Moreover, the efficiency of a biological control will often depend on the detection of such initial foci. Early trials have been performed in tomato greenhouses to develop whitefly sampling schemes that are compatible with the time constraints imposed by the grower. Eggenkamp-Rotteveel Mansveld used stratified random sampling in which absolute counts were performed on 0.6 per cent of the plants, spread evenly through the greenhouse. These data were compared with an enormous set of absolute counts obtained from the 18,000 plants in the greenhouse.

The results demonstrated that the random sampling did not accurately reflect the actual numbers and distribution of the whitefly and also that, in practice, absolute counts were not useful. The same conclusions were drawn by Ekbom. She suggested that some device should be used to detect at an early stage the presence of whiteflies. This was tried out by Guldemont and den Belder in chrysanthemum greenhouses. They simultaneously used yellow sticky traps and incidence counts (percentage of infested plants) to detect the moment and the level of the attacks by the major pests of the crop: leafminers, thrips, aphids, whiteflies, spider mites and caterpillars. They concluded that traps were still useful for monitoring the number of leafminers and thrips during the entire season and for aphids in the winter season, but that less emphasis should be placed on the use of traps and more on crop sampling.

The low density of pests and their aggregated distribution, however, makes the use of fixed sampling sites less suitable. Different approaches to aphid sampling have been tested, but unfortunately most of these experiments were performed in open-field cereal crops, and not in greenhouses. The spatial heterogeneity of populations was incorporated into the sequential sampling plans, based on the relationships of variance and the mean of density. The sample size may be adapted according to the reliability desired. More recently, incidence counts (percentage of infested/non-infested tillers) have replaced precise counts. Incidence counts will improve aphid sampling efficiency if there is a strong relationship between the incidence and the precise counts at the scale of the sampling unit, and if the loss of precision for each unit is more than compensated by the increase in the number of units observed in a given time. It is useful to combine the errors that are induced by the representativeness of the sampling scheme and by the use of incidence counts, as has been done for aphid predators in cereal crops. The monitoring of insect pests in greenhouses thus remains a complex problem. The most accurate and least expensive methods need to be developed in each situation and then adjusted to give the necessary precision for each particular biological question to be answered.

MEASURING DISEASE INTENSITY

The intensity of disease may be estimated by two distinct measurements: disease incidence and disease severity. Disease incidence is defined as the number of units infected, expressed as a proportion of the total number of units assessed, *e.g.* the percentage of infected plants, leaves, fruits, tubers, twigs, etc. This is a quintal measurement (*i.e.* the unit is infected or it is not infected). Disease severity expresses the intensity of the symptoms, *e.g.* the area of plant tissue affected by disease expressed as a proportion of the total leaf area, number of lesions per plant unit, etc. Measurement of disease intensity in a crop is fundamental for IPM. Disease incidence is generally easy to assess with considerable accuracy, but accurate estimates of the severity of many diseases are much more difficult to obtain.

Moreover, a fanner concerned about his crop readily overestimates severity. Thus for decision-making, disease incidence rather than disease severity is the preferred measurement. However, disease severity generally correlates better with yield and crop loss. Because of the relative ease of obtaining most incidence values with accuracy, many attempts have been made to correlate severity to incidence. At low disease levels, good correlation between disease severity and incidence has been found. At high disease levels, the relationship between incidence and severity becomes insufficient. When the correlation is significant, the similarity of the two measurements is confirmed and more easily measured incidence values for disease assessment may be used. When this relationship is not linear, an appropriate transformation may be employed. A square root transformation of the severity values is often used to create regression equations that predict severity from incidence. Thus, many schemes that warn against pests and diseases depend on enumeration rather than estimation procedures.

Estimating Disease Severity in Field Situations

Visual estimation of disease severity is almost exclusively used for estimating disease severity in the field. Methods for visual assessment of disease generally fall into two categories. The first category contains descriptive keys that utilize arbitrary scales, indices, ratings, grades or percentages to quantify disease. Such keys have been successfully used to estimate disease severity of host plants with differing disease resistance, or of host plants subjected to different environmental conditions or cultivation procedures. For example, disease can be described using categories of 1–5 to denote incidence (none to extreme) or severity (none to heavy). It is not appropriate to perform mathematical manipulations such as averaging on these records because values between two adjacent categories have no meaning.

The second category for visual assessment of disease involves the use of standard area diagrams. Pictorial representations of the host plant with known and graded amounts of disease are compared with diseased leaves to allow estimation of disease severity. Estimates of disease severity are proportional to the absolute area of the leaf that is diseased, and are not expressed as a percentage of an arbitrary maximum severity value. In contrast to descriptive keys, standard area diagrams allow estimation of intermediate levels of disease severity by comparing a diseased plant with diagrams that show both more and less disease.

Horsfall and Barratt, while noting the Weber-Feckner law, emphasized the limitation of the eye in the assessment of plant disease. The Weber-Feckner law states that the visual acuity of the eye is proportional to the logarithm of the intensity of the stimulus. These authors also noted that in visually estimating disease severity, the observer actually assesses the diseased proportion of leaves having <50 per cent injury and the healthy portion of leaves having >50 per cent injury. Horsfall and Barratt developed a disease-rating scale that contained 12 equal divisions of disease severity on a logarithmic scale with a median value of 50 per cent.

Thus, divisions of this scale included decreasing ranges of disease severity when either increasing or decreasing from 50 per cent disease severity. This scale and many standard diagrams constructed thereafter account for the logarithmic decrease in acuity of the eye in estimating severities approaching 50 per cent by their selection of representative keys. Estimations of disease severity intermediate between two keys are made by careful interpolation.

Accuracy, Repeatability and Reliability of Disease Assessments

Visual estimation of disease severity can differ significantly from the actual amount of disease. If the observer is not aware of the limitation in visual acuity at the midrange of disease severity, estimated disease severity and actual disease severity will be linearly related, and the variance of estimates will be independent of disease severity. However, the Weber-Feckner law indicates that the true confidence interval of estimates of disease severity will approach the expected linear relationship at both low and high disease levels, but will increasingly depart from this line with increasing disease severity, with a maximum variance at 50 per cent disease.

Inter-rater reliability has been operationally defined as the ratio of true variance to total variance, which includes a variance component for the error among raters. Although improved sampling designs and increased sample size can lower the actual and total variance, limited resources often restrict sample size. In addition, when more than one rater is involved, it is difficult to quantify the bias attributable to any one individual. Shokes proposed measuring intra-rater repeatability with the test-retest correlation procedure. The correlation coefficient (r) provides a statistical measure of the relationship between repeated assessments of the same sampling units by the same individual or instrument. However, correlation analysis between two variables cannot be used to infer a cause-and-effect relationship, nor can one variable (repeated assessments) be used to predict the value of another variable (first-time assessments). Least-squares regression can be used to determine if there is a significant linear relationship between disease assessment performed by different raters and whether there is a statistical relationship between related assessments performed by the same individual. Regression-equation parameters, such as the slope and the intercept, could be used to evaluate and compare the accuracy and precision of disease assessment raters and methods. Slopes that are significantly different from one indicate the presence of systematic bias among rates, whereas intercepts significantly different from zero indicate the presence of a constant source of error among raters.

RESISTANT CULTIVARS IN PATHOGEN RESISTANCE

The primary management technique for most plant diseases is to develop cultivars that possess genetic resistance or tolerance to infection; tolerance is the ability of the plant to be productive despite being infected. Of course, not every cultivar is resistant to every disease, but the prevalence of a disease in a particular locality often will determine the choice of a cultivar. Even when using a resistant cultivar, producers should be vigilant in watching for disease. Microorganisms, particularly those with genetic diversity derived from sexual reproduction, which occurs in many fungi, can quickly overcome resistance with a new form or race. For example, the fungus *Bremia lactucae,* which causes downy mildew of lettuce, exists in a number of pathogenic races, so lettuce growers must be aware of the races that are prevalent in their area and choose a cultivar accordingly. Recently, a new race of the tomato wilt pathogen *Fusarium oxysporum* f. sp. *lycopersici* appeared in Florida, and it is perhaps just a matter of time before it shows up in Canada and elsewhere. It is known as race 3 and it overcomes resistance to the two previously known races 1 and 2. Plant breeders work continually to produce cultivars resistant to new races of pathogens.

INSECT AND MITE RESISTANCE

The development and use of resistant or tolerant cultivars has been less successful for arthropod pests than for diseases. Some cultivars of vegetable crops are less attractive to pest insects and mites, and a few cultivars and hybrids are resistant to some insect and mite pests by virtue of chemical or physical characteristics. For example, there is a substantial difference in the frequency of tunneling by the European corn borer among cultivars of sweet corn. On the other hand, no apparent difference in feeding of the imported cabbageworm is found among cultivars of cabbage.

The development of pest-resistant cultivars of vegetable plants will be enhanced by the use of genetic manipulation techniques. These advanced biotechnological procedures permit more precise and faster addition and removal of specific factors or characters than do conventional plant breeding techniques.

EFFECTIVE BIOLOGICAL CONTROL OF VEGETABLE DISEASES

Effective biological control of vegetable diseases is rarely achieved, even on an experimental scale. Unlike predatory and parasitic insects and mites, a disease controlling organism is almost invariably a fungus or bacterium and has to be registered in the same manner as the microbial insecticide *Bacillus thuringiensis* Berliner or a chemical pesticide.

The registration process entails tests to ensure that the biological pesticide is environmentally safe, not harmful to operators or crop handlers, and safe for humans and livestock to consume with food. These tests, combined with the need to demonstrate that the biological pesticide is at least as effective as the best alternatives, as well as being economical and non-phytotoxic, require several years of research and testing, which leads in the end to a very expensive product. At present, Dygall *(Agrobacterium radiobacter* Conn) is the only biological pesticide available commercially in Canada for control of a bacterial disease, in this case crown gall in nursery stock. However, organisms to control white mold in bean and powdery mildew in greenhouse cucumber offer considerable commercial promise, and a number of patents of biological fungicides are pending. A yeast-like fungus is being developed commercially to control powdery mildew of cucumber in greenhouses, and fungi in the genera *Trichoderma* and *Gliocladium* show promise for control of root-infecting fungi. These agents actively parasitize the pathogen, produce enzymes that degrade the cell wall, or produce antibiotics. Examples are *Coniothyrium minitans* W.A. Campbell, which is a parasite of the white mold *(Sclerotinia* spp.) and gray mold *(Botrytis* spp.) pathogens, and *Sporidesmium sclerotiovorum* Uecker et al., which parasitizes sclerotia of *Sclerotium cepivorum*.

A wide variety of indigenous or naturally occurring biological agents affect populations of pest insects and mites, in most cases limiting populations to densities well below economic significance. These indigenous, beneficial organisms include predaceous and parasitic species of invertebrates, principally arthropods, in soil and on plants, and a variety of small animals, such as birds, rodents and toads, that feed on insects and mites, and microorganisms (fungi, bacteria, protozoa and viruses) that infect pest species.

The impact of indigenous beneficial organisms is enhanced by procedures that sustain or promote their activity. It is apparent that the application of a broad-spectrum toxic chemical to the crop will kill not only the target insect or mite but also the beneficial arthropods in a field. In addition, use of the chemical may repel birds but also may discourage their feeding because of the sparse numbers of surviving insects. The use of pesticides that are selective for the target pest species and that have minimum effect on non-target species is desirable in a pest management system. Also, the activity of indigenous organisms may be encouraged by such management practices as provision of hedgerows, cover crops, interplanted crops and field borders that can provide a habitat not only for birds but also for parasitic and predaceous insects and mites. Beneficial arthropods that occur naturally in small numbers may be augmented by introduction of mass-produced colonies. Releases of ladybird beetles have been found effective in controlling aphids on some field-grown crops. Also, the greenhouse whitefly on cucumber and tomato is controlled in a majority of greenhouses in Canada by release of the parasite *Encarsia formosa* Gahan propagated commercially for this purpose. The release of exotic, predaceous and parasitic arthropods has considerable potential for control of pests, particularly introduced pests, for which indigenous parasitic or predaceous species are not present or are not effective.

The augmentation of naturally occurring pathogenic microorganisms for pest insects has potential for a major role in management of pests of vegetable crops. Applications of the bacterium *Bacillus thuringiensis* Berliner represent the most widespread distribution of a masspropagated, indigenous biological agent. For example, formulations of *B. thuringiensis* are commonly used as alternatives to chemical insecticides to protect cruciferous crops against the cabbage looper, the imported cabbageworm and the diamond back moth. More recently, formulations of strains of the bacterium have been developed to control the Colorado potato beetle on potato. Applications of *B. thuringiensis* have little or no direct effect on parasitic and predaceous arthropods and other non-target animals in the field habitat and are not considered to be hazardous to humans, domestic animals or crop plants.

Several species of pest insects are susceptible to viruses. These agents, which are usually specific for one species of insect, occur naturally in the field habitat and cause high mortality of some pests. For example, a granulosis virus is a major factor in natural regulation of the imported cabbageworm, and other naturally occurring viruses cause high mortality in populations of several species of cutworms and armyworms. Some entomoviruses have been found to be very effective bio-insecticides when applied to vegetable crops; for example, applications of viruses are effective alternatives to chemical insecticides to control the cabbage looper and imported cabbageworm on cruciferous crops. These viruses are specific for the target insect, have no direct effect on beneficial species, and are not considered hazardous to higher animals and man.

Several species of naturally occurring entomopathogenic fungi cause substantial mortality of some species of pests, but fungi introduced into the field habitat to augment the natural populations generally have not caused significant predictable mortality of pests on vegetables. An exception is the effectiveness of applications of *Verticillium lecanii* (A. Zimmerm.) Viegas and *Aschersonia aleyrodis* Web. to control the greenhouse whitefly on greenhouse-grown tomato and cucumber.

4

Management of Plant-parasitic Nematodes in Vegetable Crops

MANAGING DISEASES AND PESTS IN HOME VEGETABLE GARDENS

Vegetable gardening is a rewarding and enjoyable hobby that provides a supply of fresh produce in season and that also can help to reduce a family's food costs. Vegetable crops are threatened by various disease, insect and weed problems that may reduce the yield and quality of the produce and even destroy the plants. Diseases and pests can damage vegetables from the time seeds are planted until after the crops are harvested, and gardeners who fail to follow good growing practices that minimize damage from diseases and pests in their gardens will suffer many disappointments.

The impact of plant diseases on vegetables often appears less direct than that of insects and weeds. Insects usually can be seen with the naked eye, and their activities quickly noted, while weeds are conspicuous, competing directly with vegetable plants for space, nutrients and moisture. Diseases, on the other hand, may make their presence known only when plant growth begins to slow down or when productivity is less than expected. No causal agent may be visible, only symptoms indicating that something is wrong.

Based on general cause, there are two kinds of diseases: biotic and abiotic. Biotic diseases are caused by an identifiable microorganism or infectious agent, for example, bacteria, fungi and viruses. Abiotic diseases result from unfavourable growing conditions or environmental stresses, such as extreme temperatures, too much or too little water, and nutrient imbalances. Both kinds of diseases exist widely in vegetable gardens across Canada.

Plant-feeding insects, mites and nematodes become pests if they injure crop plants sufficiently to interfere with the capacity of the plant to produce food. Some species of insects also transmit plant pathogens. Aphids, for example, transmit several viruses, and cucumber beetles can spread bacterial wilt and squash mosaic virus. The presence of some plant-feeding pests can be tolerated, but control may be necessary if the pests begin to cause significant damage. Home gardeners usually tolerate more pest injury, especially cosmetic injury, to their vegetables than do commercial producers. Many pest species are kept at a low level by natural enemies, such as predators and parasites, and additional control measures may not be required. It is important to realize that the population build-up of natural parasites and predators lags behind that of the pests, so some damage usually is evident. Special practices may have to be employed to keep the pest population at a level where damage will not be serious.

The most effective pest control strategy for the home gardener utilizes a combination of practices; these include planting resistant cultivars, providing optimum soil conditions, fertility and moisture supply, selecting pest-free seed and transplants, employing crop rotations, monitoring carefully for pests, destroying plant residue, cultivating, mulching, and using pesticides only when necessary. Such practices often keep the populations of garden pests at tolerable levels. The role of these and other techniques in helping to minimize disease, insect and weed problems in home vegetable gardens.

MONITORING

Traps and lures: Baits and lures using light, colour, sex attractants (pheromones) or food to draw a pest to a trap may be effective for some pests. Wireworms can be trapped by burying pieces of potato or carrot in the soil, then checking the bait every few days for the insect. Infested baits can be collected and destroyed. Slugs can be drawn to reservoir traps of beer or to a mixture of molasses, water and yeast.

Boards laid out in the garden can be used to trap slugs because they look for a cool, damp spot to hide during the day. These pests can then be collected and destroyed daily. Earwigs can be trapped in wooden traps or in containers filled with water containing a small amount of detergent. Growing an insect pest's favorite plant as a trap crop to lure it from the garden also works. However, the pests must be controlled on the trap crop or they will migrate back to the garden.

Cultural Practices

Healthy seeds and transplants: Many disease-causing organisms are capable of living in and on vegetable seeds; therefore, it generally is unwise for a person to save seed produced in the home garden. Rather, it should be purchased from dealers who have a reputation for producing or selling high quality, disease-free seed. When growing transplants, gardeners should use high quality seed and a pasteurized growth medium. If transplants or rootstocks are purchased from a supplier, they should be carefully checked for signs of pests and diseases, and any found to be infested should be destroyed.

Crop rotation: Moving the preferred host plant of an insect or disease organism hampers that pest's ability to feed and reproduce. Many disease-causing organisms do not survive long in the soil in which a different crop is planted.

Exceptions include fungi that cause such diseases as fusarium wilt on cabbage, potato or tomato, and clubroot of cruciferous crops; once the soil is infested with these pathogens, it can remain so for several years. Many insects and mites hibernate or lay their overwintering eggs in the soil near their preferred hosts. Moving the garden to a different location, or even switching the host crops from one to another part of the garden, has the effect that, upon emergence in the spring, the pest may find a non-host plant on which it is unable to feed and reproduce. This practice also may help to control weeds, especially if part of the garden is summer fallowed for at least one growing season. Crop rotation has the added advantage of allowing the soil to rejuvenate itself. Cabbage, turnip and potato use large amounts of nitrogen and should be preceded with legumes, for example, pea or bean, followed by fallow and the incorporation of compost or manure.

It is advisable to rotate among vegetable families because different members of a plant family often are susceptible to many of the same pest problems. For example, tomato, potato, eggplant and pepper are in the potato family and share many common diseases and insect pests, as do members of the cucurbit family (cucumber, melon and squash) and the crucifer family (cabbage, broccoli, Brussels sprouts, cauliflower, kohlrabi, turnip and radish). It is desirable to leave as much time as possible between related crops; three to six years is ideal. If rotation is not practical, gardeners should at least alternate the cultivars of vegetables that they are growing, choosing pest-resistant ones whenever possible.

Manual and mechanical methods: Hand-picking large, slow-moving pests, such as caterpillars, Colorado potato beetles and slugs, and dropping them into a container of soapy water or a 5 per cent isopropyl (rubbing) alcohol solution is effective. Shaking the plants or hosing them off with a spray of water dislodges some insects and mites. Barriers are meant to keep a pest away from the crop that it may harm.

Some examples are floating row covers, copper strips, abrasive materials and mulches. Used cans or frozenjuice containers opened at both ends and placed around transplants will protect them from subterranean cutworms. Copper strips repel slugs because their slimy coating interacts chemically with the copper.

Applying bands of abrasive materials, such as diatomaceous earth, stone dust or crushed egg shells, around the perimeter or between rows in a garden helps to deter slugs and crawling insects. Diatomaceous earth is an insecticide made up of the ground shells of tiny sea creatures called diatoms. The silica daggers pierce the skin of the insect, causing dehydration. Soap solutions smother the pests to which they are applied. Tar paper collars placed on the soil around individual crucifer seedlings will prevent root maggot flies from depositing their eggs at the base of the plants.

Mulches provide an effective means of weed control in vegetable gardens. Organic materials, such as weed-free straw or grass clippings 7 to 10 cm thick, will inhibit germination of weed seeds brought to the surface by cultivation, retard development of weed seedlings, conserve moisture, and help to maintain a uniform soil temperature. After the crop is removed, the mulch can be incorporated into the soil to enhance its organic matter content. Plastic mulches increase the soil temperature and also conserve moisture. Black plastic mulches, with openings for the crop, will prevent weed growth, except at the opening.

Clear and white plastic mulches, however, permit weed growth, as some light can penetrate through the plastic. Plastic mulches are best suited for warm-season vegetables. Mulches also may prevent soil from being splashed onto plants, thereby keeping the produce clean and reducing the risk of spreading some diseases. Weed mats made of woven fabric allow water and air to reach the soil but screen out light and act as a barrier to emerging weeds, thereby providing effective, long-term control.

Removal of infested plant material: Many fungal diseases can build up and spread rapidly through the production of millions of spores. Prompt removal or destruction of diseased plan residues retards the spread of disease-causing organisms. This is an effective practice with such diseases as gray mold, powdery mildew, and various leaf spots and fruit rots. Removing diseased plant material from the garden at the end of the season eliminates an important source of inoculum for the following spring.

The remaining residues should be rototilled or spaded into the soil to destroy disease organisms and to expose overwintering pests to the elements and to predators. Composting diseased plant material is not recommended because it may not destroy all of the disease organisms, even if the compost heats properly and is turned frequently. In some cases, leaving the remains of insect-infested plants in the garden may help to increase the population of beneficial parasites.

Good growing practices: When watering vegetable gardens, allow the soil surface to dry before another watering. A void frequent, light waterings as they tend to promote disease development and favour the germination of weed seeds. Watering with soaker hoses or in-ground furrows may reduce disease incidence because the foliage stays dry. When using overhead sprinklers, gardeners should be sure that the water is applied during the late morning or early afternoon.

This allows for rapid drying of the foliage. Prolonged leaf wetness favours the development of most foliar diseases. Wide row-spacing ensures rapid leaf drying and reduces the spread of certain pests and diseases by direct contact.

However, wide spacing also reduces shading of the soil, thus enhancing the loss of soil moisture and favoring the growth of weeds. Keeping plants as healthy as possible is an important strategy in managing insect and disease problems. Planting into a warm, moist, well-prepared, well-drained seedbed, maintaining a high organic matter content, applying fertilizers correctly, and not cultivating when either the plants or the soil is wet, are recommended practices. Monitoring the crop through the growing season may help to keep pest problems small by facilitating prompt control. Accurate record keeping is suggested to ensure success from year to year.

Companion planting: Many insects prefer to feed on plants belonging to specific families and reject others. For example, the imported cabbageworm feeds on cabbage, radish, kohlrabi and other cole crops; therefore, interplanting unrelated plants, such as onion, with cole crops will help to deter feeding and limit damage by this pest. Furthermore, garlic, onion and other aromatic plants interplanted among other garden crops are believed to repel certain insect pests. Alternating vegetable cultivars or species also reduces the chances of disease spread. Where root-knot and root-lesion nematodes are a problem, interplanting with French marigold or African marigold may help to reduce the populations of these pests.

Altered planting times: Seeding dates of vegetables can be altered to allow the plants to grow at times when they are less likely to be injured by certain pests. This requires knowing the life cycles of the common pest species in the area. In some cases, it may be feasible to plant certain vegetables before or after the pests have passed through their most active feeding period. For example, in Ontario, onion maggot eggs are laid near onion plants in May, and maggots soon attack plants and may cause them to die. However, onion sets planted after June 1 will escape most first-generation maggots.

Resistant cultivars: Some vegetable cultivars are resistant or tolerant to certain pests and diseases, and growing resistant plants is an excellent way to decrease the risk of damage. Certain disease-causing fungi can survive for many years in the soil and may not be managed by routine cultural practices; therefore, planting resistant cultivars is the best means of controlling them. Packages of tomato seed marked VFN produce plants that are resistant to the *Verticillium* and *Fusarium* fungi that cause wilt diseases, and to root-knot nematodes that may attack and cause galls on the roots. Other diseases best controlled by planting resistant cultivars include cabbage yellows, potato scab and cucumber mosaic. Certain vegetable cultivars also are less vulnerable to insect damage than others. For example, Red Pontiac potato is more resistant than other potato cultivars to tuber tlea beetle, and Champion radish seems to be somewhat resistant to the crucifer flea beetle.

Biological control: Beneficial insects can be attracted to the garden by planting some of their favorite nectar- and pollen-producing plants in the garden or nearby. Members of the carrot (dill, caraway, fennel and parsley), mint (catnip, hyssop and lemon balm), and daisy (yarrow and coneflowers) families are known to attract such insects. Providing a source of water and shelter also helps to keep beneficial insects around. Birds and toads also aid in the control of insect pests.

Care must be taken when applying pesticides because beneficial insects can be killed as well as the target pest. The bacterium *Bacillus thuringiensis,* or Bt, is a pathogen of certain insects and must be ingested by the larva of the insect to be effective. Only butterfly and moth larvae are susceptible to this product, although a strain of Bt has recently been developed that will kill the Colorado potato beetle. Various formulations of this microbial pestcontrol product are available for controlling larvae of certain moths and buttertlies on home garden vegetables. This and other treatments must be applied when the pest is most susceptible.

Chemical control: *Seed treatment* Planting fungicideor hot-water-treated seed helps to ensure good stands and may avoid the need to replant. Seed treatments kill diseasecausing organisms on the seed and help to protect the vulnerable seed and young seedlings from certain soil-borne disease organisms.

Treated seed is available to the home gardener and is marked 'treated' on the package. Fungicide-treated seed is usually red or some other easily identifiable colour. If untreated seed is used, it should be certified disease-free or be hot-water treated.

Seed can be treated by the gardener, using recommended chemicals according to the manufacturer's directions. Small packets of seed can be treated by tearing off one corner and putting about twice as much chemical in the packet as can be picked up on the first centimetre of the flat end of a toothpick. The packet should be shaken until the seed is thinly coated. *Foliar treatment* Most foliar diseases can be controlled by spraying or dusting plants with an effective fungicide as a preventative treatment. Protectant fungicides work on the plant surface to protect against infection, but they cannot cure established infections.

If a considerable amount of disease is present, it is usually too late to apply a fungicide treatment, except to protect newly emerging leaves. Fungicides should be applied at 7- to 10-day intervals, or as directed by the manufacturer, with reapplication after rain or watering has washed the material away. A thorough covering of the plants is necessary for proper disease prevention. Early detection, timely fungicide applications and removal of diseased leaves are necessary for effective control of many foliar fungal diseases. Gardeners can choose from a wide variety of organic and inorganic fungicides for use in combatting vegetable diseases in home gardens.

The home gardener also has a number of insecticides and miticides to choose from. Most botanical insecticides are derived from plant parts and will break down quickly into harmless substances. Botanical insecticides include pyrethrin and rotenone. Insecticidal soaps are effective against aphids, but they also are non-selective and may upset the natural balance of beneficial insects and their prey.

Chemical control of weeds usually is not practical for small home gardens but can be employed for larger gardens and for market gardens. If the decision is made to use a herbicide, extreme care should be taken to avoid accidental drift onto non-target species. In addition, residual soil herbicides should not be used where residues may carryover and affect succeeding crops.

Growers who apply pesticides in their home gardens should use only products that are registered and recommended for the crops being grown and for the specific diseases or pests that are prevalent in the area. It is very important to carefully follow the manufacturer's directions to ensure maximum effectiveness of a pesticide and to minimize potential adverse effects, such as poor control, crop injury and unacceptable chemical residues on the edible produce. Spot treatment is preferable to a general application because it lessens the risk of harming beneficial organisms, humans and pets.

FUTURE TRENDS IN WEED MANAGEMENT

Alternative methods of weed control will become increasingly important to vegetable growers in the future. Although these techniques will likely reduce the reliance on chemical herbicides, they will be only a partial substitute.

Reducing the use of herbicides: Methods by which herbicide use can be reduced include banding applications; using new, low-rate herbicides; timing of post-emergence herbicides to maximize effectiveness; placing fertilizer below the seed to target the crop instead of weeds; using new herbicide combinations, adjuvants, and application equipment; monitoring and determining weed thresholds as a basis for applying herbicides.

Alternatives to chemical herbicides: Alternative strategies for weed management are becoming increasingly important to vegetable growers. Reliance on chemical herbicides may be reduced but not eliminated. Methods under study and development include: biological control by weed-specific mycoherbicides; weed suppression by cover crops; rotation; the use of allelopathic plants as green-manure crops; and the use of liquid nitrogen fertilizer to control certain weeds in cruciferous crops. Chemical herbicides will likely remain the major method of weed control for some time, although specific products can be expected to change. The general trend to use less herbicide will continue, with greater reliance on cultural practices, including cover crops for weed suppression and soil conservation.

Genetically engineered crops: Biotechnology may have an impact on weed management through the development of vegetable cultivars with resistance to low rates of 'environmentally friendly' herbicides and by the incorporation of allelopathic properties into vegetable crops.

Herbicide-resistant weeds: The first occurrence of a herbicide-resistant weed in North America was documented in 1970 in a commercial nursery in Washington State, where common groundsel, which originally was susceptible to simazine, had become resistant.

This herbicide had been used for many years at this location. With prolonged use, in some cases for as few as five years, resistance to most of the herbicide groups with similar modes of action has developed in a number of weed species, although some of these resistant biotypes have only recently been detected. Methods used to delay or prevent the development and establishment of herbicideresistant weeds include: rotating crops; rotating herbicides (using herbicides with different modes of action); using short-term or non-residual herbicides; using the lowest possible effective rate; using herbicide mixtures; and practicing cultural/mechanical control where possible.

MANAGEMENT OF WEED PESTS

By managing weeds in headlands and other non-productive areas, and by preventing them from setting seed on crop land, growers can gradually decrease the reservoir of weed seeds in vegetable fields.

Monitoring: Scouting vegetable fields, particularly in the early stages of crop emergence, is essential for making decisions on the management of weeds. Although information on the economic threshold levels for specific weeds in vegetable crops grown in Canada is generally not available, some assessment of the weed population is required to decide whether a herbicide application or cultivation is necessary.

Similarly, regular monitoring is required to properly time management operations when weeds are at a susceptible stage. In a field where weeds are uniformly distributed, a zig-zag pattern for scouting is recommended. A map of the field showing dense weed patches or areas requiring special management should be prepared for immediate as well as future use. Because most weeds originate from the soil seed-bank, there usually is some consistency in weed problems from one year to the next.

Cultural practices: Weeds can be managed mechanically by cultivation, mulching and mowing. Cultivation and hand hoeing have been relied upon for centuries to control small weeds in row-cropped vegetables and for improving soil aeration.

However, if cultivation is too frequent or if the soil is too moist, compaction occurs and can adversely affect crop growth. Frequent rototilling can destroy soil aggregates and structure. Mulching with plastic has been a common practice in recent years for heat-responsive crops, such as cucumber, melon, pepper, sweet corn and tomato. Although clear plastic mulch provides the highest soil temperature, it also permits weed growth. Black plastic, which blocks sunlight, will inhibit weed growth. A newly developed plastic permits the transmission of infrared radiation and is being studied for its effectiveness in preventing weed emergence. Recent research also has focused on the use of living mulches or surface crop residues for reducing or suppressing early weed emergence. Other strategies, such as rotation and altering row widths and seeding rates, have an important role in integrated weed management.

Biological control: BioMal, the first commercial mycoherbicide in Canada, was registered in 1992 for the control of round-leaved mallow in certain field crops. The active ingredient is the pathogenic fungus *Colletotrichum gloeosporioides* f. sp. *malvae* Mortensen. In the future, this product also may be registered for use in some vegetable crops. Results from research are expected to provide new mycoherbicides for controlling other species of weeds. In contrast to most chemical herbicides, biocontrol agents (insects or plant pathogens) are very specific for individual weed species.

Biological control methods for weeds in vegetable crops will require the application of an organism to the weed to be controlled, a practice called inundative biocontrol. Classical biocontrol of weeds involves the establishment in an area of organisms from another region to provide ongoing control.

This latter method is suitable for rangeland and other noncultivated areas, but not for annual vegetable crops. Chemical control Herbicides are commonly used along with cultivation to provide overall weed control. Chemical companies rarely develop herbicides specifically for vegetable crops because of the limited market. Scientists, therefore, have had to adapt existing herbicides for vegetable production. In many instances, federal and provincial researchers in Canada provide the data on efficacy and crop tolerance required for the registration of herbicides through the "User Requested Minor Use Label Expansion" programme.

The recent development of sethoxydim and fluazifop-pbutyl has contributed significantly to the control of annual grasses and quack grass in dicotyledonous vegetable crops. These products are applied after emergence. Because they do not have significant soil residual activity, later weed flushes may reinfest vegetable crops and interfere with harvesting. These products are used in vegetable production, on both mineral and organic soils, to destroy cereal windbreaks before they compete with vegetable crops.

STATUS OF CHEMICAL WEED CONTROL

The status of chemical weed control for broadleaved weeds in vegetable crops ranges from excellent for carrot, potato, sweet corn, and tomato to unsatisfactory in cucurbits and most lowhectarage vegetable crops. Other modes of weed control need to be employed for these crops. In situations where only pre-plant incorporated or pre-emergence herbicides are specified, growers must rely on hand weeding and cultivation to remove weeds that emerge later. Crop injury from herbicide drift or soil residues Herbicides, such as 2,4-D and dicamba, which act as plant growth hormones, often injure susceptible vegetable crops by accidental drift or through the use of contaminated sprayers or watering cans. Typical symptoms are stem or petiole bending, leaf curling and cupping, and abnormal leafvein development. Picloram residues in manure from cattle fed with a treated crop often have caused problems in home gardens, and occasionally in commercial fields.

Sulphonylurea (chlorsulfuron, metsulfuron methyl) and imidazolinone (imazethapyr, imazamethabenz) herbicides can be injurious to most vegetable crops, even at very low concentrations. Soil residues of these herbicides can produce stunted, chlorotic growth and cause plant mortality. Trifluralin products are widely used for weed control in vegetable crops. However, in the following year, carry-over in soil can seriously damage beet, sweet corn and, to a lesser extent, cucurbit crops. Close monitoring is necessary to ensure that only tolerant crops are grown after triazine and urea herbicides are used.

In greenhouse crops, damage may result from the use of cereal straw or other mulches and amendments that have been contaminated with herbicides. Occasional cases of direct injury to crops from careless spraying of alleys, walkways or areas beneath benches also have been reported.

IN THE END, WE CAN SAY

Successful management of plant-parasitic nematodes in vegetable crops requires the integration of cultural practices, resistant cultivars, and chemical nematicides to reduce nematode populations at planting. Pesticide use may be economically feasible when other management strategies prove too costly or inadequate. For information on the major nematode pests of vegetable crops in Canada and their management, see Carrot, root-knot nematode; Onion, stem and bulb nematode; Potato, root-lesion nematode, potato cyst nematode, and stubby-root nematodes; and Beet, sugarbeet cyst nematode.

Monitoring There is a direct relationship between population density of plant-parasitic nematodes in the soil at planting and crop damage, and theoretically a damage threshold can be established for each parasitic nematode on its respective vegetable host. When nematode density in the soil exceeds this level at planting, losses in yield, quality or both become noticeable. Determining such thresholds is time consuming. Relatively little information of this type is available for Canadian vegetable crops. In practice, nematode numbers in the soil are used to predict nematode damage; this is also the first step in diagnosis. After determining the density of nematodes in a soil sample, the number is often extrapolated to estimate the population in the entire field; however, caution is advised if using this procedure. An average count above a certain threshold could mean only that some areas of the field will show crop injury.

Cultural practices: It is almost impossible to design effective and economical rotation schemes for farms that grow vegetables exclusively. Destruction of infested crop residue, clean fallowing between crops, and rotation with non-host crops are effective measures for nematode species that attack only a few crops. Useful rotations against the sugarbeet cyst nematode are alfalfa, cereals, and bean and potato, which are non-host vegetable crops.

The root-lesion and root-knot nematodes have very wide host ranges and are more difficult to manage. Some grasses and cereals, such as wheat, barley, oat and rye, which are non-hosts of the northern root-knot nematode, are used extensively. An ideal practice for reducing numbers of root-knot and rootlesion nematodes in small vegetable plantings is to interplant with French marigold, *Tagetes patula* L., or African marigold, *T. erecta* L. The use of these plants is more effective than fallowing or other cultural practices. The nematodes are attracted to and penetrate *Tagetes* roots but are unable to feed and multiply. Consequently, the density of nematode pests in the soil is lowered. Marigolds are not effective against *Heterodera, Ditylenchus,* or most ectoparasitic species.

Solarization is practical for small gardens. Intense summer sunlight can raise the soil temperature above 40/C beneath a transparent plastic tarp, killing many nematodes to depths of 5 to 10 cm. The soil must be well worked and moistened so that the heat will penetrate evenly. The soil should be covered for 3 to 6 weeks. Control is complete at the surface but less so at greater depths.

NEMATODE-FREE SEED AND TRANSPLANTS

Despite the extra cost, the use of certified, nematode-free seed and transplants is usually worth the expense. Transplants should be vigorous and free of root galls or lesions caused by nematodes. The use of soil-free media, or pasteurized and fumigated soil, is suggested for cucumber, tomato and other susceptible crops. To eliminate plant-parasitic nematodes from small quantities of soil to be used to germinate seeds and grow transplants, growers should moisten the soil for several hours, then heat it to 80/C for one hour. In heavily infested greenhouse soils, annual or even semi-annual pasteurization may be required to avoid severe damage to greenhouse cucumber and tomato crops, as well as to some ornamentals sometimes found in vegetable greenhouses.

Resistant cultivars: A few vegetable cultivars are resistant to species of nematodes. In most cases, there is some degree of resistance to some but not all root-knot species or other plantparasitic nematodes. There are no vegetable cultivars with specific resistance against the rootlesion nematode or the northern root-knot nematode, but some are more tolerant; for instance, some carrot cultivars have been shown to be more tolerant than others to root-knot nematode infection, and nematode-resistant rootstocks of tomato are available for grafting.

Biological control: Microbial nematicides are not yet commercially available. However, some soil microorganisms are natural enemies of nematodes and there are ways to enhance their activity. Nematode-trapping fungi, for instance, can be stimulated by heavy applications of manure or other types of organic matter.

Chemical control: Several fumigant nematicides were developed in the 1950s. Non-fumigant, non-phytotoxic nematicides that can be applied at planting have been developed, but lack of registration for use on vegetable crops prevents their use. A successful fumigation under optimal conditions and at the recommended rate should eliminate 80 to 90 per cent of nematodes to a depth of 25 cm. Activity depends on the presence of water and air in the pore spaces of the soil. Fumigant gas dissolves in water, killing the nematodes. Some of the gas adheres to organic matter and plant residues. The gas moves approximately 1000 times further in air than in water, and faster in warm than in cold soil. In wet soil in which pore spaces are filled with water, the gas dissolves but cannot diffuse; consequently, most of the spaces in the medium are not reached and not all nematodes are killed.

All fumigant nematicides are phytotoxic. It is essential to allow enough time for the gas to disperse from the medium being treated before seeding or transplanting. The aeration process usually requires cultivating or turning the medium once a week for two to three weeks between the time of fumigation and planting. A further disadvantage of fumigants is that they kill beneficial organisms, such as mycorrhizal fungi, *Rhizobium* bacteria, predatory nematodes, and fungi and bacteria that compete with or prey upon the plant pathogenic nematodes.

5

Diseases of Plantation Crops and their Management

HORTICULTURE RESEARCH

India has a wide variety of climate and soil on which a large range of horticultural crops such as, fruits; vegetables, potato and other tropical tuber crops; ornamental, medicinal and aromatic plants; plantation crops; spices, cashew and cocoa are grown. After attaining independence in 1947, major emphasis was laid on achieving self sufficiency in food production. Development of high yielding wheat varieties and high production technologies and their adoption in areas of assured irrigation paved the way towards food security ushering in green revolution in the sixties.

It, however, gradually became clear that horticultural crops for which the Indian topography and agro climates are well suited is an ideal method of achieving sustainability of small holdings, increasing employment, improving environment, providing an enormous export potential and above all achieving nutritional security. As a result, due emphasis on diversification to horticultural crops was given only during the last one decade.

RESEARCH INFRASTRUCTURE

The Indian Council of Agricultural Research is the premier agency which pioneered systematic research on agricultural crops in the country. Horticulture research in India received very little attention till the 3rd Five Year Plan.

The establishment of the Indian Institute of Horticultural Research at Bangalore and starting of eight All India Coordinated Crop Improvement Projects to cover different horticultural crops was a landmark in the history of horticulture in 4th Five Year Plan (1969-74). Rapid expansion of infrastructure took place in 7th and 8th Plans. Today, the horticultural research in the country is being carried out at eight ICAR institutes (with 26 regional stations), 10 National Research Centres (on major crops) and a Project Directorate on Vegetable crops.

Area specific, multi-disciplinary research is also being conducted under 14 All India Coordinated Research Projects each on Tropical, Subtropical and Arid Fruits; Vegetables, Potato, Tuber Crops and Mushrooms; Ornamental Crops, Medicinal and Aromatic crops; Palms, Cashew, Spices and Betel vine; and Post Harvest Technology at 215 centres located at various research Institutes, and State Agricultural Universities.

In addition, four net work projects each on hybrid research in vegetable crops, drip, irrigation in perennial horticultural crops, protected cultivation of ornamental crops and *Phytophthora* diseases of horticulture crops are now in operation. Research on horticulture is also being undertaken at several multi-crop, multi-disciplinary Institutes. Departments of Horticulture in 24 Agricultural Universities, one deemed to be University and one full fledged University of Horticulture and Forestry are also engaged in horticultural research. Besides 280 adhoc schemes supported from Agriculture Produce Cess Fund and a number of foreign-aided projects have also been in operation on specific problems of different horticulture crops. As a result, the country now has a sound research infrastructure in horticulture to meet the growing needs and expectations of the fast developing horticulture industry.

Budgetary Support

The investment in horticulture research by the ICAR in the Central sector has increased significantly in the last two Plans. The Plan allocation for horticultural crops started in 4th Plan (1969-74) with a modest allocation of ₹ 34.78 million and was enhanced to ₹ 319.56 million in the 7th Plan (1985-90) and to ₹ 1047 million in the 8th Plan (1992-97). Non-Plan expenditure also increased from ₹ 73.55 million in the 5th Plan to ₹ 768 million in 8th Plan.

Overall increase in Plan investment in 25 years has been of the order of 2775.21 per cent. The per cent budget allocation for horticulture research out of the total budget for agriculture research rose from 6.1 in 5th to 6.5, 6.67 and 7.7 in 6th, 7th and 8th five year plans, respectively. Similarly, expenditure for Central Sector Schemes of the Department of Agriculture and Cooperation for horticulture crop development also rose tremendously from ₹ 20.5 million (4th Plan) to ₹ 76.18 million (5th Plan), ₹ 146.37 million (6th Plan), ₹ 250 million (7th Plan) and ₹ 10,000 million (8th Five Years Plan).

Table: Budgetary Support (Million Rupees)

Five year plans	*Total for Agriculture Research*	*Share of Horticulture (%)*	*Total Development Support*
IV 1969-74	6105	N.A	20.50
V 1974-78	7292	6.10	76.18
VI 1980-85	10684	6.50	146.37
VII 1985-90	8445	6.67	250.00
VIII 1992-97	15165	7.70	10,000.00

Manpower

Nearly one sixth of the total strength of 5906 scientists working in ICAR is allocated for horticulture research in ICAR Institutes. Besides, approximately, 560 scientists are working in State Agricultural Universities in ICAR funded All India Coordinated Projects. In addition, a large number of scientists are working on horticultural crops in State Agricultural Universities.

Total Scientists in ICAR	5906
Scientists working in the Division of Horticulture	832
Scientists working in other ICAR Institutes	150
% of Total	15
Scientists working in SAUs in ICAR funded projects	560

Research Achievements

Several new crops have been introduced for commercial cultivation, *e.g.*:

- Kiwi fruit in sub-mountain areas of North India
- Olive in mid hills of North Western Himalayas
- Low chilling stone fruits in the North Western plains
- Oilpalm in coastal states of Karnataka, Andhra Pradesh, *etc*.
- Gherkin in south and west India
- Baby corn and sweet corn in certain specific pockets, and
- Broccoli, Brussels' sprouts, asparagus, celery, parsley near the cities.

Crop Improvement

- A large number of high yielding varieties developed in several horticultural crops *e.g.* fruits (76), vegetables (160), potato (29), other tuber crops (24), ornamental crops (300), palms (20), spices (51), cashew (33) and betel vine (1).
- First seedless variety of mango developed.
- 40 F_1 hybrids developed in brinjal, tomato, chillies, cauliflower, carrot, capsicum and muskmelon.
- Self incompatible lines in cauliflower, gynodioecious lines in cucumber and muskmelon, genetic male sterile lines in tomato and temperature tolerant strains of button mushroom developed.

Propagation of Quality Planting Material

- Standardized propagation technique for many fruits hitherto propagated by seed. *e.g., aonla, bael, ber,* black pepper, cardamom, cashew, cassia, cinnamon, clove, custard apple, jack fruit, *jamun,* nutmeg, sapota and walnut.
- Standardized Seed Plot Technique resulting in successful disease free potato seed production in the tropics and sub tropics of the country.

Standardized method of micro-propagation and *in vitro* micro-tuber production in potato.

- Identification of suitable parental lines for production of True Potato Seed and Standardized technology for raising commercial crops.
- Micropropagation protocols developed in banana, black pepper, betel vine, cardamom, ginger and turmeric.
- Production of coconut hybrids through establishment of Seed gardens of Tall (T) x Dwarf (D) and D x T hybrids.
- Standardized rootstocks in citrus, grape and apple.

Agrotechniques

- Standardized high density plantations in banana, citrus, mango and pineapple and high production technology in several crops *e.g.*, pineapple, black pepper and cardamom.
- Year round production technology in tomato and "off season" cultivation of onion and cauliflower developed.
- Standardized use of several plant growth regulators and chemicals now commercially employed in production and quality improvement of horticultural crops *e.g.*, paclobutrazol for induction of flowering in mango; gibberellic acid for improving berry size and quality in grape; Maleic hydrazide for preventing sprouting in onion and potato; Dormex for hastening bud burst in grapes; and Boron and Calcium for changing flower cycle in some cucurbits.

Crop Protection

- Developed improved disease detection techniques such as ELISA and ISEM for improving seed quality and tissue culture technique for rapid multiplication of potato.
- Developed IPM for fruit borer in brinjal, diamond back moth in cabbage, thrips in chillies, phytophthora foot rot in black pepper, "Katte", rhizome rot in cardamom, rhizome rot in ginger, late blight and bacterial wilt in potato.
- Developed biological control measures for mealy bug in grape, fruit borer in tomato and okra.

Post Harvest Management

- Preharvest treatments to control post harvest losses in citrus, mango and grape standardized.
- Chemical treatment for regulation of ripening in mango, sapota and banana standardized.
- Optimum storage temperatures worked out for several fruits, vegetables and tuber crops.

- A mango harvester, fruit peeler, hand and pedal operator cassava chipping machine, harvesting tools (5-14 times efficient); Implements for mechanization of potato cultivation *e.g.,* oscillating tray type potato grader, fertilizer application cum line marker, potato culti-ridger, soil crust breakers, potato digger and automatic potato planter/diggers developed.
- Low cost environment friendly storage system for fruits, vegetables, potato and onion developed.

TISSUE CULTURE TECHNIQUES

SOMACLONAL VARIATIONS

The production of plantlets by callus regeneration, cell suspensions, protoplast cultures, could present some deviations with regard to the mother plant. This is a way to increase the genetic variability. Associated with a selective pressure (stress to toxins, pH, salinity, cold...) it's used to obtain resistant lines. Indeed, after regeneration, plants can express new potentialities rarely obtained another way. Stable and profitable variants are sale another way. Stable and profitable variants are selected and introduced in breeding programmes.

In 1976, a Pelargonium cv Velvet Rose was created by this technique. Later, other applications were described, as in sugarcane, tomato, red pepper... Following Sibi (1994), extrachromosomic systems must be involved to explain the genetic behaviour of some new expressed tissue culture characteristics. She uses the term "epigenic" to evoke a whole of hereditary elements which belongs to the cytoplasmic or nuclear compartments, but are not transmitted by mendelian rules.

Haplodiploidisation

Many species are able to produce haploids through different in vitro techniques. The oldest was anther culture of androgenesis. To date, the results vary considerably from one species to another. Solanae (datura, tobacco, red pepper, eggplant, petunia), cereals (wheat, barley, rice, triticale, maize) or crucifers (soybean, cabbage) are species easy to regenerate by anther culture. To the contrary, tomatoes, leguminous, or Compositae are recalcitrant.

Ovule culture, or gynogenesis, was a successful technique for Gerbera, beet, courgette... However, the most common technique is pollination with irradiated pollen, in order to induce in vivo parthenogenesis. The oospheres developed in embryos without fertilisation are saved by embryo rescue.

The haploid plantlets (n chromosomes) are tree. The haploid plantlets (n chromosomes) are treated with colchicine to produce fertile homozygous lines, called "double haploid lines". This haplodiploidisation technique could give immediately new elite genotypes, hybrid parents after in vitro propagation, as asparagus supermales (MM), or useful genetic material to establish gene mapping. A good synthesis book concerning all these possibilities was published by Bajaj (1990).

PROTOPLAST CULTURE

Protoplasts are the smallest units able to regenerate a whole plant. Therefore protoplasts cultures can serve to enlarge genetic variability by introducing somaclonal variation. However, the main interest of protoplasts, is their capacity to fuse and to produce hybrids or cybrids, new organisms most often unknown in nature. Indeed, protoplasts allow to transgress the barrier of the botanical species or genus. Naked protoplasts can accept without rejection external elements: nuclei, cytoplasmic corganelles, liposomes.... containing genetic information. The first protoplast fusion application was a cytoplasmic transfer from one genotype to another to induce male sterility (CMS) from mitochondrial origin. These male sterile hybrids are interesting to produce F1 hybrids. Another cytoplasmic transfer concerns the introduction of chloroplastic DNA to induce herbicide (atrazine) resistance, from Solanum nigrun duce herbicide (atrazine) resistance, from Solanum nigrun to tomato for instance.

Many intergeneric and interspecific hybridations are possible but not always profitable because these hybrids are too often infertile. This problem does not exist if we can vegetatively propagate the hybrids, as in the case of Citrus rootstocks. In Florida, Grosser (1990) was able to analyse several thousand Citrus hybrids.

CONTRIBUTIONS OF MOLECULAR BIOLOGY IN HORTICULTURE

More and more pathologists are using molecular probes for early detection of diseases. However, the main contributions of modern biotechnology remain probably in the hands of the plant breeders. Indeed, molecular markers are of great use in detecting desirable new genes, or to identify important QTL. This marker-assisted selection is very useful for many vegetable plants. Other applications are DNA - fingerprinting and genetic engineering. This recent technology aims to insert and to express new specific genes into a selected plants.

MARKER-ASSISTED SELECTION

Molecular markers and genome mapping will reach a large extension in the next breeding programmes of tomatoes, peas, cabbages, melons, potatoes...

It will be possible to improve the speed and the incorporation efficiency of desirable new genes by utilizing of closely linked selectable molecular markers genes by utilizing of closely linked selectable molecular markers. The number of backcross will be reduced. Tanksley *et al.* (1981) show that only with 12 markers, one on each tomato chromosome, the composition of the recurrent parent genome is similar to that observed in the absence of selection after the third backcross two years later. Therefore, it's easy to understand the high improvement that we can get with the 700 tomato markers already known in 1990 for this crop.

The utilization of molecular markers could also identify important quantitative trait loci (QTL). Recently Foolad and Chen (1998) have identified 13 RAPD markers at eight genomic regions, that were associated with QTLs affecting salt tolerance during germination in tomato.

DNA Fingerprinting

The DNA polymorphism observed by RFLP, AFLP, or RAPD, allows cultivar identification of fruit trees, apple, citrus... or other vegetatively propagated plants. Although some mutants, as coloured fruits, escape these identification techniques, some people would evaluate the genetic uniformity of regenerated plants in this way.

Genetic Engineering

Genetic engineering consists of introducing a foreign gene into a plant genome to create a new function. Or at the contrary, it could be to reduce or suppress an existing gene function, as in the antisense strategy.

1. *Historic:* In this field, progress were very rapid. In 1974 the pathogenicity of Agrobacterium tumefaciens, due to a large plasmid called Ti, was demonstrated. In 1977, Chilton *et al.* demonstrate a stable incorporation of plasmid Ti-DNA into higher plant cells. In 1983, the team of Van Montagu and Schell in Gent succeeded in producing a transgenic plant, and eleven years later the first transgenic cultivars were available in the market.
2. *Transgenic Plants with New Agronomic Traits:* Two different techniques are commonly used to introduce genetic information (DNA) inside cells, protected by pectocellulosic walls. A particle bombardment process, called biolistic, is used with success for monocots gynmosperm, squash, peas...Today, this technique allows the bombardment of meristems and avoids the difficulties of regeneration. But the mediation of Agrobacterium remains till now the most routinely system to transform dicotyledonous plants.
   The main agronomic traits introduced in horticultural plants and already commercialised are Bt toxin and herbicide resistance. Other studies concern virus resistance, male sterility...
3. *Antisense Strategy:* Calgene created in 1994 the first commercial transgenic plant, a long shelf life tomato, by the suppression of polygalacturonase activity due to an antisense gene. However, t

activity due to an antisense gene. However, this Flavr Savr tomato variety was removed from the trade 3 years later, because of its disease susceptibility and its lack of productivity !

Later, other tomato varieties with long storage qualities were obtained by the utilization of an antisense RNA inhibition of ACC synthase or ACC oxydase, two ethylene precursors.

The antisense technique was also used to reduce the lignification of woody plants, by blocking the enzymes involved in the precursor of lignin biosynthesis. Another interesting application was the induction of white flowers in petunia and in different other ornamental plants by the suppression of chalcone synthase activity.

4. *Transmission of the New Traits:* These traits induced by the transferred DNA are transmitted to the progeny as a dominant Mendelian character. Nevertheless in some cases, transgenic plants with a strong gene expression can generate progenies with only a faint no gene expression. This problem is not clearly elucidate.
5. *Marker, Reporter, Promotor and Expression Genes:* To select the transformed cells, some marker genes, very often resistant to antibiotics or to herbicides, are attached to the coding sequence, as a promoter or an expression gene. Those allow the new gene to express in the whole plant or in a specific plant tissue.

   Some reporter genes are used to follow the evolution one.

   Some reporter genes are used to follow the evolution of the transformed cells. The most frequent reporter genes are the gene "gus" or the gene lux (luciferase).
6. *Horticultural Transgenic Crops already released in USA:* The Flavr Savr tomato was the first genetically engineered whole food approved for commercial sale in 1994. Four other transgenic tomatoes with a delayed ripening were approved later (1995 and 1996).

In 1995, potatoes with Bt genes and a squash cultivar resistant to two viruses were released. Two years later, another squash cultivar resistant to 3 viruses and a papaya line also resistant to viruses were approved. Two other horticultural crops are waiting the authorities approval: red hearted cichory (Radicchio) with male sterility and resistance to herbicide, and a tomato with Bt toxin.

It there are so few commercial transgenic plants in horticulture, it's probably due to the relatively low acreage of these horticultural crops in comparison with other agricultural crops. In 1997, in the world 5 million ha were planted with glyphosate resistant soybeans released by Monsanto alone.

Nowadays, it's true also that many transgenic plants exist in research laboratories, awaiting authorization for field testing, as they are very interesting model plants for learning physiology: lettuce with less nitrate by increasing nitrate reductase gene active with less nitrate by increasing nitrate reductase gene activity, a yeast ribonuclease gene to reduce viroid infection on potato, modification of lignin synthesis in plant to produce trees adapted to paper industry, or timber, or biomases production, regulation genes to control tree architecture....

Situation of Biotechnologies in the World: Development and Risks

Situation in the Young Countries or in Developing Countries: A general use of in vitro micropropagation in the developing countries is now a reality, as it was predicted by Albert Sasson (1993). African countries for instance are producing potatoe, banana, cassava...

Even the production of transgenic plants is no longer a prerogative of the Northern countries. Different experiments are now starting in the South, specially in the international laboratories from CGIAR. Although CGIAR's expenditures on biotechnology for 1997 raise only 24.2 million US$ on an annual budget of 345 million US$ this Consultative Group on International Agricultural Research plays a very important role as contributor to agricultural research for developing countries. The large international research centres IITA (Ibadan, Nigeria), CIP (Lima, Peru), CIAT (Cali, Columbia)... belong to the CGIAR. Their roles and NGO's are primordial. The genetic transformation their roles and NGO's are primordial. The genetic transformation of cassava is an excellent example. Its tuberous roots provide food for over 500 million people, mostly small-scale farmers. Unfortunately, till recently this integral plant for food security in developing countries has been recalcitrant to transformation approaches. However, thanks to the participation of international institutes located in cassava growing countries, CIAT and IITA with the help of four development associations from Swiss, UK, The Netherlands and USA, were able to initiate procotols for cassava transformation and regeneration.

Other networks, like the Cassava Biotechnology Network (CBN) previously described, exist. REDBIO, a technical cooperation network on plant biotechnology, is supported by FAO to promote a best use of scarce manpower, equipment and other resources in Latin America.

On the other hand, more and more private laboratories exist in the developing countries which have a high scientific and technical level (China, Singapore, Taiwan, India, Brasil, Mexico, Chile...). In these countries, European, American or Japanese agrochemical companies are investing. Also these countries will have their chance to develop their own technologies, and to export to the Northern markets.

Meanwhile, genetic engineering techniques are being applied mainly to crops which are important for g applied mainly to crops which are important for the industrialized world, not crops on which the world's hungry depend. Therefore, it is unrealistic when Monsanto writes "the experts said that biotechnological innovation will increase the crop productivity without occupation of new lands, saving tropical forest of best quality and animal habitat". However, the World Bank was promising in a 1997 publication Bioengineering of crops that" transgenic crops could improve food yield by up to 25 per cent in the developing countries and could help to feed an estimated additional three billion people over the next 30 years".

But, not all people are confident in this prospective. To the contrary, some are afraid that the poorest countries, where more than 700 million people are chronically undernourished will miss the biotechnological revolution.

Nevertheless, there is some hope to transfer to the developing countries high technology, which don't undermine the environmental and social network. So, the Agricultural Biotechnology for Sustainable Productivity Project (ABSP project) is managed by Michigan University. Four American universities, two research centres from France and USA, and two American private companies are also involved in this project. The ABSP's objectives are the reduction of losses due to pathogens and the reduction of losses due to pathogens and pests by using transgenic plants (sweet potatoes, potatoes, cucurbits, tomatoes) and by cloning commercial value-plants (bananas, pineapple, coffee).

Situation in the Northern Countries

1. *Mass Propagation:* In 1988, Pierik (1991) predicted a European in vitro propagation of 212 million plantlets/year. He was also predicting an important development of this activity for the future. Following Pierik this quantity accounted for only 5 per cent of the plants able to be propagated by vegetative means. However a few years later Pierik's prediction was not confirmed. The world production was estimated at 600 million. In USA, Zimmerman (1997) was recording a production of 121 million plantlets for a total of 110 laboratories, half produced in 11 laboratories, and 6 labs were propagating each more than 6 million plants per year.
   In Europe, only the Dutch production of micropropagated plants is precisely known. In 1990 the total production was 95.5 million. In 1995, the Dutch in country production reached only 53,7 million, whereas importations increased 77,3 million. (37 million from Poland and 17,1 million from India) (Pierik, data unpublished). These laboratory relocations to low-cost manpower countries are commonly observed, and put the question of the quality management.
2. *Transgenics:* The GMOs invade progressively all the US lands: 1.6 per cent in 1996, 15 per cent in 1997, 50 per cent in 2000..... It will represent an estimated market of 3,9 million US$ in 2003 ! The largestt agrochemical companies are merging to become giants of agrobusiness such as Novartis, Monsanto, Zeneca...

Following Monsanto in 1996, about 950 T insecticides were saved by introduction of resistant cotton, id est a net savings of 81 US$/ha for the grower. Glyphosate-resistant soybeans (4 million ha in 1997) allow a savomgs of 44 to 49 US$/ha. (Information from cultivar, suppl. n 436, Feb. 16, pp. 24-37, 1998).

3. *Perspectives, Limitations and Environmental Risks*: Ecological impact related with the introduction of GMOs is always an open debate.

   The application to agriculture of these new technologies certainly opens interesting perspectives, but also raises potential problems. The risk of crop transgene spreading has been demonstrated. A researcher of Clemson University in South Carolina reported "that in a population of wild strawberries growing within 50 meters of a strawberry field, more than 50 per cent of the wild plants contained marker genes from the cultivated strawberries".

A Danish team have shown a possible rapid spread of genes from oilseed rape to the weedy relative *Brassica campestris*.

There are other risks. *Brassica campestris*.

There are other risks. The introduction of Bt gene allows a drastic reduction in the use of toxic chemicals for crop protection. But a poorly controlled use of Bt-technology can destroy more effectively the predators than the pests.

Or when" many crop plants are transformed with similar effective traits, in such situation, many polyphagous pest species, which by nature are more flexible evolutionarily than those that have a narrower diet, are likely to overcome Bt resistance very quickly".

Therefore, before releasing a transgenic plant, any risk has to be weighed against the benefit of the transgenic crops. We must not forget that annually in the world 500.000 acute pesticide poisonings, with 5000 deaths, are observed.

DURING THE MICROPROPAGATION PROCESS, THE GENETIC STABILIT

During the micropropagation process, the genetic stability of the new shoots is dependent upon their origin. Axillary shoots issue from pre-existing buds and are normally true to type, indeed the meristematic cells are genetically very stable. Adventitious shoots, such as somatic embryos, are neo-formed buds developed directly on some organs, or indirectly through a callus phase formed on this organ.

So, if the mother plant presents a cell mosaic or chimaeric tissues, risks of genetic variation exists. It is similar in the case of an indirect regeneration, when the callus phase is too long.

PROPAGATION BY AXILLARY SHOOTING

This technique has proved to be the most applicable and reliable method of in vitro propagation. Axillary shoot growth is stimulated by overcoming apical meristem dominance. Commercial tissue culture laboratories are now able to propagate a large number of herbaceous ornamental species and several woody plants in this way. However, the propagation of Pelargonium, Howea, and a few other horticultural plants are always difficult to propagate by axillary branching.

Propagation by Direct or Indirect Organogenesis

Adventitious shoots could arise directly from the tissue of explants without callus formation. Several plants of the family gesneriaceae (Saintpaulia, Streptocarpus...) regenerate directly buds on leaf explants, likewise Lilium rnerate directly buds on leaf explants, likewise Lilium regenerates on scales.

However more often, like for Ficus lyrata, adventitious buds appear on callus. While coffee, cocoa trees, and many conifers are produced by somatic embryogenesis developed on callus or cell suspensions.

Constraints of in Vitro Micropropagation

The establishment of axenic culture could be difficult when the explants are coming from hot and humid countries.

Moreover, in the course of micropropagation several authors observed a sudden appearance of endogeneous bacteria. Nevertheless only Holland and Pollaco (1994) have demonstrated the presence of this kind of bacteria in more than 70 different species. For Leifert and Woodward (1997), the major source of contamination is the initial explant. They add that microbial contamination in commercial plant tissue culture laboratories is the most important cause of losses. For this reason, they support the introduction of a microbiological production control strategy.

A second important constraint during the initial phase mainly concerns woody plants. A mature tree must be rejuvenated to recover its morphogenetic competence. Franclet (1981) described various treatments, micrografts, cascade grafts, cytokinin treatments to reverse the adult phase to juvenile. The micropropagation of Acacia Senegal was only possible after rejuvenation by micrografting (Palmnly possible after rejuvenation by micrografting.

A distinction between chronological, ontogenic and physiological age is needed to understand the rejuvenation concept. This was definitively demonstrated after a complete reversion of Sequoia sempervivens.

Improvement of Axillary Branching

The cost of micropropagated plantlets is also an important limitation of the techniques. In New Zealand, where they produce 2-3 million micropropagated radiata pine per annum, the relative cost of micropropagated planting stock had dropped from 13,8 times the cost of seedlings in 1988 to 6,9 by 1993. To reduce manpower costs, several improvements have been proposed. The more simple method was in vitro layering developed by Wang (1977) to clone PVX-free potato plants. The first plantlets placed on the medium in a horizontal position developed axillary shoots. They are harvested by cutting one centimetre above the medium surface, at 3 weeks intervals. A similar technique called 'hedging system ' by Aitken Christie and Jones (1987) was later used to produce Pinus radiata.

Ziv (1990) proposed for corm plants, Gladiolus and Nerine, a very rapid propagation system. She reduces the internodes and leaves by introduction of an anti-gibberellin agent in the medium. Finally, only aggregates of buds are formed, then they finally, only aggregates of buds are formed, then they are divided and introduced in bioreactors for mass production. Similar systems were developed in Gembloux, to propagate carub trees and asparagus. Since 1988, Duhem was producing very large quantities of Eucalyptus plantlets in Petri dishes without anti-gibberellin but in complete darkness. Transfers from one Petri to another is made by a simple squashing.

Somatic Embryogenesis Propagation

For genetically stable species, somatic embryogenesis offers a very fast scaling-up system, especially when it's possible to produce embryos in bioreactors. Unfortunately this production via fermentors was not as simple as first envisaged. And today, only a few model plants are successfully produced by such technology: carrot, celery. Other applications remain at the experimental stage: coffee, oil, palms, conifers, Euphorbia pulcherrima... and several other horticultural species.

Several bottlenecks limit the use of this interesting technology. One of the main problems is genetic stability. So, despite the clonal nature of nucellar embryos, different morphological anomalies can occur among mango somatic embryos, as it was also observed in Citrus plants derived from nucellar cultures.

Another difficulty is the loss of embryogenic capacity over time, a phenomenon observed with different species. It is also a phenomenon observed with different species. It is also important to note that somatic embryogenic lines of conifers are always originated from immature embryos.

Artificial Seeds

Another very interesting possibility of the somatic embryogenesis technology has been developed during the past fifteen years by Redenbaugh and his team (1991, 1993). They were able to encapsulate somatic embryos by hydrogel coatings (sodium alginate), producing single embryo artificial seeds. To date, some improvements offer the possibility to directly plant the artificial seeds in the greenhouse on special substrates (vermiculite, sans...). This methodology will provide in the future a good technique to reduce the cost of transplants.

Gene Bank

Tissue culture methods offer the opportunity for in vitro collecting, rapid multiplication and distribution of important, elite, or rare plants that are threatened with extinction. The two major in vitro storage strategies are slow growth and cryopreservation. Since the first results of Seibert (1976), who was able to initiate shoots from carnation shoot apices frozen to -196C, this technique is now successful for many of horticultural species. Dereuddre *et al.* (1991) have provided a very simple technology to freeze encapsulated meristems in dried alginate beads. It works for pear, strawberry, eucalyptus, potato....

Some Ior Pear, Strawberry, Eucalyptus, Potato

Some international institutions have very large collections of old and current varieties, available for exchange and introduction in crop improvement programmes. The International Potato Centre (CIP) in Lima, Peru, has a large world potato collection. Germplasm of sweet potato and cassava is at the International Institute of Tropical Agriculture (IITA), Ibadan, Nigeria.

The movement of germplasm involves the risks of accidentally introducing plant quarantine pests along with the host plant material. To limit these risks, the plant material should be transferred from one country to another as in vitro cultures through a transit Centre, where it should be indexed. For bananas, in the framework of INIBAP, the transit Centre is the Catholic University of Leuven in Belgium where a very large in vitro germplasm exists.

RESEARCH PREPAREDNESS FOR ACCELERATED GROWTH OF HORTICULTURE

The horticulture scenario of the country is rapidly changing. The production and productivity of horticultural crops have increased manifold. Production of fruits and vegetables has tripled in the last 50 years. The productivity has gone up by three times in banana and by 2.5 times in potato. Today horticultural crops cover about 25 per cent of total agricultural exports of the country. The corporate sector is also showing greater interest in horticulture. A major shift in consumption pattern of fresh and processed fruits and vegetables is expected in the coming century. There will be greater technology adoption both in traditional horticultural enterprise as well as in commercial horticulture sectors. Diversification and value addition will be the key words in the Indian horticulture in the 21st Century.

Horticulture research in India is about four decades old. Systematic research on fruit, vegetable and ornamental crops began in 1954 with the initiation of independent institutions and programmes. The research agenda is designed relevant to national plans and priorities for the horticulture development. Today, eight ICAR institutes with 27 regional stations, 1 project directorate, 10 national research centres, 16 all India coordinated research projects (AICRPS) with 223 research stations, 1 full-fledged university of horticulture, 25 state agricultural universities and 7 multi-disciplinary institutes of the ICAR are engaged in horticulture research. In addition, a few R&D establishments of crop/commodity boards and private sectors are providing research support to Indian horticulture. Research system in horticulture is now geared to provide necessary technological support to the expanding horticultural industry. The research efforts in the past were mainly concentrated on crop improvement, propagation of seed/planting material, agrotechniques, crop protection and post harvest management.

VARIETAL DEVELOPMENT

Among the fruit crops, improved high yielding mango varieties, Mallika, Amrapali, Ratna, Sindhu, Arka Aruna, Arka Puneet, Dashehari-51 and hybrids namely CISH-M-1 and CISH-M-2 have been developed. Mallika is coming up in southern states like Karnataka and Amrapali is performing well in eastern India. Dashehari-51, a regular bearing cultivar with about 38 per cent higher productivity than the normal Dashehari, has been identified after 14 years of rigorous selection. In guava, three selections, namely Lalit, CISH-G-1 and CISH-G-2 have been developed for domestic and export markets.

The fruits of Lalit are of medium size weighing about 150 g each and suitable for both table and processing purposes. In banana, high yielding hybrids like FHIA-01 and FHIA-03 are promising for replacing varieties, Panchananda and Bluggoe, respectively. Cultivar, Saba is found promising under sodic soils. In addition, high yielding varieties like Col, Hl and H2 have been developed. In grape, superior and high yielding varieties have been developed *e.g.* Beauty Seedless and Pusa Seedless and early ripening variety, Perlette for cultivation under North Indian conditions and Anab-e-Shahi, Dilkhush, Thompson Seedless, Tas-A-Ganesh, Sonaka, Bangalore Blue and Pachadraksha for south Indian conditions.

In citrus, high yielding and cluster bearing varieties of acid lime have been developed. *e.g.*, Rough Lemon, Rangpur Lime, Pramaini, Vikram PKM-1. Trifoliate oranges, namely Flying Dragon and Rich 16-6 are dwarfing types. In papaya, high yielding superior varieties both for table purpose and papain production have been developed. *e.g.*, Co-1 to Co-7, Coorg Honey Dew, Pusa Delicious, Pusa Majesty, Pusa Giant and Pusa Nanha. In apple, superior hybrids have been developed. *e.g.*, Lal Ambari, Sunehari. Red Spur, Star Crimson, Golden Spur, Red Chief, Oregon Spur, Skyline Supreme and Vance Delicious have been identified. Tissue culture protocols for micro-propagation of two commercial varieties have been developed.

Among the vegetable crops, more than 130 open pollinated varieties, 36 hybrids, 3 synthetics and 29 resistant varieties of 20 vegetable crops have been developed and released for cultivation in different agro-climatic regions. These include 40 in tomato, 45 in brinjal, 13 in cauliflower, 12 in chillies, 20 in pea, 9 in musk melon, 16 in onion and 44 in other crops.

In potato, 33 high yielding varieties have been developed indigenously for large scale cultivation in different regions. Kufri Ashoka and Kufri Pukhraj mature in 75 days. Kufri Jawahar and Kufri Satluj are field resistant to late blight. Kufri Jawahar has most ideal plant type for inter-cropping. Kufri Chipsona-1 and Kufri Chipsona-2 have been developed with excellent processing attributes, comparable to exotic varieties. Kufri Swarna resistant to golden nematode is ideal for Nilgiri Hills.

In tuber crops, improved varieties of different tuber crops have been recommended/released for commercial cultivation. These includes 9 varieties of cassava, 15 varieties of sweet potato, 6 varieties of colocasia, 3 varieties each of greater yam and lesser yam, 1 each of *Amorphophallus,* taro and yambean. Cassava varieties, Sree Visakham and Sree Prakash are popular in Kerala. Triploid clone, Sree Harsha with high dry matter and starch content is suitable for industrial belt of Tamil Nadu. Two early maturing varieties Sree Jaya and Sree Vijya have been released for culinary purposes. Elephant foot yam variety, Am-15 has been released with high yield potential of 41 t/ha.

Among the plantation and spice crops, India is the first country to exploit hybrid vigour in coconut. Twelve hybrids involving tall and dwarf parents and 4 varieties have been released for commercial cultivation. These varieties yield 21 to 89 per cent more than the local cultivars. Some of the released varieties like Chandra Kalpa and Pratap (Banawali Green Round) are receiving wide acceptance of farmers. Chowghat Green Dwarf variety is good for tender coconut purpose. In arecanut, 4 high yielding varieties, namely, Mangala, Sumangala, Sreemangala and Mohitnagar have been developed, giving about 30 per cent higher yield than the local cultivars.

In oil palm, first efforts for improvement were made by producing Tenera hybrids using Pisifera pollen imported from Nigeria. Dura x Pisifera hybrids are field tested in East and West Godavari districts, Khammam and Krishna districts of Andhra Pradesh, with yield potential of 20-25 tonnes/ha of FFB from the fifth year. In cashew, 22 region specific selections and 12 hybrids with yield potential of 1.5-2 tonnes of raw nuts/ha have been produced and released for commercial cultivation.

The present standards fixed for cashew varieties include export grade kernels of W-210 to W-240 and at least one tonne per ha yield with 30 per cent shelling. In black pepper, 6 varieties, namely, Sreekara, Subhakara, Palode-2, Panniyur-2, Panniyur-4, Panchami and Pournami, and 2 hybrids *viz.* Panniyur-1 and Panniyur-4 have been developed. In cardamom, a number of improved varieties have been developed and released for commercial cultivation *e.g.,* CCS-1, Mudigere-1, PV-1, ICRI-1 and ICRI-2. In ginger, varieties like Suprabha, Suruchi, Suravi and Varada have been developed. In turmeric, several varieties *viz.*, Co-1, Krishna, Sugandham, BSR-1, Suvarna, Roma, Suroma, Rajendra Sonia, Sugana, Sudarshana, Ranga, Rasmi, Prabha, Prathiba, Mega Turmeric and RCT-1 have been developed with yield potential of up to 44 tonnes of fresh rhizomes per ha. Three high yielding cinnamon lines, namely, Navashree, Nithyashree and Konkan Tej have been released for cultivation.

Agrotechniques

In fruit crops, improved agrotechniques developed have helped the farmers in improving the productivity and quality of produce. Soil application of paclobutrazol (4 g/tree) increase flowering and fruiting in mango on commercial scale in coastal Maharashtra. It also controls irregular bearing in cultivar Dashehari. Spray of NAA @ 200 ppm in October is recommended for control of malformation. Heavy fruit drop at maturity in cultivar Langra can be controlled by spraying NAA (20 ppm). In guava, double spray of 10 and 20 per cent urea on cultivars, Allahabad Safeda and Sardar twice at bloom eliminate poor quality rainy season crop and increases winter season yield by 3 and 4 times, respectively. Application of neem coated urea (800 g/plants) yields 98 kg fruits/plant in variety, Sardar compared to 37 kg from untreated plants. In banana, high density planting (4550 plants/ha) yield up to 174 t/ha.

Adoption of improved technology in Maharashtra has resulted in fruit yield increase up to 52 t/ha. In citrus, two grafting methods using inverted 'T' cut and apical triangle cut have been developed with overall success of around 36 per cent using either method. For accelerating the survival of growth, shoot tip grafts, the successful grafts are double grafted (side grafted) on vigorous greenhouse grown Rough Lemon and Rangpur Lime seedlings. Rangpur Lime rootstock is found superior for sweet oranges and mandarins. In papaya, closer spacing of 1.4 x 1.4 m is recommended for high yield. Drip irrigation techniques have been standardized. In banana, it has resulted in production gain (60-70 per cent) and early harvesting (40-50 days), besides improved water efficiency. Likewise, in grapes higher yields have been obtained with better water use efficiency (11 per cent).

In vegetable crops, improved production technology has been developed for major crops. Drip irrigation is economical in tomato and brinjal. In cucumber, replenishment of evaporation loss through irrigation resulted in maximization of yield of quality fruits. Drip irrigation in watermelon provided 33 per cent higher yield with a water saving of 40 per cent. Nutrient requirements and fertilizer schedules have been worked out crop-wise and recommended for different agro-climatic regions. In leguminous vegetables, high N depresses nodulation. The VAM fungi increases P availability to plants.

In all leguminous vegetables, inoculation of the VAM fungi along with *Rhizobium* culture is beneficial. Production technologies for *kharif* onion in northern India and long day type onions for high altitudes have been standardized. Pendimethalin (Stomp) has been found effective in controlling weeds in tomato, brinjal, chilli, bell pepper and okra. In potato, a number of potato-based multiple and inter cropping systems have been developed for different potato growing regions. Intercrop combinations with sugarcane in Maharashtra, wheat in Chhota Nagpur area and linseed in central Uttar Pradesh are found remunerative. A suitable method of urea application has also been worked out.

In tuber crops, short duration legumes *viz.*, groundnut and French bean and cowpea can be successfully inter-cropped with cassava. Short-duration cassava, Sree Prakash is ideal in double cropped rice fields. Studies on cassava-based multiple cropping systems involving banana, coconut, *Leucaena* and Eucalyptus, have shown banana-cassava combination to give maximum root yield. Banana and coconut combination reduces soil loss and surface run-off considerably. *Dioscorea* and elephant foot yam with banana, Nendran can generate an additional income of ₹ 20,000/- over the sole crop of banana. For Inter-cropping *Dioscorea* in coconut garden, the ideal planting density is 9000 plants/ha. When *Amorphophallus* is raised as an inter-crop in coconut garden, one third dose of recommended fertilizers is sufficient. Inoculation with VAM fungi in cassava give about 15-20 per cent increase in yield.

Among the plantation and spice crops, density of 175 coconut palms/ha (7.5 m x 7.5 m spacing) is found ideal. In general, NPK application of 500:320:1200 g/palm/year is found optimum. A multi-storied cropping system involving black pepper trained on coconut trees, and cocoa in between the rows of coconut and pineapple in the ground floor has been found ideal for exploiting light, soil and air spaces. In arecanut, application of NPK (100:40:140 g) and green leaf (14 kg) per palm per year is recommended for coastal regions of Kerala and Karnataka and for plains of West Bengal, Karnataka and Assam.

In oil palm, application of NPK (1200:600:1200 g/palm) is found to give 17.1 tonne of FFB per ha. In black pepper, rapid methods for production of rooted cuttings of pepper have been developed and a commercial protocol has been standardised for micropropagation of black pepper. Ginger yield could be increased up to 33 per cent by application of neem cake at the rate of two tonnes per ha and the fertilizer schedule of 75 kg each of N, P_20_5 and K_20. Technology for storage of ginger seed rhizome is standardised and recommended.

Protection Technologies

For major fruit crops, plant protection schedules have been developed for the control of significant insect-pests for wider adoption. In the recent years, research efforts are directed to devise eco-friendly, economical and long lasting control measures. Success has been achieved in biological control of mealy bugs in mango and guava. The *Beauveria bassiana* has been found killing mango mealy bug and hopper. In grapes, integrated management of *Spodoptera* caterpillar involving light and pheromone traps, NPV and neem based insecticides and biological control of mealy bug by the beetle *Cryptolaemus montrouzieri* have been standardized. Studies on pesticide residues have resulted in working out of safe-waiting periods for harvesting and consumption of fruits.

In vegetable crops, about 50 improved measures for efficient management of diseases and 23 for insect-pests have been worked out and popularized in different agro-climatic regions in the country. Integrated pest management (IPM) for controlling diamond back moth on cabbage through a trap crop like mustard has been demonstrated. Fruit borer *(H. armigera)* on tomato can be controlled by the release of *Trichogramma pretiosum* alone and in combination with HaNPV. In potato, integrated management schedules for control of bacterial wilt and tuber moth have been developed. A late blight forecasting system has been developed for the hills.

Among the tuber crops, the major diseases affecting tuber crops are cassava mosaic and brown leaf spot in cassava, *Phytophthora* leaf blight in colocasia, *Fusarium* wilt in elephant yam and virus diseases of sweet potato.

Foliar sprays of Bavistin (0.1 per cent) combined with disodium and dipotassium phosphates (100 ppm) and calcium sulphate at 15-day interval was found to check foliar diseases in sweet potato. The major pests include spider mites, scale insects and white fly on cassava, weevil on sweet potato, defoliators, aphids and mites on colocasia, and scales and mealy bugs on yams and elephant yam. Cultural methods for weevil control include clean cultivation, destruction of alternate hosts and timely harvest. An effective IPM package using synthetic sex pheromone has been developed. Control measures involving insecticides have been evolved for the control of pests of other tuber crops.

Among the plantation and spice crops, bud rot of coconut caused by *Phytophthora palmivora* can be effectively controlled by spraying Bordeaux mixture. Calyxin root feeding and drenching of soil with 1 per cent Bordeaux mixture along with neem cake application @ 5 kg per palm per year is recommended for controlling Thanjavur wilt disease reported in Tamil Nadu, Andhra Pradesh and Karnataka.

A package of practices has been developed for managing mycoplasma like organisms (MLOs) in root wilt affected coconut palms in Kerala and Thatipaka disease affected palms in Andhra Pradesh.

Eradication of all root wilt affected palms is recommended. In cashew, tea mosquito bug (TMB) can be effectively controlled through a schedule of spray coinciding with flushing, flowering and fruiting. For effective control of stem and root borer infestation, constant monitoring and adoption of strict sanitation in the plantations coupled with prophylactic application of coal tar and kerosene in the ratio of 1:2 on trunks are recommended. In black pepper, spraying Bordeaux mixture (1 per cent) and drenching the soil with copper oxychloride (0.2 per cent) is found effective in managing *Phytophthora* foot rot.

Seed/planting Material Propagation

In most of the fruit crops, vegetative propagation techniques have been standardized. Soft wood grafting has been standardized for mango, sapota, custard-apple and jackfruit. Other vegetative propagation techniques have been developed for *ber, aonla,* jackfruit, custard-apple and *bael.* In mango, veneer grafting and stone-grafting is practised commercially. Mango variety, Vellaikolumban is suitable semi-dwarfing rootstock for Alphonso. Old unproductive mango trees can be rejuvenated successfully by pruning the 4th order branches during December-January. Flowering and fruiting are regular in pruned trees. For mandarin orange, Rangpur lime is a drought hardy rootstock. In grapes, Dogridge and Salt Creek (Ramsey) are suitable for minimizing adverse effects of soil salinity on Thompson Seedless.

A tissue culture technique for mass multiplication of Dogridge has been standardised. In banana, sword suckers of 700-1000 g are optimum. Rhizomes with active lateral buds and dead central buds are preferred for distant transportation in western India. Double paring and shade drying followed by dipping in Monocrotophos (0.5 per cent) and Bavistin (0.2 per cent) is recommended to disinfect nematodes and soil borne fungi. Tissue cultured banana plants are now commercially adopted for their uniformity in flowering and produce. Shoot-tip grafting technique in citrus has been considerably advanced.

In vegetable crops, seed production of over 120 open pollinated high yielding varieties of different vegetables has been well established in the country. Hybrid seed production has become easier with the development of male sterile lines in tomato, self incompatible lines in cauliflower and gynoecious lines in cucumber and muskmelon. In brinjal, functional male sterility controlled by a single recessive gene has been identified.

Temperature barrier in cole crops (cabbage and cauliflower) has been overcome by developing heat tolerant hybrids. It is now possible to cultivate cabbage and cauliflower in southern India. Development of tomato varieties resistant to bacterial wilt has made their cultivation successful in non-traditional areas. Onion seed production technology for cultivation in kharif season has been developed for north Indian states especially, Haryana, Punjab and western Uttar Pradesh. Seed Plot Technique has been developed for production of disease-free potato seed in plains. It is widely adopted by farmers. A new technology for raising commercial crop of potato using 'True Potato Seed' (TPS) has been developed and standardized as supplementary technology to the traditional tuber grown crop. Two TPS populations, TPS-C-3 and HPS-113 are recommended for commercial production in Bihar, Gujarat, Tripura and West Bengal. Micropropagation protocols have been developed in banana, oil palm, cashew, black pepper, ginger, *etc*. Seed gardens of Tall (T) x Dwarf (D) and D x T hybrids have been established for production of coconut hybrids.

Post Harvest Management Technologies

The post harvest handling of fruits and vegetables accounts for 20-30 per cent of losses at different stages of storage, grading, packing, transport and finally at marketing as a fresh produce or in processed form. A number of improved technologies have been developed for commercial exploitation. An on-farm, low cost, environment friendly cool chamber, Zero Energy Cool Chamber has been developed using locally available material.

The principle of evaporative cooling reduces the inside temperature by as much as 17-18 °C and keeps the relative humidity above 90 per cent during peak summer. The chamber increases the shelf life and reduces PLW of banana, mango, orange lime, grape fruit, tomato and potato in different situations in India. Maturity standards for mango, guava, grape, litchi and *ber* and chemical treatments for regulation of ripening in mango, sapota and banana have been standardised.

Optimum storage temperatures worked out for several fruits, vegetables and tuber crops. A mango harvester, fruit peeler, hand and pedal operator cassava chipping machines, harvesting tools (5-14 times efficient), coconut dehusking machine, implements for mechanization of potato cultivation and other crops have been developed. A number of improved technologies have been developed for commercial exploitation *viz.*, tent type foldable solar dryer, packaging boxes for distant transportation of apple, mango, citrus and plum, production of value added products-pectin from peel and flour from mango fruit kernel, production of fruit post carbonated beverages *etc*.

Production Constraints

In spite of great strides made, the productivity of horticultural crops, in general, is still quite low and the post harvest losses particularly of perishable commodities, are considerable. Improvement in quality standards of the produce and their marketing are essential to increase our share in the global market. The research agenda in horticulture is by design relevant to national plans and priorities and research programmes are normally formulated keeping in view the thrust areas in development. The major technology related constraints contributing to low productivity of horticultural crops and inferior quality of produce are:

- Vast majority of holdings are small and un-irrigated.
- Large tracts of low and unproductive plantations needing replacement/ rejuvenation.
- Low productivity of crops due to inferior genetic stocks and poor management.
- Inadequate supply of quality planting materials of improved varieties.
- High incidence of pests and diseases.
- Heavy post harvest losses and low utilization in processing sector.

For addressing the above constraints, research institutions are engaged in both basic and applied research. While formulating research strategies some of the inherent weaknesses associated with perennial tree crops and certain perpetual problems in Indian horticulture must be kept in mind. They are:

- Long period required for development of improved genotypes. Application of biotechnological tools/methods in horticultural crops is still in its early stage of development in the country.
- Chronic production problems due to major disorders like alternate bearing, malformation and spongy tissue in mango, guava wilt, citrus decline, root wilt in coconut, viral disease in vegetables, *Phytophthora* diseases in large number of crops *etc*. remained largely unresolved.
- Lack of advanced technologies for post harvest handling, processing and marketing of produce.

Losses caused by biotic stresses are very high and due to pesticide residue problems development of eco-friendly IPM strategy is more relevant in horticulture. There is a threat for loss of valuable genetic resources, if measures are not taken for their conservation.

Wastelands and hilly terrains being the potential future expansion areas, matching technologies for dry land and hill horticulture need to be developed. Counter seasonal advantages from diverse agro-climatic situations provide strength for extended availability of horticultural crops round the year and such potentials can be harnessed only with relevant research support.

PLANT TISSUE SYSTEMS

STRUCTURE AND FUNCTION

Although the architecture varies considerably among roots, stems, and leaves, all plant organs are comprised of 3 basic tissue types: dermal tissue, vascular tissue, and ground tissue. A tissue is simply an assemblage of cells that work together as a structural or functional unit. In this lab, we will examine the general arrangements of those tissues among different plant organs. You will find that the distribution of tissues varies somewhat among different plants as well as among different organ systems. For example, one of the key differences between monocots and dicots has to do with the arrangement of vascular tissues in leaves and stems. In next week's lab, we will continue our examination of plant tissues by exploring some of the different cell types that form these three plant tissue sytems.

DERMAL TISSUE

The dermal tissue system is derived from the protoderm and consists of 2 tissue types*:

- *Periderm*: Periderm includes cork and cork-producing tissues and is only found in plants with secondary growth.

- *Epidermis*: The epidermis constitutes the outmost layer of cells of roots, stems and leaves. Epidermal cells are usually covered by a waxy layer, called a cuticle that serves for waterproofing the cells.

VASCULAR TISSUE

Vascular tissues, derived from the procambium*, are the tissues responsible for transporting substances from one point in the plant body to another. As is the case with animals, the vascular tissues are responsible for the movement of dissolved gases, sugars, and minerals throughout the organism.

In plants, there are 2 types of conducting tissues:

- *Xylem*: Xylem is the tissue specialized for the conduction water and dissolved minerals. Cells of the xylem are dead at maturity and act as the structural medium for the movement of water. Individual xylem cells have multiple holes in the walls to allow for the passage of water between adjacent cells. The bulk of wood consists of xylem tissue.
- *Phloem*: Phloem conducts dissolved sugars throughout the plant. The main conducting cells of phloem, called"sieve elements", are alive at maturity, but are highly modified for carrying out the function of transport. For example, many of the organelles you might expect to find in a living plant cell, such as nuclei, ribosomes, Golgi apparati, cytoskeleton, and vacuoles are absent from the sieve elements in order to make more space for fluid transport. The small amount of protoplast in sieve elements is distributed in a thin layer along the periphery of the cell wall. In some cases, you may be able to observe smaller cells called"companion cells" distributed among the sieve elements. Companion cells contain all the organelles you would expect to find in a typical plant cell and carry out many of the necessary life functions for the sieve elements which lack organelles.

GROUND TISSUE

Derived from the ground meristem*, ground tissues are distributed between and among the dermal and vascular tissues of roots, stems, and leaves.

Ground tissues include a wide variety of tissue types that perform a multitude of functions within the plant.

Some of the more familiar ground tissues include the following:

- *Cortex*: the ground tissue distributed between the epidermis and vascular tissues in roots and stems. Cortex is usually modified for storage, especially of starches. In herbaceous green stems, cortex often contains chloroplasts and thus is further modified for carrying out photosynthesis.
- *Pith*: like cortex, pith is generally modified for starch storage, but is only found to the interior of the vascular tissues in stems and roots.

- *Mesophyll*: the ground tissue of leaves found between the two layers of epidermis. Leaf mesophyll is the tissue most adapted for photosynthesis and thus is generally rich with chloroplasts.

In addition to the functions indicated above, ground tissues also have an important role in wound healing and the regeneration of damaged or dead cells throughout the plant.

PLANT TISSUES ARE DERIVED FROM MERISTEMS

Unlike animals, most plants grow (increase the number and size of their cells) throughout their lifetime. This ability to grow indeterminately is due to a special form of tissue known as a"meristem".

A meristem is a region of tissue that is perpetually embryonic - that is, it is relatively undifferentiated and divides continuously, giving rise to all new cells and tissues that make up the plant body. Can you think of an analogy to the plant's meristem in animals/humans? There are several meristems distributed throughout the plant body which are named for their location and/ or for the tissue systems that are derived from them.

APICAL (PRIMARY) MERISTEMS

Are located in the tips (or, apex) of the growing roots and shoots of all plants. Primary meristems produce primary tissues and account for all the primary growth of plants. Primary growth is growth that contributes to elongation of the plant (height and root lengthening) and for the establishment of its basic organ structure (roots, stems, leaves).

LATERAL (SECONDARY) MERISTEMS

are organized as concentric cylinders (think of a can inside a can inside a can) within the plant body and are responsible for secondary growth. Secondary growth contributes to the thickening of a plant structure, or increases in girth. With few exceptions, lateral meristems tend to be present only in woody plants, such as trees, shrubs, and some vines.

As cells in the apical meristem divide and give rise to new cells, the tips of roots and shoots extend outward, away from the central plant body. As a consequence of this elongation due to the addition of new cells, the slightly older meristematic cells become distributed into different regions within the root and shoot. These slightly more mature cells still retain their meristematic properties, but are somewhat differentiated from the true apical meristem. These regions of meristematic tissue derived from the apical meristem are referred to as"primary meristems" and are named for the tissues that they will give rise to:

- *Protoderm*: Gives rise to dermal tissues,
- *Ground Meristem*: Gives rise to ground tissues, and
- *Procambium*: Gives rise to vascular tissue (aka,"cambial tissue").

The organization of primary meristems also varies among plant types and organs. Like other organisms, the cells in a plant are grouped together into various tissues. These tissues can be simple, consisting of a single cell type, or complex, consisting of more than one cell type.

DERMAL TISSUE SYSTEM

The dermal tissue system consists of the epidermis and the periderm. The epidermis is a single layer of closely packed cells. It both covers and protects the plant. It can be thought of as the plant's"skin." Depending on the part of the plant that it covers, the dermal tissue system can be specialized to a certain extent. For instance, the epidermis of a plant's leaves secretes a coating called the cuticle that helps the plant retain water.

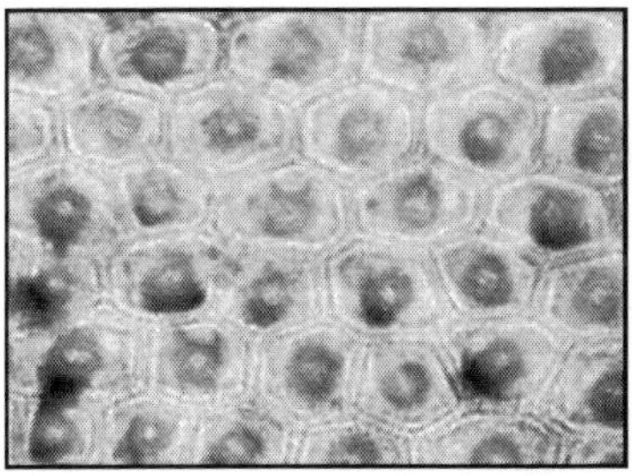

EPIDERMIS CELLS

The periderm, also called bark, replaces the epidermis in plants that undergo secondary growth. The periderm consists of cork cells and protects the plant from pathogens, prevents excessive water loss and provides insulation for the plant. Xylem and phloem throughout the plant make up the vascular tissue system. It allows water and other nutrients to be transported throughout the plant.

GROUND TISSUE SYSTEM

The ground tissue system synthesizes organic compounds, supports the plant and provides storage for the plant. It is mostly made up of parenchyma cells but can also include some collenchyma and sclerenchyma cells as well.

PLANT TISSUE CULTURE

Practically any plant transformation experiment relies at some point on tissue culture. There are some exceptions to this generalization, but the ability to regenerate plants from isolated cells or tissues *in vitro* underpins most plant transformation systems.

PLASTICITY AND TOTIPOTENCY

Two concepts, plasticity and totipotency, are central to understanding plant cell culture and regeneration. Plants, due to their long life span, have developed a greater ability to endure extreme conditions and predation than have animals. Many of the processes involved in plant growth and development adapt to environmental conditions. This plasticity allows plants to alter their metabolism, growth and development to best suit their environment.

Particularly important aspects of this adaptation, as far as plant tissue culture and regeneration are concerned, are the abilities to initiate cell division from almost any tissue of the plant and to regenerate lost organs or undergo different developmental pathways in response to particular stimuli. When plant cells and tissues are cultured *in vitro* they generally exhibit a very high degree of plasticity, which allows one type of tissue or organ to be initiated from another type. In this way, whole plants can be subsequently regenerated. This regeneration of whole organisms depends upon the concept that all plant cells can, given the correct stimuli, express the total genetic potential of the parent plant. This maintenance of genetic potential is called 'totipotency'. Plant cell culture and regeneration do, in fact, provide the most compelling evidence for totipotency.

The Culture Environment

When cultured *in vitro*, all the needs, both chemical and physical, of the plant cells have to met by the culture vessel, the growth medium and the external environment (light, temperature, etc.). The growth medium has to supply all the essential mineral ions required for growth and development. In many cases (as the biosynthetic capability of cells cultured *in vitro* may not replicate that of the parent plant), it must also supply additional organic supplements such as amino acids and vitamins.

Many plant cell cultures, as they are not photosynthetic, also require the addition of a fixed carbon source in the form of a sugar (most often sucrose). One other vital component that must also be supplied is water, the principal biological solvent. Physical factors, such as temperature, pH, the gaseous environment, light (quality and duration) and osmotic pressure, also have to be maintained within acceptable limits.

Plant Cell Culture Media

Culture media used for the *in vitro* cultivation of plant cells are composed of three basic components:

(1) essential elements, or mineral ions, supplied as a complex mixture of salts;
(2) an organic supplement supplying vitamins and/or amino acids; and
(3) a source of fixed carbon; usually supplied as the sugar sucrose.

For practical purposes, the essential elements are further divided into the following categories:

(1) macroelements (or macronutrients);

(2) microelements (or micronutrients); and
(3) an iron source.

Media Components

It is useful to briefly consider some of the individual components of the stock solutions.

Macroelements

As is implied by the name, the stock solution supplies those elements required in large amounts for plant growth and development. Nitrogen, phosphorus, potassium, magnesium, calcium and sulphur (and carbon, which is added separately) are usually regarded as macroelements. These elements usually comprise at least 0.1 per cent of the dry weight of plants.

Nitrogen is most commonly supplied as a mixture of nitrate ions (from the KNO_3) and ammonium ions (from the NH_4NO_3). Theoretically, there is an advantage in supplying nitrogen in the form of ammonium ions, as nitrogen must be in the reduced form to be incorporated into macromolecules. Nitrate ions therefore need to be reduced before incorporation.

However, at high concentrations, ammonium ions can be toxic to plant cell cultures and uptake of ammonium ions from the medium causes acidification of the medium.

In order to use ammonium ions as the sole nitrogen source, the medium needs to be buffered. High concentrations of ammonium ions can also cause culture problems by increasing the frequency of vitrification (the culture appears pale and 'glassy' and is usually unsuitable for further culture). Using a mixture of nitrate and ammonium ions has the advantage of weakly buffering the medium as the uptake of nitrate ions causes OH^- ions to be excreted.

Phosphorus is usually supplied as the phosphate ion of ammonium, sodium or potassium salts. High concentrations of phosphate can lead to the precipitation of medium elements as insoluble phosphates.

Microelements

These elements are required in trace amounts for plant growth and development, and have many and diverse roles. Manganese, iodine, copper, cobalt, boron, molybdenum, iron and zinc usually comprise the microelements, although other elements such as nickel and aluminium are frequently found in some formulations.

Iron is usually added as iron sulphate, although iron citrate can also be used. Ethylene Diamine Tetra acetic Acid (EDTA) is usually used in conjunction with the iron sulphate. The EDTA complexes with the iron so as to allow the slow and continuous release of iron into the medium. Uncomplexed iron can precipitate out of the medium as ferric oxide.

Organic Supplements

Only two vitamins, thiamine (vitamin B1) and myo-inositol (considered a B vitamin) are considered essential for the culture of plant cells *in vitro*. However, other vitamins are also added to plant cell culture media.

Amino acids are also commonly included in the organic supplement. The most frequently used is glycine (arginine, asparagine, aspartic acid, alanine, glutamic acid, glutamine and proline are also used), but in many cases its inclusion is not essential.

Amino acids provide a source of reduced nitrogen and, like ammonium ions, uptake causes acidification of the medium. Casein hydrolysate can be used as a relatively cheap source of a mix of amino acids.

Carbon Source

Sucrose is cheap, easily available, readily assimilated and relatively stable and is therefore the most commonly used carbon source. Other carbohydrates (such as glucose, maltose, galactose and sorbitol) can also be used, and in specialized circumstances may prove superior to sucrose.

Gelling Agents

Media for plant cell culture *in vitro* can be used in either liquid or 'solid' forms, depending on the type of culture being grown. For any culture types that require the plant cells or tissues to be grown on the surface of the medium, it must be solidified (more correctly termed 'gelled').

Agar, produced from seaweed, is the most common type of gelling agent, and is ideal for routine applications.

However, because it is a natural product, the agar quality can vary from supplier to supplier and from batch to batch. For more demanding applications, a range of purer (and in some cases, considerably more expensive) gelling agents are available.

These components, then, are the basic 'chemical' necessities for plant cell culture media. However, other additions are made in order to manipulate the pattern of growth and development of the plant cell culture.

PLANT GROWTH REGULATORS

We have already briefly considered the concepts of plasticity and totipotency. The essential point as far as plant cell culture is concerned is that, due to this plasticity and totipotency, specific media manipulations can be used to direct the development of plant cells in culture.

Plant growth regulators are the critical media components in determining the developmental pathway of the plant cells.

The plant growth regulators used most commonly are plant hormones or their synthetic analogues.

Classes of Plant Growth Regulators

There are five main classes of plant growth regulator used in plant cell culture, namely:

1. Auxins;
2. Cytokinins;
3. Gibberellins;
4. Abscisic acid;
5. Ethylene.

Each class of plant growth regulator will be briefly looked at.

Auxins

Auxins promote both cell division and cell growth The most important naturally occurring auxin is IAA (indole-3-acetic acid), but its use in plant cell culture media is limited because it is unstable to both heat and light. Occasionally, amino acid conjugates of IAA (such as indole-acetyl-L-alanine and indole-acetyl-L-glycine), which are more stable, are used to partially alleviate the problems associated with the use of IAA. It is more common, though, to use stable chemical analogues of IAA as a source of auxin in plant cell culture media. 2,4-Dichlorophenoxyacetic acid (2,4-D) is the most commonly used auxin and is extremely effective in most circumstances. Other auxins are available, and some may be more effective or 'potent' than 2,4-D in some instances.

CYTOKININS

Cytokinins promote cell division. Naturally occurring cytokinins are a large group of structurally related (they are purine derivatives) compounds. Of the naturally occurring cytokinins, two have some use in plant tissue culture media. These are zeatin and 2iP (2-isopentyl adenine). Their use is not widespread as they are expensive (particularly zeatin) and relatively unstable. The synthetic analogues, kinetin and BAP (benzylaminopurine), are therefore used more frequently. Non-purine-based chemicals, such as substituted phenylureas, are also used as cytokinins in plant cell culture media. These substituted phenylureas can also substitute for auxin in some culture systems.

Gibberellins

There are numerous, naturally occurring, structurally related compounds termed 'gibberellins'. They are involved in regulating cell elongation, and are agronomically important in determining plant height and fruit-set. Only a few of the gibberellins are used in plant tissue culture media, Gibberelic Acid 3 (GA3) being the most common.

Abscisic Acid

Abscisic acid (ABA) inhibits cell division. It is most commonly used in plant tissue culture to promote distinct developmental pathways such as somatic embryogenesis.

Ethylene

Ethylene is a gaseous, naturally occurring, plant growth regulator most commonly associated with controlling fruit ripening, and its use in plant tissue culture is not widespread. It does, though, present a particular problem for plant tissue culture.

Some plant cell cultures produce ethylene, which, if it builds up sufficiently, can inhibit the growth and development of the culture. The type of culture vessel used and its means of closure affect the gaseous exchange between the culture vessel and the outside atmosphere and thus the levels of ethylene present in the culture.

Plant Growth Regulators and Tissue Culture

Generalisations about plant growth regulators and their use in plant cell culture media have been developed from initial observations made in the 1950s. There is, however, some considerable difficulty in predicting the effects of plant growth regulators: this is because of the great differences in culture response between species, cultivars and even plants of the same cultivar grown under different conditions.

However, some principles do hold true and have become the paradigm on which most plant tissue culture regimes are based. Auxins and cytokinins are the most widely used plant growth regulators in plant tissue culture and are usually used together, the ratio of the auxin to the cytokinin determining the type of culture established or regenerated.

A high auxin to cytokinin ratio generally favours root formation, whereas a high cytokinin to auxin ratio favours shoot formation. An intermediate ratio favours callus production.

CULTURE TYPES

Cultures are generally initiated from sterile pieces of a whole plant. These pieces are termed 'explants', and may consist of pieces of organs, such as leaves or roots, or may be specific cell types, such as pollen or endosperm. Many features of the explant are known to affect the efficiency of culture initiation. Generally, younger, more rapidly growing tissue (or tissue at an early stage of development) is most effective. Several different culture types most commonly used in plant transformation studies will now be examined in bit detail.

CALLUS

Explants, when cultured on the appropriate medium, usually with both an auxin and a cytokinin, can give rise to an unorganised, growing and dividing mass of cells. It is thought that any plant tissue can be used as an explant, if the correct conditions are found. In culture, this proliferation can be maintained more or less indefinitely, provided that the callus is subcultured on to fresh medium periodically.

During callus formation there is some degree of dedifferentiation (*i.e.* the changes that occur during development and specialization are, to some extent, reversed), both in morphology (callus is usually composed of unspecialised parenchyma cells) and metabolism. One major consequence of this dedifferentiation is that most plant cultures lose the ability to photosynthesise. This has important consequences for the culture of callus tissue, as the metabolic profile will probably not match that of the donor plant. This necessitates the addition of other components—such as vitamins and, most importantly, a carbon source—to the culture medium, in addition to the usual mineral nutrients.

Callus culture is often performed in the dark (the lack of photosynthetic capability being no drawback) as light can encourage differentiation of the callus. During long-term culture, the culture may lose the requirement for auxin and/ or cytokinin. This process, known as 'habituation', is common in callus cultures from some plant species (such as sugar beet). Callus cultures are extremely important in plant biotechnology. Manipulation of the auxin to cytokinin ratio in the medium can lead to the development of shoots, roots or somatic embryos from which whole plants can subsequently be produced. Callus cultures can also be used to initiate cell suspensions, which are used in a variety of ways in plant transformation studies.

CELL-SUSPENSION CULTURES

Callus cultures, broadly speaking, fall into one of two categories: compact or friable. In compact callus the cells are densely aggregated, whereas in friable callus the cells are only loosely associated with each other and the callus becomes soft and breaks apart easily. Friable callus provides the inoculum to form cell-suspension cultures. Explants from some plant species or particular cell types tend not to form friable callus, making cell-suspension initiation a difficult task.

The friability of callus can sometimes be improved by manipulating the medium components or by repeated subculturing. The friability of the callus can also sometimes be improved by culturing it on 'semi-solid' medium (medium with a low concentration of gelling agent). When friable callus is placed into a liquid medium (usually the same composition as the solid medium used for the callus culture) and then agitated, single cells and/or small clumps of cells are released into the medium. Under the correct conditions, these released cells continue to grow and divide, eventually producing a cell-suspension culture. A relatively large inoculum should be used when initiating cell suspensions so that the released cell numbers build up quickly. The inoculum should not be too large though, as toxic products released from damaged or stressed cells can build up to lethal levels. Large cell clumps can be removed during subculture of the cell suspension. Cell suspensions can be maintained relatively simply as batch cultures in conical flasks. They are continually cultured by repeated subculturing into fresh medium. This results in dilution of the suspension and the initiation of another batch growth cycle. The degree of dilution during subculture should be determined empirically for each culture. Too great a degree of dilution will result in a greatly extended lag period or, in extreme cases, death of the transferred cells.

After subculture, the cells divide and the biomass of the culture increases in a characteristic fashion, until nutrients in the medium are exhausted and/or toxic by-products build up to inhibitory levels—this is called the 'stationary phase'. If cells are left in the stationary phase for too long, they will die and the culture will be lost. Therefore, cells should be transferred as they enter the stationary phase. It is therefore important that the batch growth-cycle parameters are determined for each cell-suspension culture.

Protoplasts

Protoplasts are plant cells with the cell wall removed. Protoplasts are most commonly isolated from either leaf mesophyll cells or cell suspensions, although other sources can be used to advantage. Two general approaches to removing the cell wall (a difficult task without damaging the protoplast) can be taken—mechanical or enzymatic isolation.

Mechanical isolation, although possible, often results in low yields, poor quality and poor performance in culture due to substances released from damaged cells. Enzymatic isolation is usually carried out in a simple salt solution with a high osmoticum, plus the cell wall degrading enzymes. It is usual to use a mix of both cellulase and pectinase enzymes, which must be of high quality and purity.

Protoplasts are fragile and easily damaged, and therefore must be cultured carefully. Liquid medium is not agitated and a high osmotic potential is maintained, at least in the initial stages. The liquid medium must be shallow enough to allow aeration in the absence of agitation. Protoplasts can be plated out on to solid medium and callus produced. Whole plants can be regenerated by organogenesis or somatic embryogenesis from this callus. Protoplasts are ideal targets for transformation by a variety of means.

ROOT CULTURES

Root cultures can be established *in vitro* from explants of the root tip of either primary or lateral roots and can be cultured on fairly simple media. The growth of roots *in vitro* is potentially unlimited, as roots are indeterminate organs. Although the establishment of root cultures was one of the first achievements of modern plant tissue culture, they are not widely used in plant transformation studies.

SHOOT TIP AND MERISTEM CULTURE

The tips of shoots (which contain the shoot apical meristem) can be cultured *in vitro*, producing clumps of shoots from either axillary or adventitious buds. This method can be used for clonal propagation. Shoot meristem cultures are potential alternatives to the more commonly used methods for cereal regeneration as they are less genotype-dependent and more efficient (seedlings can be used as donor material).

Embryo Culture

Embryos can be used as explants to generate callus cultures or somatic embryos. Both immature and mature embryos can be used as explants. Immature, embryo-derived embryogenic callus is the most popular method of monocot plant regeneration.

Microspore Culture

Haploid tissue can be cultured *in vitro* by using pollen or anthers as an explant. Pollen contains the male gametophyte, which is termed the 'microspore'. Both callus and embryos can be produced from pollen. Two main approaches can be taken to produce *in vitro* cultures from haploid tissue.

The first method depends on using the anther as the explant. Anthers (somatic tissue that surrounds and contains the pollen) can be cultured on solid medium (agar should not be used to solidify the medium as it containsinhibitory substances). Pollen-derived embryos are subsequently produced via dehiscence of the mature anthers. The dehiscence of the anther depends both on its isolation at the correct stage and on the correct culture conditions. In some species, the reliance on natural dehiscence can be circumvented by cutting the wall of the anther, although this does, of course, take a considerable amount of time. Anthers can also be cultured in liquid medium, and pollen released from the anthers can be induced to form embryos, although the effi-ciency of plant regeneration is often very low. Immature pollen can also be extracted from developing anthers and cultured directly, although this is a very time-consuming process.

Both methods have advantages and disadvantages. Some beneficial effects to the culture are observed when anthers are used as the explant material. There is, however, the danger that some of the embryos produced from anther culture will originate from the somatic anther tissue rather than the haploid microspore cells. If isolated pollen is used there is no danger of mixed embryo formation, but the efficiency is low and the process is time-consuming. In microspore culture, the condition of the donor plant is of critical importance, as is the timing of isolation. Pretreatments, such as a cold treatment, are often found to increase the efficiency. These pretreatments can be applied before culture, or, in some species, after placing the anthers in culture.

Plant species can be divided into two groups, depending on whether they require the addition of plant growth regulators to the medium for pollen/anther culture; those that do also often require organic supplements, *e.g.* amino acids. Many of the cereals (rice, wheat, barley and maize) require medium supplemented with plant growth regulators for pollen/anther culture.

Regeneration from microspore explants can be obtained by direct embryogenesis, or via a callus stage and subsequent embryogenesis.

Haploid tissue cultures can also be initiated from the female gametophyte (the ovule). In some cases, this is a more efficient method than using pollen or anthers.

The ploidy of the plants obtained from haploid cultures may not be haploid. This can be a consequence of chromosome doubling during the culture period. Chromosome doubling (which often has to be induced by treatment with chemicals such as colchicine) may be an advantage, as in many cases haploid plants are not the desired outcome of regeneration from haploid tissues. Such plants are often referred to as 'di-haploids', because they contain two copies of the same haploid genome.

PLANT REGENERATION

Having looked at the main types of plant culture that can be established *in vitro*, we can now look at how whole plants can be regenerated from these cultures.

In broad terms, two methods of plant regeneration are widely used in plant transformation studies, *i.e.* somatic embryogenesis and organogenesis.

SOMATIC EMBRYOGENESIS

In somatic (asexual) embryogenesis, embryo-like structures, which can develop into whole plants in a way analogous to zygotic embryos, are formed from somatic tissues. These somatic embryos can be produced either directly or indirectly. In direct somatic embryogenesis, the embryo is formed directly from a cell or small group of cells without the production of an intervening callus. Though common from some tissues (usually reproductive tissues such as the nucellus, styles or pollen), direct somatic embryogenesis is generally rare in comparison with indirect somatic embryogenesis.

In indirect somatic embryogenesis, callus is first produced from the explant. Embryos can then be produced from the callus tissue or from a cell suspension produced from that callus. Somatic embryogenesis usually proceeds in two distinct stages. In the initial stage (embryo initiation), a high concentration of

2,4-D is used. In the second stage (embryo production) embryos are produced in a medium with no or very low levels of 2,4-D.

In many systems it has been found that somatic embryogenesis is improved by supplying a source of reduced nitrogen, such as specific amino acids or casein hydrolysate.

Organogenesis

Somatic embryogenesis relies on plant regeneration through a process analogous to zygotic embryo germination.

Organogenesis relies on the production of organs, either directly from an explant or from a callus culture. There are three methods of plant regeneration via organogenesis. The first two methods depend on adventitious organs arising either from a callus culture or directly from an explant. Alternatively, axillary bud formation and growth can also be used to regenerate whole plants from some types of tissue culture. Organogenesis relies on the inherent plasticity of plant tissues, and is regulated by altering the components of the medium. In particular, it is the auxin to cytokinin ratio of the medium that determines which developmental pathway the regenerating tissue will take. It is usual to induce shoot formation by increasing the cytokinin to auxin ratio of the culture medium. These shoots can then be rooted relatively simply.

PLANT TISSUE ANALYSIS

Crop nutrient uptake is influenced by many factors other than the soil test. Soil testing and plant analysis are designed to work together. Soil testing identifies the soils nutrient reserves and predicts the nutrient needs, while plant analysis identifies the actual nutrient uptake.

Plant tissue analysis is a useful diagnostic tool and can greatly help with fertility management. From emergence through the first six weeks of growth, plant analysis is helpful in identifying nutrient uptake. Testing corn leaf samples at this early stage may indicate where additional nitrogen should be applied. This is a way to determine the cost-effectiveness of the additional application of fertilizer.

Plant analysis may be utilized at the reproductive stage (full grown) to determine nutrient uptake for that season. This allows for determination of fertility requirements for the upcoming growing season. Tissue sampling and analysis is a key tool in determining the soil's ability to meet the sufficient nutrient requirements of the crop, thereby reducing any possible nutrient stress that may result in a potential yield drag or loss. A variety of tests are available, including phosphorus, potassium, calcium, magnesium, sulfur, sodium, zinc, iron, manganese, copper, nitrogen, and boron.

Remember: test results are only as good as the sample. The most common mistake in plant sampling is not obtaining enough tissue for an accurate analysis.

Several leaves from several plants will ensure a more representative sampling and ensure enough plant material. Fertility management samples should be collected from representative areas of the field. Always take composite samples. One plant is not a sample. Generally, select only upper mature leaves for your tissue sample. When samples are being collected for nutrient review, specific leaves are normally selected depending upon the plant. Here are a few examples:

- Corn-at tasseling, collect the leaf at the ear node
- Soybeans-at flowering collect the most mature trifoliant from the top of the plant
- Wheat and small grain-at beginning of head set collect the flag leaf
- Alfalfa-at flowering collect the top 6 inches of the plant

When troubleshooting is the issue, it is extremely important to take leaves from several plants experiencing the same problem and use this composite for one sample. Compare this analysis with a composite sample taken from the non-affected area of the field. Sample #2 is the control sample. This gives the grower an idea of the sufficient nutrient level for that crop in that field at that sampling period.

For example: a portion of a corn field is exhibiting chlorosis (yellowing) and stunted growth. Plant tissue analysis is the best way to determine the reason for the chlorosis and abnormal growth. A composite sample of upper mature leaves from 6 to 10 chlorotic plants and a composite sample taken from the upper mature leaves from 6 to 10 plants located in the non-affected area of the field are submitted for analysis.

Where to Sample

Avoid leaves that were heavily preyed upon by insects or mechanically damaged. Avoid taking samples from areas located in border rows or heavy shade. Do not collect tissue that is already dead. Do not take samples from contaminated or chemically altered plants. Take diagnostic samples from the affected site for one composite sample and one from the non-affected area for a second composite sample. If the entire field is distressed, collect a composite sample over the entire area and note that the entire field is affected.

When to Sample

Timing of diagnostic sampling depends on the first sign of trouble. The earlier the problem is detected, the easier it is to treat with fertilizer or lime application. If nutrient application is not feasible due to plant growth, these results, along with soil analysis, can be used in fertilizer recommendations for the following growing season. Fertility management sampling is done at the time of flowering or pollination (reproductive stage). Avoid taking samples too late in the growing season. Take a single composite sample at the onset of the reproductive stage to measure nutrient uptake throughout the growing season.

Take samples when the plant is still viable and growing. It is too late once it has started to produce seed. These values determine the following growing season's fertility needs.

SUBMITTING THE SAMPLE

Place each dried composite sample in a paper bag. Do not use plastic bags for plant samples because the plants mold. Label each bag with the field name and sample number. Please label them good, bad, affected or non-affected. Please include information that contains: name, address, telephone number, field names, number of samples per field and the type of test.

A plant's ability to take up adequate nutrients is influenced by many factors such as soil fertility level, soil temperature, soil compaction, soil moisture, nutrient balance in the soil, fertilizer applications, plant genetics, and many other factors. A chemical soil test cannot accurately account for all of these factors. A producer must use plant analysis if he expects to identify and understand the nutrient efficiency and needs of his crops and soils. Only plant analysis can tell us the actual nutrient status of the crop. Plant analysis can be used to detect low nutrient levels in plants before they are seen by the human eye. Once visual symptoms are visible, top yields and quality have already been compromised. Plant analysis is an important part of planning future fertility programmes.

Sampling Guide

Our plant sampling guide contains information on most plant species. If you need furthur assistance, contact the lab for specific information on your crop.

USES OF PLANT ANALYSIS

- Plant analysis can be used by turf managers to:
- Confirm suspected nutrient deficiency symptoms;
- Verify toxicities;
- Reveal hidden hunger (*i.e.* plants show no visible symptoms, but the nutrient content is low enough to reduce growth or affect quality characteristics);
- Aid in evaluating the efficiency of applied fertilizers;
- Assist in formulating fertilisation practices; and,
- Monitor the nutrient status of plants throughout the growing season to determine whether each nutrient is present in sufficient concentration for optimum growth characteristics.

Plant analysis is a proven and effective means of predicting fertilizer needs for many plant species. However, it does not completely replace a soil test. Soil and plant analysis serve different purposes and when properly used they compliment each other in providing detailed information for maximizing the efficiency of fertility programmes. Soil testing is based on the concept that the concentration of a particular nutrient in a given volume of soil reflects whether or not the nutritional level of that soil is adequate for optimum plant growth or production. Plant analysis is based in part on the concept that the amount of a specific nutrient in the plant tissue is related to the plant availability of that element in the soil. Plant analysis also reflects nutrient uptake conditions in the soil. Soil properties such as compaction, root disease, impervious layers and/or poor drainage may inhibit the uptake of nutrients by plants. Or a low concentration of one nutrient in the plant may result from the excessive application of another nutrient. Conversely, favourable soil physical properties and optimum soil moisture may accentuate nutrient uptake even though the soil may not have an abundant supply of nutrients.

As a result of these soil-plant interactions, there are certain instances when contradictions occur between soil and plant analysis results. For example, assume turf is growing on a soil in which the soil tests revealed a medium level of extractable Magnesium. A plant analysis from the area a few weeks later indicates that Magnesium is low. Immediately, the validity of the test results are questioned, which is an absolutely normal response. However, a closer examination of the plant analysis results revealed that the Calcium and Potassium concentrations of the turf were high.

Upon checking the information accompanying the plant analysis results, it was noted that Lime ($CaCO_3$) had been applied at 500g/m2 and a high rate of Potassium was applied in the fertilizer programme. As a result of these two management practices, the level of Calcium and Potassium in the soil was sufficient to reduce the uptake of Magnesium. This is one example of how soil testing and plant analysis can be used together for making better nutrient management decisions. Plant analysis can also be used to supplement a soil testing programme. It is particularly useful in distinguishing between Nitrogen and Sulphur deficiencies in turf, as deficiency symptoms of the two elements are similar. Plant analysis offers an excellent means of delineating which element is deficient (which cannot be ascertained through soil testing). If this distinction is not made properly and the wrong corrective treatment is applied, plant growth can be affected appreciably.

In the case of most turf grass species, a soil analysis prior to active growth in spring makes it possible to determine whether lime, dolomite, gypsum and/or specific fertilizer applications will be needed. Plant analysis of the turf during the growing season will indicate if the applied materials were effective and whether the predictions based on the soil analysis were correct.

A&L EASTERN LABS-SAMPLING FOR PLANT ANALYSIS

Plant analysis assesses nutrient uptake while soil testing predicts nutrient availability. The two tests are complementary as crop management tools. Plant analysis will detect unseen hidden hunger and confirm visual deficiency symptoms. Toxic levels may also be detected. If it is done early, plant analysis will allow a corrective fertilizer application in the same season. A basic knowledge of plant structure is necessary before collecting samples. A leaf is made up of a leaf "blade" and a "petiole". The petiole is the stalk attached to the blade. A compound leaf may have several "leaflets" attached to it. In some cases, only terminal "leaflets" may be sampled, as in the case of walnuts and pistachios. A common error in tomatoes is when only leaflets are sampled instead of the whole compound leaf. This shows the importance of understanding proper sampling technique.

The most recent mature leaf (MRML) is the first fully expanded leaf below the growing point. It is neither dull from age nor shiny green from immaturity. For some crops, the most recent mature leaf is a compound leaf. The most recent mature leaf on soybean and strawberry, for example, is a trifoliate compound leaf: three leaflets comprising one leaf.

For cotton, grape, potato and strawberry, petioles provide an additional indication of nitrogen status. When sampling these crops, collect most recent mature leaves and their petioles. Detach leaves from petioles in the field to stop the translocation of nutrients. Put petioles in a separate bag. "Midribs" are the middle ribs to large leaves such as corn, lettuce, and cabbage, and would equate to a petiole sampling.

Deciding When to Sample

To monitor plant nutrient status most effectively, sample during the recommended growth stages for your specific crop Take samples weekly or biweekly during critical periods, depending on management intensity and crop value. However, to identify a specific plant growth problem, take samples whenever you suspect the problem.

The best time to collect samples is between mid-morning and mid-afternoon. Nitrate nitrogen varies with time of day and prevailing conditions but generally not enough to alter interpretation. Sampling during damp conditions is okay but requires extra care to prevent tissue from decomposing during shipping. Keep samples free of soil and other contaminants that can alter results.

The appropriate part of the plant to sample varies with crop, stage of growth, and purpose of sampling. When sampling seedlings less than 4 inches tall, take whole plants from 1 inch above the soil line. For larger plants, the most recent mature leaf is the best indicator sample.

Taking A Representative Sample

Proper sampling is the key to reliable plant analysis results. A sample can represent the status of one plant or 20 acres of plants. In general, a common-sense approach works well. When problem solving, take samples from both "good" and "bad" areas. Comparison between the two groups of samples helps pinpoint the limiting element. Comparative sampling also helps factor out the influence of drought stress, disease, or injury. Take matching soil samples from the root zones of both "good" and "bad" plants for the most complete evaluation.

When monitoring the status of healthy plants, take samples from a uniform area. If the entire field is uniform, one sample can represent a number of acres. If there are variations in soil type, topography, or crop history, take multiple samples so that each unique area is represented by its own sample.

Choosing Sample Size

The actual laboratory analysis requires less than one gram of tissue. However, a good sample contains enough leaves to represent the area sampled. Therefore, the larger the area is, the larger the sample size needs to be.

Sample size also varies with crop. For crops with large leaves, like tobacco, a sample of three or four leaves is adequate. For crops with small leaves, like azalea, a sample of 25 to 30 leaves is more appropriate. For most crops, 8 to 15 leaves is adequate. For crops requiring petiole analysis, collect at least 15 to 20 leaves.

Submitting the Sample

Send the completed information sheet and proper fee with each sample. Use permanent ink or pencil on sample forms and bags Avoid numbering samples simply as 1,2,3 ... as it may lead to confusion later. Give each sample a unique identifier that will help you remember the plants or area it corresponds to-such as HOUSE1, 15B, GOOD, or BAD. You can use up to six letters and/or numbers. Put the identifier on both the information sheet and the sample envelope. Pay attention to detail when filling out the information sheet. Note any conditions-drought, disease, injury, pesticide or foliar nutrient applications-that might be relevant. Indicate the analysis desired and provide very specific information on stage of growth and plant part if an interpretation is required. The laboratory does not automatically provide an interpretation, as some clients prefer to make their own.

Diagnostic interpretations require more details than predictive. When sending matching soil, solution, or waste samples, indicate the matching sample ID in the designated areas on the information sheet. Be sure the grower name and address are exactly the same on all matching information sheets. Ship all matching samples in the same container. Ship the tissue sample in a paper envelope or cardboard box so it can begin drying during transport. Samples put in plastic bags will rot, and decomposition may alter test results.

If samples are very wet, air-dry to a workable condition before packaging. Otherwise, decomposition or molding will occur. Include a completed plant analysis information sheet or cover letter with instructions within the same package. Processing will be delayed if sent separately. Also, include payment if you do not have an established account. Samples should be shipped by a carrier such as UPS or FEDEX, or by first class mail.

Interpreting the Report

Samples are analyzed the next day of their arrival. The prompt turnaround makes it possible for growers to take any corrective action needed to salvage the current crop. The report can be e-mailed or post mailed to the grower

PLANT ANALYSIS AS A GUIDE FOR MINERAL NUTRITION OF SUGAR BEETS

Nitrogen is the nutrient most likely to limit growth and sucrose production in sugar beets. Profitable responses to P fertilization are becoming common. Potassium fertilization is necessary in certain soils of the Sacramento-San Joaquin Delta. Sulfur is deficient in some soils of the upper Sacramento Valley. Micronutrient deficiencies of sugar beets are uncommon in California.

Research and practical experience have shown that plant analysis can serve as an effective guide in the fertilization of sugar beets. A plant analysis programme, properly carried out, will help optimize income for the producer, provide quality beets for processing, conserve fertilizer, and minimize the leaching of nutrients to underground water supplies.

CRITICAL CONCENTRATIONS

Critical concentrations of most of the essential nutrients have been determined for various parts of the sugar beet plant. These values along with concentration ranges found in healthy plants and levels in plants with deficiency symptoms.

The critical concentration for any nutrient is a relatively constant value which changds little with sugar beet variety, climate, or soil conditions. Nutrient concentrations, on the other hand, are changed by conditions that affect plant growth. For example, petioles of mature leaves of plants infected with the beet yellows virus may be higher in N03-N concentration than are healthy plants. This is because diseased plants grow more slowly, and nitrate accumulates in their petioles.

Potential users of sugar beet leaf analysis sometimes express concern that differences in plant moisture content at different times of the day may significantly alter the concentrations of nutrients in plant tissues and thus affect the interpretation of analytical results. In field experiments we observed little or no change in dry-weight N03-N concentration in petioles samples hourly from sunrise to sunset. Nor were there important differences with moderate (slight wilting) compared to no moisture stress. Only when plants were severely stressed was there a significant decrease in petiole nitrate. Plants moderately stressed compared to those severely stressed (loss of many leaves) had N03-N concentrations of 5, 1 00 and 3,600 ppm, respectively.

Differences in the nutrient concentration due to variety, weather, moisture stress, disease and the like, are largely related to the kind and extent of root development or to other growth factors, and not to differences in internal requirements.

SAMPLING AND EVALUATION

For the sugar beet, recently matured leaves (the youngest leaves that have just attained maximum size) are usually the most appropriate part to sample. Petioles are most satisfactory for evaluating the plant's nutritional status for N, P, and Cl, while the blade is most satisfactory for the other essential elements.

A series of four samplings should be made throughout the growing season: one at thinning time, the second at early mid-season, the third at late rnid-season, and the fourth just before harvest. By comparing the analysis of these samples to critical nutrient concentrations, it is possible to detect and correct impending nutrient deficiencies before they occur, and to change the timing or method of fertilization to assure that fertilizer needs of the crop are fully met.

Large amounts of N are required to insure adequate top and root growth, but if storage roots are to be high in sucrose concentration the plants must be N deficient for 4 to 8 weeks before harvest. An estimate of the amount of fertilizer N needed for a specific crop should be made according to past history of the field or by soil tests.

This amount should be applied to the crop early in the growing season, not later than thinning time. If an error is made, it is best to make it on the side of under-fertilization, because such a mistake can be corrected. If, however, the initital fertilizer application is more than the crop can use, fertilizer will be wasted, sucrose concentration (and, consequently, crop value) will be depressed, and excess nitrate will possibly be leached to the ground water.

Adequacy of the initial fertilizer application can be ascertained by analyzing petiole samples taken at 4 and 2 weeks before mid-season-the date midway between the dates of emergence and harvest. (To determine midseason for a crop to be overwintered, determine the date midway between emergence and November 1.) Indication of an impending deficiency of more than 10 weeks before harvest can be corrected by an addition of from 40 to 80 pounds of N per acre. These two early-season petiole samples can also contribute to the overall evaluation of the fertilizer programme as outlined above.

Abnormal Plant Growth

Frequently, plants in a field or part of a field show abnormal growth. A knowledge of nutrient deficiency symptoms gained from previous experience or a printed visual guide may lead to a tentative diagnosis which can be confirmed or discarded by analyses of comparable petioles and blades of abnormal and normal appearing plants. A useful visual guide is: Sugar Beet Nutrient Deficiency Symptoms, A Colour Atlas and Chemical Guide (Ulrich and Hills, 1969). To confirm a diagnosis of deficiency symptoms by chemical analysis, plants should be sampled soon after the symptoms appear. After a prolonged period of stress, the concentration of certain nutrients increases in physiologically mature tissue, especially if new supplies of the nutrient become available to the plant.

Guide to Harvesting

When there is a choice of fields to be harvested, the results of a systematic plant analysis programme can serve as a guide to scheduling of beet harvest. Those fields depleted of N first would be scheduled for an early harvest; those high in N would not be harvested until the N is depleted or would be held as long as possible before harvesting. Root pulp as well as petioles may be analyzed to determine depletion of N before harvest. Pulp from preharvest root samples may be analyzed for nitrate by a semi-quantitative spot test using the diphenylamine reagent or, more accurately, by other quantitative methods. To maximize the sucrose concentration of beet roots, petioles and/or roots should be below critical N03-N levels about 4 to 8 weeks before harvest.

Nutrient Survey

A nutrient survey of sugar beet fields in an area can result in detection of deficiencies and identification of fields on which future fertilizer trials should be located. Surveys conducted by Farm Advisors or other interested agricultural groups can do much to improve fertilizer practices in an area.

Sampling Procedure

To serve as a measure of variability within a field, a minimum of two, but preferably four, samples should be taken from each field. From 25 to 50 leaves are usually sufficient to reasonably estimate the true mean of the nutrient concentration in the plants represented by the sample. A convenient sampling system is to divide a field into imaginary quarters.

The sampler walks across the centre of each quarter at right angles to the plant rows and collects from 25 to 50 leaves per quarter of the field. Samples are placed in a paper bag of convenient size and taken to the laboratory for processing as soon as possible. Delays of more than 48 hours at ordinary temperatures should be avoided, and storage in closed compartments of automobiles during warm weather should not be allowed. If samples must be stored before drying they should be kept at 5'C; at this temperature, sugar beet petioles change very little in nitrate, phosphate, and K concentration.

For microelement analyses, blades must be free of dust and should be washed for 30 seconds in a 0. I N HC1 bath followed by two successive rinses in distilled water.

A sample is reduced in size by cutting it into small pieces (approximately 1/8 inch), thoroughly mixing the pieces and then placing a sub-sample of about 100 grams (a large handful) of fresh material in a cut-down paper bag or other suitable container and drying it overnight in an oven, preferably with a forced draft, at 70'C to 80'C (158'- 1 76'F). After drying, samples are ground to pass a 40-mesh screen and stored. Samples treated in this way have been kept for many years without significant change in their mineral composition.

INTERPRETATION AND LIMITATIONS

With plants in the field, the likelihood of a growth response from the addition of a nutrient to the soil will depend upon whether the nutrient concentration of the plants is above or below the critical level. When the nutrient concentration is above the critical level and remains there throughout the growing season, there is little chance of a response in growth from the addition of more of the nutrient. When the nutrient concentration of the plants falls below the critical level, the chance of a growth response becomes greater the longer the plants remain deficient, provided other growth factors are favourable (*e.g.*, climate is favourable for rapid growth, and growth is not seriously limited by disease, water stress, and the like). If the deficiency appears late in the growing season, there is little chance for a significant yield increase with the addition of more of the nutrient as time is required for any deficiency to cause a measurable decline in crop yield.

When plant material for a sample is taken from a large area, the sample may differ considerably in nutrient concentration within itself. When the average nutrient concentration of the sample is at or near the critical level, the material collected will represent both deficient and non-deficient plants. If there are areas in a field differing in plant-growth characteristics, these areas should be sampled separately.

Analysis of plant tissue usually reveals only one deficiency at a time. A second nutrient, or even a third, may be in short supply, but due to reduced growth caused by the primary nutrient deficiency, other nutrients will accumulate in the tissue. However, when the primary deficiency is corrected, the increased growth will decrease the concentration of the second nutrient, and most likely the second nutrient will become deficient rather soon. An exception to this general rule is with regard to P, deficient surgar beets which, particularly as seedlings, are also occasionally deficient in N, because P deficiency retards nitrate uptake. In the interpretation of the results for multiple sampling dates, the sharper the drop in nutrient concentrations and the earlier in the season it occurs, the greater the degree of deficiency. Even when the results of several sampling dates indicate the plants have been deficient for a considerable period, the exact degree of deficiency is not known. At harvest, however, the degree of deficiency for a given field can be estimated in terms of the maximum yield for the area, and fertilization for the next crop is then adjusted to this finding. For example, a field that produced only 25 per cent of the potential yield is much more deficient and will require more fertilizer than one that produced 90 per cent of its potential.

One of the greatest values of plant analysis is the prevention of deficiencies rather than their correction. Trends in nutrient content of leaves followed over a period of years can be a valuable guide in delineating the kinds, amounts, frequency, and methods of fertilization for more efficient crop production on a given field. Nutrient deficiencies can be predicted or at least detected as soon as they occur, and corrective measures can be taken before serious crop loss takes place.

6

Management of Horticultural Crops Under Protected Cultivation

CROP PRODUCTION

Attempts to adapt crop production to the environment with protective devices or practices date back to ancient times. Structures for crop production were first used in the early period of the Roman Empire, under Emperor Tiberius Caesar, 14–37 AD. Such structures consisted of mobile beds of cucumber placed outside on favourable days and inside during bad weather. Covers were slate-like plates or sheets of mica or alabaster. Greenhouses in the UK and The Netherlands developed from glass structures built to protect plants imported from tropical Asia and America in the 16th and 17th century during the winter period. However, such methods of cultivation ceased with the decline of the Roman Empire and it was not until the 15th to 18th centuries that simple forms of greenhouses appeared, primarily in England, The Netherlands, France, Japan and China.

By the end of the 19th century, commercial greenhouse crop production was well-established. The purpose of growing crops under greenhouse conditions is to extend their cropping season and to protect them from adverse environmental conditions, such as extreme temperatures and precipitation, and from diseases and pests. Greenhouse structures are essentially light scaffolding covered by sheet glass, fibreglass or plastic. Such materials have a range of energy-capturing characteristics, all designed to maximize light transmission and heat retention.

Crops may be grown in groundbed soil, usually amended with peat or farmyard manure, in benches, in pots containing soil or soil mixtures or soil substitutes, and in hydroponic systems, such as sand or rock wool cultures and flowing nutrient systems, without a matrix for the roots. Modern technology has given the grower some powerful management tools for production. Generally, added-value crops are grown under protection. Most of them are labour-intensive and energy-demanding during cold weather. Greenhouse production therefore normally requires a high level of technology to obtain adequate economic returns on investments. Quality is a high priority for greenhouse crops, requiring much care in pest and disease management, not only to secure yields but also to obtain a high cosmetic standard. Although technological changes are ultimately intended to reduce production costs and maximize profits, precise environmental and nutritional control push plants to new limits of growth and productivity. This can generate chronic stress conditions, which are difficult to measure, but apparently conducive to some pests and diseases. Historically, not enough attention has been paid to exploiting and amending production technology for the control of pests and diseases. This makes the control of pests and diseases in protected crops even more challenging, with many important problems being unresolved and new ones arising as the industry undergoes more changes in production systems.

Additionally, the international trade in ornamental and flower plants facilitates the spread of pests and diseases around the world and their establishment in new areas. In Europe, for example, at least 40 new pests have been recorded in protected crops in the last 25 years. The increasing complexity of pest and disease problems and the high cosmetic standards of vegetable, ornamental and flower products have led growers to apply intensive preventive chemical programmes, which result in pests and pathogens becoming resistant to the most frequently used pesticides in a few years, which, in turn, increases control costs. In southern Spain, the average cost of pesticide application in 1992 in protected vegetables was estimated as USD 0.14/m^2 (16.5 per cent of the total production cost), and several whitefly, thrips, aphid and fungus species are suspected to be resistant to several active ingredients.

A similar figure is valid for Italy, where the most sophisticated structures are located in the northern part of the country: pesticides are widely applied and pest and disease resistance is quite widespread. In The Netherlands, pest and disease control costs for vegetables are still limited and are normally below 3 per cent of the total costs to produce a crop. As control costs increase, pesticide-resistance spreads and consumers become aware of the risks of pesticide-residues in fresh vegetables, a strong demand for non-chemical control methods is emerging in many countries. Integrated systems for greenhouse pest and disease control have been developed and implemented in northern Europe and Canada, but implementation is still cumbersome in other parts of the world.

FORMS OF CROP PROTECTION IN PROTECTED CULTIVATION

Protected cultivation is an extremely high-input procedure to obtain food and other agricultural products per unit of land, although inputs are the lowest when related to the yield per area. Crop protection activities contribute to the total input in variable proportions mainly through the application of pesticides.

Several features of protected cultivation are delaying the adoption of more sustainable ways to control pests and diseases. In areas where protected cultivation is most intensive, crop protection costs rarely exceed 5 per cent of the total production costs. In these circumstances, growers are not stimulated to make decisions based on economically founded criteria, and chemicals are frequently applied to prevent pest occurrence rather than to control real pest problems. This is particularly true in ornamental and flower crops, which can lose their value at extremely low pest densities. In addition, pesticides may be applied easily and little expertise is needed to spray or to recommend pesticides so that no specialized advisory personnel is usually employed by growers who rely on this "simple" technology. Consequently, innovative crop protection methods become difficult to implement in practice. From a general point of view, vegetable crops, due to their limited diversity, are most suitable for IPM. In the case of ornamentals, the enormous crop diversity and the many cultivars per species grown make the development of IPM strategies more complicated. Several stimuli are pushing growers to use less pesticides and to adopt more sustainable ways to protect crops from noxious organisms as world marketing becomes more global.

Among the factors stimulating sustainable forms of crop protection are the following:

CONSUMER CONCERN ABOUT CHEMICAL RESIDUES

This is a general stimulus for growers wishing to adopt IPM systems, but it is particularly relevant in fresh-consumed products like the majority of vegetables grown in greenhouses. Consumers not only demand high quality products, but are also concerned with how they are grown to judge them from the environmental aspects. Food marketers and European regional administrations are developing auditing procedures to sell vegetables under IPM or Integrated Production (IP) labels. In some cases, a surplus price is achieved by growers who produce vegetables under established IPM/IP technology.

Pesticide-Resistance in Pests and Pathogens

As protected cultivation allows pest and pathogen populations to increase faster than in the open air, and as protected crops receive a great number of pesticide treatments, pesticide-resistance develops rapidly. Dozens of greenhouse pests or pathogens are suspected to have developed resistance to the most common active ingredients and this has been observed in many pests (aphids, whiteflies) and pathogens such as *B. cinerea*.

Chemical Application

Because society in general and governments in particular are aware of the impact of chemicals on soil, water and air, several initiatives to restrict the use of chemicals in Europe and North America are being undertaken.

Efficacy

Some pests and diseases are difficult – sometimes impossible – to control if an integrated approach is not adopted. On the other hand, natural control can prevent several pests from building-up high populations under the action of predators, parasitoids and entomopathogens that naturally establish on greenhouse crops if chemicals are not intensively applied, and several cultural practices allow enhancement of their effectiveness.

A first step towards sustainability in greenhouse crop protection is to analyse why and which phytophages and pathogens are able to increase their population densities until reaching damaging levels. Methods to improve the accuracy and speed of diagnosis are needed, particularly for diseases, and may be one of the most useful applications of biotechnology. Once the pest or disease is correctly diagnosed, environmental factors that allow or prevent such a pest or pathogen to reach economic injury levels should be identified. Such knowledge may help us to design integrated methods to take advantage of the whole environment. If an action threshold is determined, accurate techniques for pest and disease sampling and monitoring should permit intervention at the best moment and prevent unnecessary treatments.

The identification of key factors governing pest or pathogen population dynamics may allow modification of the greenhouse and crop environment – including greenhouse-surroundings – to adversely affect a pest or pathogen or to favour the effectiveness of the natural enemies or antagonists. Sometimes this can be achieved cheaply – in both economic and energetic terms – by means of correct crop and management practices. As mentioned before, the most damaging pests and many pathogens in greenhouses are polyphagous; although they are able to develop on many host plants, their negative effect on yield varies with host plant species and cultivar. The development of cultivars which are less susceptible to pests and diseases or that favour the activity of pest natural enemies is undoubtedly one of the most sustainable ways to control diseases in greenhouses and its potential for pests has been shown in a few but significant cases.

Many of the arthropod pests and diseases that affect greenhouse crops are exotic and became established in greenhouse growing areas from accidental importation of infested crops, mainly ornamentals. In some cases, as for *Liriomyza trifolii* (Burgess) and *Liriomyza huidobrensis* (Blanchard), native natural enemies have been able to greatly contribute to the natural control of these pests, but in other cases exotic parasitoids or predators have to be released in the environment to control them, as is done for *Trialeurodes vaporariorum* (Westwood) by means of *Encarsia formosa* Gahan. Natural and biological control is nowadays the basis of most of the integrated pest management strategies adopted in northern Europe and its practical achievements are particularly emphasized in this book. The history of biological control of diseases in greenhouses is more recent, but significant advances have also been achieved here in the last few years. Given the very high cosmetic demands and the low pest and disease thresholds applied by greenhouse growers, the progress in application of Integrated Pest and Disease Management is remarkable. Until recently, biological and integrated control was seen as a cost factor. Nowadays, however, it is considered as a beneficial marketing factor.

The greenhouse industry faces many new crop protection problems as a consequence of modification of production procedures and crops. The major changes will include more widely adopted mechanization and automation systems for improved crop management and the use of biotechnology in plant production. These modifications will affect the severity of pests and diseases. Strong cooperation among plant pathologists, entomologists and horticulturists is necessary in order to assure that new management practices have a beneficial effect on plant health.

Methods to improve the accuracy and speed of diagnosis are needed and may be one of the best applications of biotechnology. Improved and widely used monitoring and diagnosis systems to determine the degree of infestation and economic thresholds of pathogens and pests will enable rational management decisions. A high priority should be given to the production of pathogen and pest-free propagation material, obtained through sanitation. The use of pest and pathogen-free material, and growing media disinfested with steam or naturally suppressive to soilborne pathogens will help to reduce the impact of important pests and diseases considerably. When all such measures are integrated with the use of resistant germplasm, with modern techniques for applying pesticides and with biological control of several diseases and pests, a greatly reduced input of chemicals becomes realistic for protected cultivation.

CROP AND CROP ENVIRONMENT IN GREENHOUSE

Greenhouses vary in structural complexity from simple plastic film-covered tunnels, with no assisted ventilation, to tall, multispan, glass or plastic-covered structures covering several hectares and having sophisticated, computer-controlled environments.

Essentially, however, all have climates inside that are rain-free, warm, humid and windless, ideal for raising crops but at the same time also ideal for many diseases and arthropod pests. Though it is restricted, the climate within the greenhouse forms a continuum with the climate outside the greenhouse, and there are gradients in temperature, humidity, light and carbon dioxide. Depending on the needs of the crop, the need to exclude pests and pathogens, and the need to implement biological control programmes, these gradients can be manipulated to certain extents by such devices as screening, shading, cooling, heating and ventilation.

At the other end of the scale, the climate at the immediate plant surface, the so-called boundary layer, whether of shoots or roots, is of paramount importance in the avoidance of pests and diseases. It extends 1–2 mm for arthropod pests, about 30 μm for fungi and even less for bacteria. Its climate, the true microclimate, forms a continuum with the climate within the intercellular spaces of leaves on the one hand, and with the macroclimate of the greenhouse and its environs on the other hand.

While most stages of most arthropod pests and beneficial insects are free to enter and leave the boundary layer if it is inimical to their activity, most micro-organisms enter passively and leave as wind-dispersed or water-splashed secondary propagules. In order to escape arthropod pests and pathogens, the microclimates of phyllosphere and rhizosphere must be made inimical to their activity but at the same time biological control organisms have to be encouraged with appropriate microclimates. It is often overlooked that biological control organisms have their own hyperparasite and predator chains extending theoretically indefinitely and acting alternately counter to effective biological control on the crop or beneficially with it. They also have their own adverse environments. It is apparently an insoluble task to manage boundary layer microclimates without detriment to the crop or to biological control, at the same time not permitting primary pests and diseases to become established.

MANAGING THE GREENHOUSE OF LOCAL CLIMATE

The local climate, the external disease and insect pressures, the greenhouse structural design, the climate-control equipment available, and the skill level of greenhouse workers have a major bearing on how a greenhouse is managed to control insects and diseases. From the outset, it is important to have the input of a greenhouse manager to ensure that the physical facilities are properly designed for IPM when building a new greenhouse operation. Once a greenhouse is in operation, greenhouse managers have to be forever mindful of how activities in and around a greenhouse will affect IPM.

Structures and Equipment

The structural complexity of successful greenhouse operations tends to increase with time as older structures are replaced with more advanced designs, as the operations increase in size, as profits are reinvested, and as the need for improved climate-control becomes apparent.

The low cost, low height, plastic film-covered structures that are often fust built by growers provide some protection from outdoor weather and pests, but without any means for climate-control, conditions inside are often more favourable for diseases and pests than outside. Higher structures with more substantial framing members are required to accommodate climate-control equipment. The trend in greenhouse structural design in recent years has been towards large gutterconnected complexes with high (4–5 m) gutter heights. As the size of operations under one roof has increased, increased gutter heights have become necessary to create the chimney effect needed to ventilate these structures naturally.

With increased air space between the crop and the greenhouse cover, the uniformity of horizontal and vertical air movement has improved, temperature gradients in the crop canopy have been reduced and the uniformity of lighting of the crop has improved because shadows cast by higher overhead structural members move around more throughout the day. Increased gutter heights have also been beneficial for IPM because they increase the height that insects and pathogen propagules must be transported by wind to find their way into greenhouses through vents. With larger complexes and the economies of scale they provide, it is feasible to incorporate features in a greenhouse design that favour IPM. With large-scale operations, it is practical to build header-house facilities that restrict access to the greenhouse.

Separate shower and lunch room facilities, foot baths, refuse handling facilities, concrete floors, etc., mat reduce the transport of insects and pathogen propagules into the growing areas can be justified. The costs of pressure washing equipment and specialized potting and growing medium sterilizing equipment are easier to justify. Also, for large scale operations, it is feasible to have separate propagation facilities specially designed for the production of disease-free transplants. On the other hand, because of the increased number of nooks and crannies, it is more difficult to eradicate insects and disease propagules from large complexes once they have gained a foothold.

Radiation Transmission Characteristics

The radiation transmission characteristics and the air tightness of greenhouse cover materials have a major effect on the climate for IPM inside a greenhouse. Ideally a cover material should have a high photosynthetically active radiation (PAR) transmission to maximize productivity and solar gain, low infra-red (IR) transmission to minimize radiation heat loss, and low ultraviolet (UV) transmission to inhibit sporulation of fungi. Unfortunately, no material has all these radiation transmission characteristics. Depending on latitude and local climate, some cover materials have been found better than others for IPM. Glass is the preferred greenhouse cover material at high latitudes, where winter light levels are limiting and outdoor temperatures are low, because of its high PAR and low IR transmission characteristics.

Glass, however, does transmit the UV radiation necessary for the sporulation of fungi and has relatively high air leakage which can lead to very low humidity during cold periods with high heat demand. During these periods it is necessary to humidify glass greenhouses to ensure the continued activity of biological control agents. Polyethylene is the preferred greenhouse cover material at lower latitudes where high PAR transmission is not as critical and where retention of humidity for IPM is important. Some manufacturers include admixtures in their polyethylene films to block the UV wavelengths necessary for sporulation of fungi.

The effectiveness of these blockers decreases as the films age. Polyethylene-film covered greenhouses are tighter than glass houses and therefore are better at retaining humidity during hot dry periods. During cool wet periods, high humidity and condensation on the underside of polyethylene films is a problem that can lead to indiscriminate dripping and spread of diseases in the crop. Surfactant sprays have been developed for polyethylene films that cause a film-wise condensation and run-off at the gutter. In recent years, roof arches used for polyethylene greenhouses have been modified from a semi-circular shape to a gothic shape to enhance film-wise condensation and run-off at the gutter.

Siting and Orientation

On a world-wide basis, commercial greenhouse production is concentrated in regions between 25° and 65° latitude where the climate is moderate and local weather patterns are favourable. At high latitudes solar irradiance is low, day length is short and temperatures are low during the winter months resulting in poor growth and increased susceptibility to disease. Under such conditions, diapause of predatory insects may make biological control difficult. Large inputs of energy are required to maintain greenhouse temperatures, and humidification is often necessary to overcome the drying effect of continual heating. At low latitudes, high solar irradiance stresses crops making them more susceptible to disease. More outside ventilation air is required which brings with it more pathogen propagules and insect pests. Within the most favourable latitudes, greenhouse production is concentrated in maritime areas where large bodies of water moderate the local climate. In continental areas, large swings in outdoor temperature and maximum solar-irradiance levels on a day-to-day basis create crop stresses that make greenhouse management more difficult. In summer, cooling of greenhouses is difficult if ambient air temperatures are above the desired greenhouse temperature, and if the relative humidity is so high that evaporative cooling is not effective.

Within any given region, the siting of a particular greenhouse operation makes a significant difference in the management of disease and insect problems. Field crops and natural vegetation growing in close proximity to a greenhouse create disease and insect pressure, especially if those crops and the vegetation are susceptible to the same disease and insect pests as the greenhouse crop. This pressure is intensified when pathogen propagules are stirred up by field operations, or when the outdoor crop is harvested or senesces and insects are forced to find a new host. Low temperatures force insects to seek out warmer climates indoors. On the other hand, freezing outdoor temperatures reduce pest pressures by inactivating pathogens and arthropod pests. Insects and pathogen propagules are carried into greenhouses through vents and doors by wind. By locating a greenhouse away from and/or upwind of outdoor crops, many pest problems can be reduced to manageable levels.

Out of concern for maximizing productivity and crop uniformity, greenhouses are oriented for maximum light penetration. This usually means an east-west orientation for free standing greenhouses and gutter-connected complexes. Achieving good lighting uniformity over the course of a day is also important for IPM because insects and diseases proliferate in shaded areas and on stunted plants. In addition to orientation for optimal lighting, greenhouses should be oriented to take advantage of the prevailing winds. High wind speeds, if not reduced by windbreaks, increase heat loss and increase static pressures against which ventilation fans must operate. Moderate wind velocity at right angles to ridge, gutter and side vents is optimal for natural ventilation air movement through vents. As said before, the environs of the greenhouse may be reservoirs of pathogens and pests. Greenhouses are often in an arable area, with trash piles, weeds and crops botanically related to the crop being grown in the greenhouse to provide ample inoculum and infestations of pathogen vectors. Entry into the greenhouse can be rapid and on a massive scale: wind-blown dust carries spores and bacteria, air currents with or without forces ventilation carry spores and viruliferous insects from trash piles and weeds, water run-off into the greenhouse can carry soilbome pathogens such as *Pythium* and *Phytophthora* species and chytrid vectors of viruses, and dirt on feet and machinery carries pathogens.

A foot bath containing a disinfectant reduces this latter risk when placed at the doorway. To surround greenhouses by a 10-m band of weed-free lawn and to eliminate trash piles may prevent or delay pest and pathogen inoculum entrance into greenhouses. Though whitefly-proof screens can keep out most insects (and keep in pollinator insects) fungal spores and bacteria cannot be excluded. Diseases of tomato such as VerticiUium wilt, Fusarium crown and root rot, and bacterial canker are often first noticed directly beneath root vents or just inside doorways, as is the *Diabrotica*-borne bacterial wilt of cucumber. Overlapping of cropping, *i.e.* raising seedlings and transplants alongside production crops, is unsound hygiene, inviting infection and infestation of the new crop from large reservoirs in the old crop.

Carefully Designed of Heating Systems

A carefully designed heating system to maintain air and root zone temperatures close to recommended levels is essential for an effective IPM programme in greenhouses. In the northern hemisphere greenhouse heating systems should be designed to maintain the desired indoor temperature when the outdoor temperature is at the 2.5 per cent January design temperature (*i.e.* the temperature below which 2.5 per cent of the hours in January occur on average) for a given location. If it is expected the greenhouse will be heated from a cold start in January, then it is common practice to add another 25 per cent of pick-up capacity to the calculated 2.5 per cent January design heating load so that the greenhouse can be fully wanned up before plants are transplanted.

Centralized hot-water or steam pipe heating systems are the most practical for commercial greenhouses. Fan-forced unit heaters are practical for small greenhouses or in greenhouses where it is only desirable to maintain temperatures above freezing, but heat delivery from fan-forced units is too costly and very non-uniform on a large scale. With hot-water or steam heating systems, heat is delivered to the base of the plants via radiation pipes running between the crop rows approximately 15 cm above floor level. Low-level positioning of heat pipes is important to provide heat to the root zone and to induce vertical air movement via natural convection. The temperature of water circulating in hotwater heating pipes is adjusted from 40 to 90 °C depending on heating demand, thus heat is always applied at the base of the plants for a uniform temperature distribution.

The flow of steam at 100 °C through steam pipes is cycled on and off as required to maintain air temperature. This cycling leads to a non-uniform heating of the base of the plants and more temperature variability in steam-heated greenhouses. During very cold weather, operation of additional heating pipes around the perimeter and under gutters in hot-water and steam heated greenhouses is required to prevent cold spots where diseases are prone to develop. In hot-water heated greenhouses, especially those with tomato crops, an additional small-bore heating pipe is often used to apply heat at the growing tip of the plants to enhance growth and to prevent condensation on developing fruit.

FAILURE OF BIOLOGICAL DISEASE AND INSECT CONTROLS

A common reason for failure of biological disease and insect controls early in the greenhouse growing season, and later on when outdoor conditions become hot and dry, is very low humidity levels in the greenhouse air. Under these conditions, transpiration of the crop is not adequate to maintain humidity levels in the optimum range for biological controls and it is necessary to add humidity to the air. Under hot and dry conditions, addition of humidity to the greenhouse has the added benefit of evaporatively cooling the greenhouse air.

The theoretical and practical management of greenhouse humidity has been discussed by Stanghellini and Stanghellini and de Jong. The best humidification systems for greenhouses are those that create small water droplets that evaporate before they have a chance to settle out on leaves where they could provide the moisture necessary for germination of fungal spores. High-pressure (4–7 MPa) misting systems with 10 μm diameter nozzles and sonic misting systems that require a compressed air supply have been developed to create 10 μm diameter water droplets for greenhouse humidification. When properly maintained, these systems create a fog that gradually disperses as the water droplets evaporate in the air.

VENTILATION SYSTEMS

Intake of outdoor air and exhaust of indoor air is necessary to prevent excessive solar-heat gain or humidity build-up inside greenhouses. Most large scale greenhouse operations are passively naturally ventilated through vents in the roofs and side walls. Small greenhouses, and polyethylene covered structures that are not equipped with roof vents, are actively ventilated with fans. Gutter vent systems have recently been developed for polyethylene covered greenhouses that allow them to be ventilated passively. Ventilation rates required for summertime temperature control are 0.75–1.0 air changes per hour. Winter ventilation rate requirements are typically 10–15 per cent of summer requirements. The relationships between greenhouse geometry, vent geometry, wind speed, wind direction, temperature and natural ventilation rates have been established by Kittas. When greenhouse vents are closed, natural convection air movement inside greenhouses is often not sufficient for good air mixing and mass transport in the crop canopy.

At low wind speeds leaf boundary layer resistance increases, resulting in decreased transpiration and increased relative humidity at the leaf surface. In large greenhouse complexes overhead fans strategically placed above the crop are required to bring horizontal air velocities up to approximately 0.5 m/s for good air mixing and to minimize boundary layer effects. Air pressure differentials between inside and outside are necessary to move air actively through greenhouses. In actively and passively ventilated greenhouses, the pressure differential between inside and outside is usually negative, and it is easy for airborne pathogens and insects to enter the greenhouse, particularly if doors and ventilators are left open in hot weather.

In special circumstances where it is essential to exclude pests and disease propagules, it may be necessary to maintain a positive pressure differential. With such a ventilation system, air can be filtered as it is drawn into the greenhouse to remove insects but removing airborne fungal spores and bacteria is impracticable. With a positive pressure differential, there is less tendency for infiltration of insects and disease propagules from outside through cracks in the greenhouse cover. Regardless of type of ventilation system, any obstructions that reduce the vent openings increase the pressure differential and/or reduce the air flow through vents. If screens are placed over vent openings then the area of the vent openings must be increased by a factor equal to the reciprocal of the per cent free area of the screen material to maintain the same pressure differential.

If screens are used in established greenhouses, it would be necessary to build boxes over vents, add screened-in bays or screen the entire head space of a greenhouse to provide adequate intake air for good ventilation.

THERMAL/SHADE CURTAINS

Thermal curtains and shade curtains are generally beneficial for IPM because they reduce the extremes in climate that stress the crop and biological controls. Thermal curtains, aside from saving energy in the winter, reduce the net radiation from leaves through a greenhouse cover to a clear sky. For this reason leaf temperatures are higher and condensation on leaves is less under thermal curtains. Shading of greenhouses is necessary in hot climates to reduce solar radiation and heat stress on crops. Paints can be applied on the exterior surface of the greenhouse cover or shade curtains can be deployed inside or outside to attenuate the radiation reaching the crop. Moveable shading systems are also useful for acclimatizing crops and biological controls to rapidly changing solar radiation conditions.

Control Systems

The climate inside a greenhouse at any given time is determined by a complex interaction between outside climate variables, status of the crop and operating state of the climatecontrol equipment. Because of highly variable solar energy fluxes, the climate can change rapidly and climate-control equipment has to be manipulated quickly and frequently to maintain optimum conditions. The complex climate-control requirements of modem greenhouses can realistically only be met with computer-control systems. Climate-control computers have been specially developed to meet the demanding requirements of greenhouse operations.

The hardware used in greenhouses has been specially designed to withstand the high humidity and high levels of electrical noise. Special temperature and humidity sensing systems have been designed to monitor the inside and outside climate for control purposes.

These sensors are shielded from the sun and are aspirated so that control is based on measurements of true ambient air temperature and relative humidity. The software in commercial greenhouse computers has been specially developed to be fault-tolerant and to integrate the operation of climate-control equipment. In most cases the software has to be configured and control loops for each piece of climate-control equipment have to be tuned by the installer to give satisfactory performance.

Currently available greenhouse control software enables greenhouse operators to schedule climate setpoints for the conditions that they believe are best for production and IPM. The actual climate-control achieved is limited by the capabilities of the climate-control equipment and the operator's skill and knowledge.

PROTECTING CROPS FROM ARTHROPODS

In the Mediterranean basin, protecting crops from arthropods is regarded as more important than protecting them from the weather, so the physical exclusion of insects from the greenhouse should help in reducing the incidence of direct crop damage and also of insect-transmitted virus diseases, theoretically this exclusion can be done by fitting fabric screens of mesh aperture smaller than the insects' body width over ventilators and doorways, or by insect-repellent fabrics, but in practice there still can be significant insect penetration.

Moreover, screens impede ventilation and reduce light transmission, so compromises in the management of light, temperature and humidity are necessary to avoid adverse effects on crops and their susceptibility to diseases. Screens do not suppress or eradicate pests, they merely exclude most of them; therefore, they must be installed prior to their appearance, and supplementary pest control measures, such as biocontrol, are still required.

Insect parasitoids and predators that are smaller than their prey can still immigrate through pest screens into the greenhouse but larger ones have to be introduced. Since they offer an economical method of biological control of pests, they must be preserved, and destructive insecticides should be avoided. Screens impede ventilation, resulting in overheating and increased humidity. Increased humidity necessitates more frequent fungicide sprays than were required previously in an unscreened greenhouse. In Israel, 5–6 sprays per season (as opposed to 2–3 previously) are required in screened greenhouses.

To minimize these harmful effects, growers add forced ventilation but this only helps to pull whiteflies through the screen, while exhausting air from the screenhouse increases the intake of small insects. Application of positive air pressure, pushing air into the structure through an insect-proof filter, reduces whitefly influx.

Thus, while screens can reduce immigrant populations of pests, they also reduce the immigration of beneficial arthropods. In neither case is exclusion total. Screens are disadvantageous in that temperatures and humidities tend to rise, promoting plant stress and susceptibility to diseases, and they also reduce light. Access to the greenhouse by workers and machinery is more difficult.

TYPES OF SCREENS

Various types of screens and plastic covers have been developed to protect crops from insects; the challenge for the grower is to match the proper type of screen to local insect populations.

Woven Screens

The conventional woven screens are made from plain woven plastic yarns. Weaving leaves gaps (slots) between the yarns both in the warp and in the weft. In commercial screens the slot is rectangular whose width must be smaller than the whitefly's body size, about 0.2 nun, but it must allow maximum air and light transmission. Elongating the slot to improve ventilation is not feasible, since the threads slide apart, allowing insect penetration. Bethke and Pain found that screens designed to exclude *Bemisia tabaci* (Gennadius) still permitted some to penetrate, and they failed to exclude *Frankliniella occidentalis* (Pergande). They did, however, exclude most larger insects such as moths, beetles, leafminers, aphids and leafhoppers, and they retained bumble bee pollinators.

Unwoven Sheets

These are made of porous, unwoven polyester and polypropylene or of clear, microperforated, polyethylene fabric. All are very light materials which can be applied loosely and directly over transplants or seeded soil, without the need of mechanical support. They have been used primarily in the open field, in early spring, as spun-bonded row covers, to enhance plant growth and to increase yield. At the same time they also proved to protect plants from insects. A polypropylene perforated sheet protected tomatoes from Tomato Yellow Leaf Curl Virus (TYLCV) transmission by *B. tabaci*.

Thrips Exclusion

Whitefly-proof (50 mesh) woven screens are by far the most widely used covers for the exclusion of whiteflies and bigger insects. In laboratory tests, thrips, with a body width of only 245 μm moved freely through this screen. However, in the field, a high proportion (50 per cent) are excluded, possibly because of the optical features of the plastic. Western flower thrips are strongly affected by colour.

A loose shading net of aluminium colour, through which even whiteflies penetrated freely in the laboratory test, was tested in the field and in a walk-in tunnel. The aluminium screen reduced thrips penetration by 55 per cent over an identically shading net but white in colour. The closer aluminium fabric is placed around the entrance the more effectively it works.

Knitted-Screens

Because of irregularity in the shape of the holes, whiteflies are not excluded

(Berlinger, unpublished). Reducing slot size to block whiteflies reduced ventilation to an impractical level. However, knitted screens can exclude larger insects.

Knitted-Woven Screen

This plastic screen is produced by a technique that combines knitting and weaving. The slot is almost 3 times longer than in the commercial woven screen, while the width is smaller than the whitefly body size. The insect cannot pass, but ventilation is improved. A laboratory test confirmed the screen's high blockage capacity for whiteflies, which was similar to that of a conventional screen (0.1 per cent vs. 0.5 per cent penetration, respectively; Berlinger, unpublished).

UV-Absorbing Plastic Sheets

These are claimed to protect crops from insect pests and from virus diseases vectored by insects, by modifying insect behaviour but Berlinger (unpublished) was unable to confirm those claims. Nevertheless, these UV-absorbing plastic sheets have become available for commercial use.

Whitefly Exclusion

The sweetpotato whitefly (*B. tabaci*) is a small insect, about 0.2 mm wide, which transmits TYLCV, and has become the limiting factor in vegetable and flower production in Israel. Its physical exclusion from greenhouses is crucial, and accordingly whitefly-proof screens were developed. While the rate of whitefly exclusion is generally proportional to the screen's mesh ($R^2 = 0.85$), the insect's ability to pass through any barrier could not be predicted solely from thoracic width and mesh size.

There is an unexpectedly high rate of whitefly penetration resulting from a great variability among the samples of the same screen resulting from uneven and slipping weave

PROPER OPERATION AND MAINTENANCE OF CLIMATE-CONTROL EQUIPMENT

Proper operation and maintenance of climate-control equipment is essential for healthy crops and avoidance of disease and insect problems in greenhouses. Mistakes in climatecontrol settings or failures of key pieces of equipment can lead to devastating losses in a matter of minutes. Even if such events do not cause immediate crop losses, physiological, disease and insect problems often show up some time later.

The key to avoiding such problems is skilled operators and preventive maintenance programmes. Regardless of the level of equipment sophistication and maintenance, alarm systems together with backup power units and fuel supplies are essential to guard against losses during equipment breakdowns or service interruptions. Computer-control systems have taken much of the manual labour out of operating greenhouse climate-control equipment. A greenhouse manager should review climate data collected by the computer on a daily basis and make adjustments to setpoints to keep the climate conditions within desired ranges. It is critical that the temperature and humidity sensors used as the basis for control in each greenhouse compartment be cleaned and checked on at least a monthly basis. Greenhouse boiler systems need to be kept on line and in peak operating condition, not only during the winter heating months, but also in the summer months when it may be necessary to provide heat in the morning hours to avoid condensation on the crops. Vents and vent drives have to be kept in good working order to ensure they open when needed or close under high wind conditions when they could be damaged. Misting systems require stringent water treatment programmes to prevent nozzle blockages. The mechanisms for thermal and/or shade curtains have to be kept in alignment so that the curtains can be deployed quickly without snags or tears of the material. Insect screens have to be repaired if damaged. Also, insect screens have to be cleaned periodically to prevent blockages of light and air flow.

WORKER EDUCATION

For an effective IPM programme, greenhouse workers have to be trained to recognize nutrient deficiencies and disease and insect problems, and to take appropriate action. Personal protective gear, disinfectants, disposal bins, markers, etc. have to be made available to workers so that they can play their part in an IPM programme. In large operations, it is necessary to have a large site map of the greenhouses and a good recordkeeping system so that disease and pest outbreaks as well as control actions that have been taken can be noted for the information of all greenhouse staff. New decision-support software programmes offer great potential for education of workers and record-keeping of all greenhouse activities, including IPM.

MANAGING THE CROP

Crop Scheduling

Seeding, pricking-out and sticking cuttings should all be done in a greenhouse separate from the main production areas, and on mesh or slatted benches allowing through-the-bench ventilation. The benches should be well above the level of soilsplash and there should be no overhead pots from which contaminated soil and drainagewater fall. Where there is risk of diseases more destructive in cool soils, for example, Fusarium crown and root rot and corky root rot of tomatoes, transplanting should be delayed until the root zone has warmed up, and insulating mulch materials put down later. Where two or more monocrops are grown each year, overlapping of transplant production and marketable crop production means that pest and pathogen populations are perpetuated unless special care is taken to keep the young and cropping plants entirely separate. There is further risk if adjacent field crops constitute a reservoir of pathogens and pests.

Spacing

Close horizontal and vertical spacing of plants both on the bench and in the ground bed invites rapid plant-to-plant spread of walking insects, and of pathogens as diverse as *Pythium* spp., tomato mosaic virus, *Clavibacter michiganensis* Davis ssp. *michiganensis* Davis, the downy mildews and *Botrytis cinerea* Pers.: Fr. The agents of virus spread are mainly water and soil splash, insects, and workers handling plants with contaminated tools and fingers. Since air movement is restricted in dense plantings, the movement of airborne propagules is restricted, giving patchy distribution of diseases and insects. Moreover, close spacing results in undue inteiplant competition for water, nutrients, light and and undue damage by workers.

Sanitation

After genetic resistance, prophylaxis is by far the most effective and cheapest way of escaping major disease epidemics and pest infestations. It reduces the need for multiple applications of pesticides (which stress the crop), the risks of pesticide resistance, and pesticide contamination of the produce, the operator and the environment. Physical screening against immigrant pests has already been discussed, which, coupled with aggressive control of insects in the environs of the greenhouse and in adjacent weeds and field crops, is very effective prophylaxis against both direct damage and insect-transmitted diseases. Some growers rely on old crop prunings to perpetuate populations of biocontrol insects. This is not a good practice because they constitute a reservoir of pathogens and non-parasitized pests. New introductions of biocontrol insects are a better practice. Reducing inoculum is also important in early crop management, with such tactics as quarantine, seed disinfestation, the use of healthy mother plants for cuttings, micropropagation, removing and properly disposing of all previous crop debris, pasteurizing or solarizing soil and soilless media, and disinfesting the greenhouse structure, benches, trays, stakes and other materials. Disinfestants include formaldehyde (as formalin) and hypochlorites but both materials are hazardous to humans and residues are phytotoxic. A persulphate oxidising agent, however, destroys viruses and micro-organisms without such side effects.

The Growing Medium

Growing media cover a wide spectrum of substrates: soil and soil-mix composts, organic materials such as sawdust and coconut fibre, inorganic materials such as rockwool and synthetic foams and aggregates, and the nutrient film technique (NFT).

Soilborne diseases are no less prevalent in soilless substrates than in soil. All substrates must be substantially free of insects and pathogens at planting and must be kept so throughout the life of the crop, thus demanding a high standard of hygiene.

Soils are usually heavily amended with peat, farmyard manure, straw or crop residues. Ploughing or rotovating the soil should be done in order to comminute plant root debris and other organic matter, and so expose pathogen propagules to natural biological control. Getting the soil into good tilth with optimum temperature, water content and aeration promotes this microbial activity. Soils also harbour several insects, such as pupae of leaf miners and thrips, as well as fungus gnat and shore fly larvae, both of which vector *Pythium* and *Fusarium* spp.

Their populations, as well as populations of predatory microarthropods, are determined by soil organic matter, soil type and pore size. Populations of omnivorous collembola and non-cryptostigmatic mites, for example, are enhanced by the organic matter usually plentifully added to greenhouse soils. Fungal parasites of insects and nematodes are also encouraged in soils of good tilth. The root-knot nematode *Meloidogyne incognita* Chitwood, however, survives at 1–2 m, well below soil disturbance levels. Most substrates can be fumigated or heat-sterilized but pasteurization to about 70 °C or serialization to about 40–55 °C is preferred over total steam sterilization to 100 °C because it preserves thermophilic biocontrol organisms. The whole greenhouse can be closed in sunny conditions for solarization of both substrate and superstructure.

High temperature and vapour pressure deficit in closed greenhouses can kill the western flower thrips (*F. occidentals* but unfortunately also its predator *Neo-seiulus* (=*Amblyseius*) *cucumeris* (Oudemans). As with the original ideas that soilless cultivation would eliminate soilbome pathogens, crops in rockwool or other inert substrate, or in NFT are no less free of soilbome arthropods. Fungus gnats, leafminers and thrips are numerous in rockwool and shore flies are always present in pools of water on plastic sheets. Even if soil is covered with plastic sheet, there are always gaps around stems, and tears and displacement of the cover readily permit insect access.

Fruit Load

Closely related to the management of pruning is the distribution of photosynthates in heavily cropping plants in relation to the susceptibility of tissues to fungal and bacterial pathogens. As Jarvis pointed out, modern technology has increased yields of greenhouse vegetables several-fold in the last two decades, with accompanying source-sink stresses on cultivars that have not changed very much. Thus, diseases such as Fusarium crown and root rot of tomatoes and Penicillium stem and fruit rot of cucumbers have become serious in that same period. Both have been shown to be stressrelated and there has been a resurgence in the incidence of corky root rot of tomatoes that might be related to a diminished flow of photosynthates to roots. Grainger referred the "plunderable" carbohydrates available to certain pathogens – the so-called high-sugar pathogens– which include *B. cinerea,* whereas other pathogens, notably *Fusarium* spp., are classed as low-sugar pathogens principally attacking tissues starved of photosynthates. It is therefore incumbent on the grower to manage the nutrition, light and pruning of fruit and foliage so that a balanced partition of assimilates is attained without unduly compromising yield.

Managing Pesticides

Pesticides are a component of integrated pest management systems but are used too freely as insurance applications rather than judiciously as almost agents of last resort. Pesticides are significant agents of stress whose over-use leads to problems of resistance, to interference with microbial, insect biocontrol organisms and bee pollinators, and so to an increase in iatrogenic diseases, diseases normally held in check by indigenous biological controls. Unlike the pesticides on crops outdoors, pesticides in the greenhouse remain unweathered and persist longer, thus putting edible produce at risk of exceeding legallytolerated residues, and exposing workers to higher concentrations for longer. There are no well-established economic threshold populations of insect pests and pathogens and the grower must thus rely largely on his own experience and on the experience of his advisors. It is at present difficult, if not impossible, to predict the course of disease epidemics in the greenhouse because the complex sequence of events in the life cycles of pathogens is dependent on a succession of different microclimates occurring in the correct order.

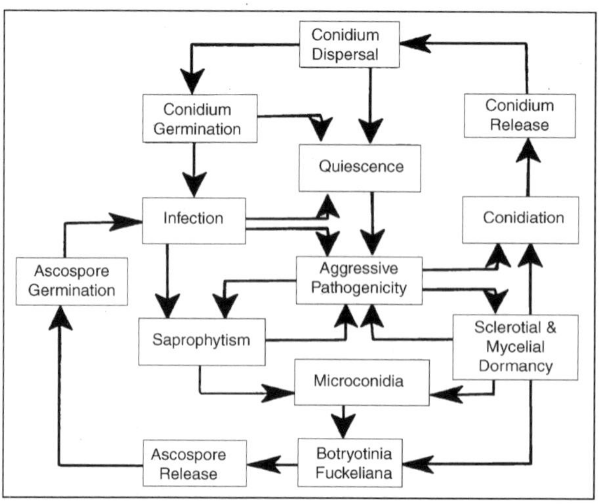

Nutrition

Deficiencies and excesses of macro- and micronutrients, and imbalances in relative amounts of fertilizers can predispose plants to most diseases. In addition, fertilizers that increase foliage density at the expense of flowers and fruit not only reduce yield but tend to lower the vapour pressure deficit (VPD) in the boundary layer by restricting transpiration and wind-assisted evaporation, and consequently increase the risks of infection. High nitrogen rates in fertilizers generally increase foliage density and softness, with increasing susceptibility to leaf and flower pathogens. For example, Hobbs and Waters found a quadratic increase in grey mould (*B. cinerea*) in chrysanthemum flowers (*Dendranthema grandiflora* Tzvelev) with nitrogen supplied with 1.5, 3.8 and 60 g/m^2. Nitrate nitrogen combined with liming gives excellent control of Fusarium wilt of several crops. Because of its role in the. integrity of cell walls, calcium imparts resistance if balanced with potassium in a high ratio. A low Ca:K ratio permits susceptibility to *B. cinerea* in tomato. The K:N ratio is important in the susceptibility of tomato stems to the soft rot bacterium *Erwinia carotovora* (Jones) Bergey ssp. *carotovora* (Jones) Bergey. The incidence of soft rot was low at a K:N ratio of 4:1, increasing at 2:1 and 1:1. Verhoeff noted similar trends in tomato stems infected by *B. cinerea.* Paradoxically Verhoeff noted that high soil nitrogen can delay the development of latent lesions of *B. cinerea* in tomato, possibly because stem senescence is delayed. Over-luxuriant foliage is conducive to greater damage by sap-sucking insects such as aphids.

Pruning and Training

Pruning and training tall staked and wire-supported crops like peppers, tomatoes and cucumbers not only modify the microclimate by altering spacing but pruning alters the fruit:foliage ratio and hence source-sink relationships in photosynthates and the disease-susceptibility of various tissues. Removal of leaves bearing prepupal and pupal stages of pests can reduce their populations, but premature removal of leaves bearing parasitized stages can result in loss of biocontrol.

MANAGING THE CROP ENVIRONMENT

DISPERSAL AND DEVELOPMENT

The effects of humidity on greenhouse crops have been reviewed by Grange and Hand, and their direct and indirect effects on diseases by Jarvis. Uncertainty about VPD and temperatures in the boundary layer raises considerable suspicion about the validity of countless experiments on the infective abilities of fungal spores and disease prediction systems at low VPDs and inadequately measured or inadequately controlled temperatures. Fungal spores and bacteria require a wet substrate in which to initiate infection, and the water on leaves and fruits is provided by dew, guttation or overhead irrigation. This last can be discounted in well-managed greenhouses as an invitation to pathogens. Fogging systems cooling the air by evaporation are permissible if all the droplets evaporate before they land on plants.

Measuring the onset and disappearance of dew is very difficult without the sensors themselves altering the boundary layer microclimate by heat conduction, shading, etc. Wei however, developed a copper-coated polyamide film sensor that could be wrapped around a tomato fruit and which had a response time of only a few seconds from dry to wet, and a response of less than 2 minutes to Peltier cooling of the surface to dewpoint. Connected to microclimate modifiers (heating, ventilation), this device could obviate much of the risk of infection.

Predicting the onset of condensation and its evaporation is even more difficult using atmospheric variables such as relative humidity, temperature, airspeed and radiation. Most predictions have errors in excess of 0.8 h and as much as ±3h. Clearly this is unacceptable in a cucumber house where infection of flowers by *Didymella bryoniae* (Auersw.) Rehm can occur in 1–2 h. Modelling the duration of dew in situations other than greenhouses has been done but with wide differences between predicted and observed durations of wetness. When the dewpoint temperature of the air falls below the temperature of the plants in a greenhouse, they become covered with water droplets and films, perhaps with hydrophilic fungal spores as nuclei, especially in still air at low VPD.

Wei developed a model from heat transfer theory that accurately simulated condensation and evaporation from tomato fruits still attached to the plant:

$$E_L = \frac{\rho\alpha Cp\left[e_a - e_s(T)\right]}{r_d\gamma}$$

where E_L is the latent heat flux, $ñ_a$ is the density of air, *CP* is the specific heat of air, *e*, and e_s (T) are vapour pressures of air and saturated vapour pressure of air, respectively at T°, r_d is the boundary layer resistance to vapour transfer between the wet surface and the air, and g is a psychometric constant. Using the wetness sensor of Wei obtained excellent agreement between simulated and measured fruit surface temperatures during condensation and evaporation, within 0.3–0.5 °C (standard deviation 0.4 °C). The model predicted wetness within 5 minutes of its detection, and dryness came as predicted. Clearly, this precision gives ample time for preventive action to be taken against most fungal infections. While free water and low VPD are to be avoided if pathogens are present, those very conditions are needed to establish epidemics of fungal pathogens of insects, such as *Verticillium lecanii* (A. Zimmerm.) Viégas, *Beauveria bassiana* (Balsamo) Vuillemin and *Paecilomyces fumosoroseus* (Wize) Brown and Smith. Similar contrary indications have been obtained for arthropod pests and their predators. While spider mites are most active at relatively high temperatures and low VPDs, their predator *P. persimilis* is inhibited in those same conditions. Optimum humidity conditions for the predatory activity of *N. cucumeris* has been established by Shipp and van Houten.

Temperature

In very general terms, diseases as well as arthropods can be said to have optimum temperatures for their dispersal and development but these cardinal points are the integral of the optima of several growth phases of the pathogen as well as of different defence reactions of the host Jarvis cited different temperature optima for different growth processes in the grey mould pathogen *B. cinerea:* mycelium growth, sporulation, conidium germination, germ tube growth, appressorium formation, sclerotium formation and sclerotium germination. All have different temperature optima, most of which lie above the general optimum range for grey mould development, 15–20 °C. In most of its hundreds of hosts, resistance to *B. cinerea* is probably least within that range. The temperatures of leaves and fruit can vary markedly from ambient air temperatures as determined by conventional greenhouse instruments, and so the temperature within the boundary layer can be assumed also to be different.

At night, energy lost by radiation from leaves can result in temperatures 1–3 °C cooler than ambient air and temperatures frequently reach the dew point. In crops transpiring well, evaporative cooling can also reduce leaf temperature but insolated leaves not transpiring can become considerably warmer, by as much as 2–8 °C, than ambient air. Similarly, Schroeder found that the temperature of red tomato fruits rose from about 20 to over 50 °C in air that rose from 26 to 37 °C in the same period. On the other hand, green fruits exposed to the same conditions remained 4–8 °C cooler than the red ones.

Temperatures of leaves, flowers and fruit can be considerably decreased by shading from direct sun and by increasing evaporative cooling by adequate ventilation and forced air flow. Eden *et al.* discussed the possibilities of raising flower truss temperatures in tomato crops to avoid grey mould. Whereas higher temperatures resulted in increased numbers of flowers infected by *B. cinerea,* the fungus was less likely to grow proximally to the main stem where the damage would be far more severe than one infected flower.

On the other hand, higher temperatures (20–25 °C) resulted in fewer infections of stem wounds than at 15 °C. Eden interpreted these results in terms of changing balances between fungal aggression and host defence reactions. Just as with diseases of shoots, temperatures can be to some extent selected to minimize diseases of roots; for example corky root rot *(P. lycopersisci)* of tomato can be largely avoided by transplanting into warm media at 20 °C, as can Fusarium crown and root rot (F. *oxysporum* f. sp. *radicis-lycopersici)*. By contrast, the optimum temperature for the expression of Fusarium wilt is 27

°C. Similarly, *Pythium aphanidermatum* (Edson) Fitzp. is most pathogenic to spinach in hydroponic culture at 27 °C, whereas *Pythium dissotocum* Drechs. is most pathogenic at 17–22 °C. It is therefore important to know exactly which of closely related pathogens is present. Insects and mites, like diseases, have also an optimum temperature for their activity, dispersal and development Generally, greenhouse pests are thermophilic and perform best within 20–30 °C night-day ambient temperatures. The preferred temperature for aphids and the greenhouse whitefly is somewhat lower, 15–25 °C. The interaction between temperature and VPD on the survival of western flower thrips was determined by Shipp and Gillespie.

Of course, temperature affects not only arthropod pests but also their natural enemies. Natural enemies may perform poorly if temperatures are too high or too low which may occur during summer and winter respectively in the Mediterranean area. Then, the more temperature-tolerant *Diglyphus isaea* (Walker) or *Dacnusa sibirica* Telenga can be used according to thermal regimes expected in greenhouses. Excessive heat, combined with high VPD is a serious constraint for *Phytoseiulus persimilis* Athias-Henriot in warmer Mediterranean areas. Shipp and van Houten determined optimum temperatures and VPD for the use of *N. cucumeris* in Canadian cucumber houses, and these types of studies serve as guides to more intelligent biological control.

Water Stress

Guttation results when the rate of water supply osmotically pumped by the roots exceeds the rate of water lost by transpiration and used in growth. To prevent guttation, the osmotic potential of the root xylem must be more negative than that of the nutrient solution. In poorly-managed greenhouses, guttation frequently happens at night when VPDs are low and root temperatures maintain high metabolic activity and root pressure.

Tissues become waterlogged (oedema) and water guttates from stomata and from hydathodes at leaf margins with profound effects on the phylloplane micro-organisms.

Water continuous with the surface and substomatal vesicles facilitates the entry of bacteria into leaves of for example *Pelargonium* spp., particularly when resumed transpiration leads to resorption of the water. Wilson described how reversal of transpiration flow permits conidia of *B. cinerea* to enter tomato stem xylem, there to remain a latent inoculum. Water alternately accumulating and evaporating from hydrothodes leaves toxic deposits of salts, a ready entry point for necrotrophic pathogens. Lesions of gummy stem blight (*D. bryoniae*) are frequently seen originating from such points on cucumber leaves. Guttation damage can easily be eliminated by regulating atmospheric humidity, ventilating effectively, reducing evening watering and adjusting the osmoticum of nutrient solutions.

Light

Setting aside the effects of daylength on flowering in florists' crops, photosynthetically active radiation (PAR) (400–700 nm) is the part of the spectrum with the greatest effect on crop growth and productivity. Low and high light intensities are important agents of stress in crops that induce physiologic strains predisposing the crops to disease. Particularly important is the effect of light combined with crop management procedures, such as plant spacing, row orientation, training and pruning systems, irrigation and nutrition, on the partition of assimilates, and the relative susceptibility of different tissues and organs to disease. Daylength, however, is important in determining diapause in both arthropod pests and their predators.

Early diapause may be a major constraint in their use. Non-diapausing strains can, to some extent, overcome this problem. Light also has direct effects on fungal sporulation, germination and sclerotium formation. In *B. cinerea,* most isolates are stimulated to form conidia by light in the near-UV band (320–380 nm), an effect temporarily reversed by blue light.

Some isolates, however, form conidia in the dark. All fungi grow mycelium in the dark, and *B. cinerea* forms its sclerotia in darkness, or in yellow or red light, or when irradiated for less than 30 min with near-UV light. The requirement of B. *cinerea* and some other fungi for near-UV light for sporulation has led to the development of greenhouse covering materials that screen out that band as a means of disease control. Tuller and Peterson found fibreglass to transmit much less light of 315–400 nm than did polyethylene but in a comparative assessment of grey mould in Douglas fir seedlings [*Pseudotsuga menziesii* (Mirb.) Franco] it was concluded that the principal effect of low irradiance transmitted by fibreglass was in inducing needle senescence in dense canopies and thus susceptibility to grey mould, rather than on any direct effect on fungal sporulation. In both types of greenhouse, the mean intensity of light that inhibited sporulation (430–490 nm) exceeded that that promoted sporulation (300– 420 nm). In those greenhouses, too, predisposing conditions of temperature (15–20 °C) and humidity (>90 per cent RH) persisted 14.5 times longer in fibreglass than in polyethylenecovered houses.

Humidity effects also seem to have outweighed effects of light wavelength in a series of trials with coloured cloches covering strawberries. Grey mould was most severe under pink and blue plastic covers, where VPDs were lower (0.41 and 0.64 kPa, respectively) than under clear plastic (1.14 kPa), or under glass (1.74 kPa). The effects of light are evidently not simple. Nevertheless, attempts have been made to filter out the near-UV light that induces sporulation in some fungi. Reuveni incorporated hydroxybenzophenone into polyethylene, which increased the ratio of inhibitory blue light (480 nm) to UV (310 nm), and reduced the sporulation of *B. cinerea* in polystyrene petri dishes. Under the treated plastic, grey mould lesions were fewer in tomato and cucumber (17 and 15, respectively) than under untreated plastic (41 and 29, respectively). Similarly, plastic coverings absorbing light at 340 nm inhibited the sporulation and reduced the incidence of grey mould lesions on cucumber and tomato as well as white mould lesions caused by *Scerotinia sclerotiorum* (Lib.) de Bary. Many isolates of *Altenaria solani* Sorauer also depend on near-UV light for sporulation, and Vakalounakis used vinyl films filtering out light of <385 nm to reduce the incidence of early blight in tomato greenhouses to less than 50 per cent of that under unamended vinyl film. Except as an agent of stress on the host, light has little direct effect on the rhizosphere microflora.

Integration of Environmental Factors

Epidemics of diseases are the result of a complex sequence of biological events each with a different set of permissive environments that have to occur in sequence, and coupled with hosts in a receptive state. Jarvis outlined the complexity of those events in the case of grey mould epidemics. Beginning with sporulation, conidia are formed at temperatures around 15 °C and in moderate VPD; they are liberated by hygroscopic movements of the conidiophore in rapidly changing conditions of humidity, and are dispersed on air currents or by water-splash; infection occurs on wet surfaces at 15–20 °C; and colonization of the host is fastest at 25–30 °C.

Marois found that epidemics of grey mould on rose depend as well on inoculum concentration, a relationship that was different in winter and summer, and affected by temperature, relative humidity and VPD, the latter the far more meaningful parameter for describing epidemiology of *B. cinerea* in roses. It has been possible to construct working models of grey mould epidemics in cucumber; tomatoes; gerbera and rose; and conifer seedlings. The value of epidemic models such as BOTMAN, an integrated chemical and biological control programme, in predicting the onset and course of epidemics, however, is severely compromised by the rapidity with which infection occurs – 9–10 h for grey mould and only 1 h for gummy stem blight in cucumber flowers – and by the wide variability of the greenhouse climate typically served by only one psychrometer in several hundred cubic metres of space. Shtienberg and Elad found that over three years, a rain forecasting system did not enable BOTMAN to perform significantly better than a weekly fungicide insurance programme in unheated tomato and cucumber crops.

However, a 4-day weather forecast proved more useful than immediate past records of weekly averages of surface wetness (calculated from dewpoint) of 7 h/d and 9.5 h/d at night temperatures between 9 and 21 °C. By the time the requisite data have been collected and analysed, infection has already begun, and is an irreversible action even with the use of fungicides, which act mostly on germinating spores and thus too late to stop infection. Surface wetness is the key factor in all infections, and so its prediction from rates of change in surface and ambient air temperatures combined, by data processor, with simultaneous rates of change in VPD would be more timely in the immediate application of environmental control measures. Powdery mildew epidemics have a somewhat less complicated sequence of events prior to infection than grey mould epidemics but they, too, are ultimately dependent on the deposition of dew.

Control of any fungus-incited disease is achieved by breaking any of the pathways in life cycles similar but the denial of water to germinable spores is the most important. Computer models can be used to optimize greenhouse climate for both crop production and pest and disease control. For example, in The Netherlands a climate management programme was developed for optimal production of tomatoes and is linked to a model for biological control of greenhouse whitefly by *Encarsia formosa* Gahan. Further, the model can be extended with a humidity management module which prevents the development of fungal diseases. Integration of pest and disease control primarily by manipulating the environment is a highly complex problem. Clarke, in describing a computer-managed system, considered the holistic production system as a six-hierarchy complex of factors in which any change at one level affected the other five levels. Thus, any change in greenhouse climate, whether engineered or not, effects changes in pesticide efficacy, biological control agents, pests and disease vectors, diseases, and ultimately productivity and profit. There are a number of electronic decision support systems for various facets of greenhouse pest and disease control and production strategy.

Jones described an expert system with grower selection of climate set points based on his experience; Jacobson further developed an expert system with pre-set points for tomato production; and Dayan developed TOMGRO that modelled physiological processes in tomato. Only Martin-Clouaire considered disease escape in their model for tomato. Van Roermund, however, described the apposition of a whitefly control model to a production model, to which can be added a disease-avoidance model. Clarke and Jewett described a holistic Harrow Greenhouse Crop Management System (HGCMS) for both greenhouse tomato and cucumber.

In addition to providing blueprints for production in which the grower has his own input, HGCMS provides user-friendly diagnoses for diseases, pests, biological controls and physiological disorders. It accepts climate monitoring. In addition, HGCMS allows the grower to enter economic data, and will analyse it for him. Conflict resolution, as far as can be agreed among experts, is a feature of HGCMS but ultimately the grower can accept or reject the advice of HGCMS. The use and analysis of computer models and controls depends, of course, on a reasonable degree of computer literacy among growers, together with a basic understanding of plant growth and pest and disease biology. Otherwise reliance on expert advisory services is obligatory.

Environments for Microbial Controls

In general, the microclimates for the successful deployment of fungal antagonists and parasites are close to those that promote plant infection by pathogens. Ideally, then, pre-emptive colonization of the phylloplane, as it is for rhizosphere, is the preferred strategy. Adaptation to that microenvironment is a prerequisite. This colonization can also be achieved by enhancing indigenous populations of phylloplane antagonists. Similarly, the use of green manures and composts can achieve control in the rhizosphere without the necessity of isolating, registering and redeploying specific antagonists. McPherson and Harriman have suggested that in recirculating hydroponic systems, antagonist populations build up naturally in a disease-suppressive system that is reminiscent of take-all decline in wheat.

The primary objective of the commercial greenhouse grower is to obtain maximum yield per unit area of space with the least financial input. However, in order to achieve this, certain minimum standards in environment management have to be maintained in such matters as crop spacing, pruning and training, irrigation, fertilization, Co_2 supplies, and temperature and humidity regimes. While much is known about disease epidemiology and insect behaviour, scant attention, however, has been paid to the manipulation of greenhouse environments expressly to avoid disease epidemics and insect infestations, which together can easily account for 30 per cent crop losses. This is a significant factor in a grower's balance sheet which is often overlooked, and usually dealt with simplistically by indiscriminate pesticide applications.

Careful analyses of epidemiological and epizootic data can indicate environments to be avoided or encouraged in greenhouse operations but integrating the desired environments into those wanted by the grower solely to maximize yields by physiological means is extremely difficult. The solution of these problems requires the consensus of several specialized experts, experienced crop advisors and, not least, good growers, whose experience and intuition are not to be ignored.

The construction of predictive models can provide valuable insight into how environments affect diseases and insects, but experts can differ widely on which environment is best to escape, for example, lettuce downy mildew, or grey mould, or whitefiies or thrips. Resolution of these apparent conflicts can now be attained, or at least reasonable compromises achieved, by the inference engine in a computer expert system. One developed by Clarke and Jewett is a decision support system for greenhouse tomatoes and cucumbers that collates expert opinions on all aspects of crop production, including disease and pest management, the grower's own input, and internal and external environmental parameters. It can also provide the financial consequences of various actions, as well as of no action. Ultimately, the grower, whose brain no-one can replace, has the final decision.

Carbon Dioxide and Oxygen

Carbon dioxide enrichment is a standard procedure in many commercial greenhouses but because it necessarily involves some restriction in ventilation to achieve the concentrations of CO_2 required, of the order of 1000 vpm, there is often increased danger of unmanageable low VPD. The concentrations of CO_2 that impair the growth of *B. cinerea* are 2–3 orders of magnitude greater than those found even in CO_2^-enriched greenhouses and so reports, for example, of Winspear, of increased incidences of grey mould in CO_2^- enriched greenhouses, can be interpreted in terms of enhanced levels of assimilates, or a denser canopy, with its increased risks of disease-susceptible wet plants. While CO_2 is a prominent component of the rhizosphere atmosphere as a product of root and microbial respiration, it has little direct effect on pathogens.

Oxygen deficiency stress readily occurs in compacted and waterlogged soils and in over-warm hydroponic solutions in which both increasing temperature and increasing solute concentration decrease oxygen solubility. Further, increased temperatures lead to higher root and microbial respiration rates which further deplete oxygen tensions. Low oxygen tension has been advanced as an explanation for physiological root death as well as decreased host resistance to root pathogens.

SAFETY OF CROPS TO NON-TARGET INVERTEBRATES

Cry proteins produced by transgenic plants are not easily extractable in the amounts that would be required for studies designed to test the effects on non-target organisms. Thus, the effects are currently assessed by feeding test species Cry proteins produced in either Escherichia coli, a Bacillus species, or on various Bt crop tissues such as leaves or pollen. Tests of Cry proteins produced in E. coli are similar to those used to evaluate these proteins when produced by B. thuringiensis, except that in many cases an activated form of the toxin is used to produce what could be considered a 'worst case' risk assessment. To complement labouratory studies, several field studies have been conducted in which non-target insect populations were monitored on Bt crops, mainly Bt maize and Bt cotton, throughout the growing season. Most of the labouratory studies have been performed in the United States, where a complex of non-target organisms serves as a standard group for which results are accepted by the US Environmental Protection Agency.

The standard test invertebrates have included a range of terrestrial and freshwater aquatic organisms generally considered beneficial. These typically are larval and/or adults of one or more of the following organisms: the honeybee, parasitic wasps, predatory ladybird beetles and lacewings, 'soil-dwelling' springtails, earthworms, and as a representative of a fresh-water aquatic crustacean, a daphnid. The non-mammalian non-target vertebrates have typically been a bird and a fish, respectively, the bobwhite quail and the channel catfish. In these tests, the non-target organisms were typically exposed to or fed amounts of toxin that were in the range of at least a hundred to several thousand times the amount they would be exposed to or consume under natural conditions.

In such tests, when no effects are observed at the highest dose or rate tested, this amount is referred to as the no-observed-effect-level (NOEL). For a crop like Bt maize, the amount of Cry protein in a maturing field is estimated to be about 500 g per hectare, and thus the test levels are adjusted to ensure a dose at least 1000 times this level.

Most of the studies conducted so far have been short-term studies, lasting from several days to a few weeks. In addition to these studies aimed at testing high levels of exposure to Cry proteins under situations that resemble a natural situation, there have been tests where non-target invertebrates have been forced to feed on Bt crops. In Italy populations of the aphid Rhodopalosiphum padi were force-reared on Bt maize (Cry1Ab).

Then its natural predator, the green lacewing, Chrysoperla carnea, was fed on these aphids. In comparison to control populations, no adverse effects were noted on either the aphid or lacewing populations with respect to survival and fecundity. In contrast to these results, other labouratory studies have shown that force-feeding can result in mortality to certain non-target insect species.

For example, it was shown that immature C. carnea fed on prey that had been fed Bt maize (Cry1Ab) suffered greater mortality than control lacewings fed on non-Bt maize-fed prey. Only 37 per cent of the C. carnea fed Bt maize-fed larvae of the cotton leafroller, Spodoptera littoralis, or the European corn borer, Ostrinia nubilalis, survived, whereas 62 per cent of the control group fed on non-Bt maize-fed caterpillars survived. In a subsequent study, using an artificial liquid diet it was determined that immature C. carnea were sensitive to the Cry1Ab toxin at a level of 100 μg/ml of diet.

However, the level of Cry1Ab in maize is about 4 μg/g fresh weight, which is considerably less than 100 μg/ml. At present there appears to be no explanation for the discrepancy between the artifical diet and exposure via prey. The apparent sensitivity of C. carnea to the Cry1Ab is an interesting labouratory observation, and should be validated with histological and receptor-binding studies.

The most widely publicized study of the potential impact of Bt maize on a non-target insect is the finding that larvae of the monarch butterfly, Danaus plexippus, are sensitive to certain doses of Bt maize pollen. In this labouratory study, milkweed leaves were covered with Bt-maize pollen and then fed to larvae. Control larvae were fed on milkweed leaves covered with non-Bt pollen or untreated milkweed. The key finding of the study was that the larvae fed milkweed leaves treated with unquantified amounts of Bt pollen had a lower survival rate (56 per cent) than controls (100 per cent).

The authors stated that their results had 'potentially profound implications for the conservation of monarch butterflies' because the central corn belt where Bt maize adoption by farmers continues to grow is also an important habitat for monarchs. In assessing the relevance of these findings on the green lacewing and monarch larvae, or other non-target organisms for that matter, it should be kept in mind that bacterial insecticides based on Bt should be just as toxic if not more so because they contain more Cry proteins than Bt crops.

In other words, monarch larvae that feed under field conditions on milkweed leaves treated with a product that contains B. thuringiensis subsp. kurstaki, from which the Cry1Ab protein gene was derived, will be equally if not more sensitive to the bacterial insecticide. Similarly, C. carnea immatures that feed on caterpillars intoxicated as a result of feeding on a Bt insecticide will be equally sensitive to the activated toxins in these larvae. So the issue here is not so much one of Bt crops, but whether Cry proteins will impact beneficial insects regardless of the source.

Despite the sensationalistic attention given to the monarch study by the popular and even the scientific press, the design of this study has been criticized by scientists familiar with monarch biology and maize production. The principal criticisms are that monarch larvae were not fed well-quantified amounts of Bt maize pollen, that larvae given a choice may find milkweed contaminated with Bt pollen unpalatable and thus should have been given a choice, that Bt pollen is only present for 7-10 days of the growing season, and that the most Bt pollen dispersed remains within several meters of the dispersal in the field. Thus, maximum monarch exposure would normally be limited to larvae developing within or at the edges of corn fields, but not in other habitats.

As a result, only monarch larvae developing in very close to Bt corn fields during pollen-shed would typically encounter quantities of Bt pollen grains on the surface of milkweed leaves. Extraordinary attention was given to the preliminary findings in the scientific and popular press on the potential negative effects of Bt pollen on monarch populations.

A benefit of this attention was that it resulted in a series of collabourative studies in 1999 and 2000 devoted to a much more rigorous assessment of these potential negative effects under field conditions throughout the US corn belt and Canada, as well as in the labouratory. The overall conclusion of these studies is that the effects of Bt maize on monarch populations will be negligible, especially in comparison with the effects of using chemical insecticides to control maize pests.

The basis for this conclusion was that most of the registered Bt maize varieties produced only a low level of Bt protein, and even moderately high levels of consumption of this pollen by larvae did not affect their development or survival. Other 'mitigating' factors that limited exposure of larvae to Bt proteins in pollen included limited overlap between larval feeding periods thoughout the corn belt and the period of pollen shed, the relatively limited adoption of Bt maize (19 per cent), and the widespread availability of milkweed outside areas where maize is grown.

Significantly, larvae in maize plots treated with the chemical insecticide (-cyhalothrin) showed no weight gain because all were killed by the insecticide within 24 hours of treatment. The experimental design used to assess risk in these studies provides a good model for other studies in that the critical aspects of the assessment focused on experiments under field conditions, further supported by labouratory experiments.

Moreover, the relative risks to a non-target organism of using Bt maize or a chemical insecticide were compared, thereby eliminating the 'double-standard' that is often applied to assessing risks of Bt crops, in other words, applying standards to Bt crops that are much higher than those applied to chemical insecticides. Regardless of whether the results obtained against non-target organisms in labouratory studies show favourable, unfavourable or neutral effects, these must be followed by long-term studies under field conditions. The reason is that labouratory studies are designed to reveal any potential adverse effects by exposing non-target organisms to excessively high levels of Bt proteins, levels that would not be encountered under field conditions. Moreover, field studies should include comparisons to current agricultural practices, which often will include the use of chemical insecticides.

At present, there have only been a few studies so far that have evaluated the effects of Bt crops on non-target organisms under field conditions over the length of the growing season. These studies are nevertheless important because non-target organisms and their prey were exposed to Bt Cry proteins in the form synthesized in the crop and over a continuous period at an operational level. The non-target organisms studied under field conditions have all been insects and the test crops either Bt maize or Bt cotton. The insects consisted of a plant bug and four beneficial insects, specifically two parasites and two predators, one of which was the lacewing, C. carnea.

In these season-long studies, no adverse effects were observed on any of the non-targets under field conditions, even on C. carnea, is apparently sensitive to Cry1Ab. Other non-target organisms that have recently received attention with respect to the environmental effects of Bt crops are soil microorganisms.

Some in the labouratory, others under quasi-field conditions, it has been shown that Bt proteins such as Cry1Ab released into soil can bind to various types of soil particles and remain active for periods of at least several months. The consequences of this accumulation remain unknown, but recent studies by Stotzky and his colleagues indicate there were no significant effects on soil microorganisms or earthworms.

Evidence from other studies indicates that Bt Cry proteins are rapidly degraded after incorporation of Bt crop tissue into soil, although low levels can persist for at least several months. While these studies indicate that the temporary accumulation of Bt Cry proteins in soil have no adverse effects, longer term studies are needed to further evaluate these initial results.

SAFETY OF BT CROPS TO NON-MAMMALIAN VERTEBRATES

With respect to non-mammalian vertebrates, tests of Bt maize have been carried out using bobwhite quail, channel catfish and broiler chickens, the latter in a 38-day study in which the chickens were fed on Bt maize grain. In the chicken studies, chickens fed Bt maize grain were compared with chickens fed non-Bt grain. The variables examined were survival, weight gain and feed efficiency. No significant differences were found between the chickens fed Bt maize grain versus non-Bt maize grain. Similar results were obtained in other studies in which the effects of feeding Bt maize versus non-Bt maize to chickens were studied.

Safety of Crops to Mammals

There have been very few studies published in the scientific literature in which mammals were fed on Bt crops and then monitored for effects. Most of the safety data used to register Bt crops, is based on studies of Cry proteins tested as bacterial insecticides or produced in E. coli. But there has been at least one comprehensive study in which the effect of feeding rats on Cry1Ab tomatoes for 91 days was examined, and amplified with other studies using Cry1Ab produced in E. coli. In this study, weanling Wistar rats were fed for 91 days on a lyophilized powder of Cry1Ab tomatoes (40.6 ng Cry1Ab/mg protein) that made up 10 per cent of the diet. Control rats were fed a similar diet consisting of a 10 per cent non-Bt lyophilized tomato powder. In terms of human consumption, this amount was calculated to be the equivalent of a human eating 13 kg of tomatoes per day. No significant differences were observed between the treated and control rats in terms of survival, food intake, body and organ weights, or other gross macroscopic examinations of tissues.

The lack of any apparent adverse effects resulting from feeding rats on a high dose of Cry1Ab tomato corresponds to results obtained in feeding rats bacterial toxins containing Cry1Ab. In related studies, these investigators used Cry1Ab produced in E. coli to study the effects and fate of this protein on various mammals and mammalian tissues.

Histological binding assays were conducted to determine whether the Cry1Ab could bind to the intestinal mucosa of mice, rats, rhesus monkeys and humans. Cry1Ab did not bind to microvilli in these mammals in any of the intestinal regions, including the esophagus, stomach, duodenum, jejunum, ileum and colon, though the toxin did bind to microvilli along the entire length of the larval midgut in the tobacco hormworm, Manduca sexta, the lepidopteran used as the insect control. In a separate study, gastrointestinal tissues were examined

7 hours after rats were fed an amount of Cry1Ab equivalent to a human consuming 2000 kg of tomatoes in a single day.

No binding was detected in any of the above intestinal regions. These results indicate that none of the mammalian species tested had receptors for Cry1Ab on their intestines. In other studies, it was found that the active toxin form of Cry1Ab (66-68 kDa) was degraded to peptides of 15 kDa or less within 2 hours when incubated in simulated human gastric juices at pH 2. In studies using Cry1Ab consumed ad libidum in drinking water for 31 days by New Zealand white rabbits in an amount equivalent to humans consuming 60 kg of tomatoes per day, no differences were found between treated and control groups.

Histological studies of the stomach and intestines showed no abnormalities, and no antibodies were induced against Cry1Ab in the treated rabbits. In another study of a Bt crop, Bt potato producing an unidentified Cry1 protein was fed to one-month-old mice for two weeks, after which the mice were sacrificed and sections of the ileum were examined by light and electron microscopy.

The control mice were fed on non-Bt potato. No statistically significant differences were observed between the mice fed Bt potato and the control mice. In another Bt potato study, raw potatoes containing the Cry3A protein (used to control the Colourado potato beetle) were fed to rats for 28 days. The control was the same potato variety that did not produce Cry3A. The rats were fed levels of Cry3A corresponding to a human consuming several kilograms of potato per day for 28 days. No adverse effects were observed. The treated and control rat groups were similar in weight gain and behaviour. In other studies on the effects of Bt grain on the health and perfomance of other animals, no significant differences were found between groups fed Bt versus non-Bt grains in dairy cows, beef cows or steers. This combination of studies using a variety of methods and feeding studies that include Bt crops and candidate Cry proteins validate the use of these methods for assessing safety and risk, and support the use of data developed for the registration of bacterial insecticides to support the safety of Bt crops.

Most studies of Bt crop safety have focused on tests designed to detect potential harmful effects. There have been few studies of potential improvements in safety, though it is often noted that Bt crops should be safer because they reduce chemical insecticide usage. Yet one notable improvement in safety that has not received much attention is that Bt maize grain contains lower levels of certain fungal toxins known as fumonisins. These toxins can cause illness and death in horses and pigs, and have been implicated in certain forms of liver and esophageal cancer in humans.

Fumonisins result from ear rot infections of maize by Fusarium species, which invade tissues damaged by the European corn borer, O. nubilalis. In Bt maize grain, the levels of fumonisins are reduced by as much as 93 per cent in comparison with non-Bt maize grown under similar conditions. Future of Bt crop safety in perspective Insecticides based on endotoxin proteins of B. thuringiensis have been in use for 40 years, and have a safety record for non-target invertebrates and vertebrates including mammals that far surpasses that of any synthetic chemical insecticide.

This safety record, combined with the efficacy of certain Bt Cry proteins and the advent of recombinant DNA technology, led to the development of transgenic insect-protected Bt crops that are being adopted rapidly, especially Bt cotton and Bt maize, by farmers in the United States and a few other countries. The safety of these crops for non-target organisms is based primarily on data from numerous toxicological studies carried out on bacterial isolates used to register Bt insecticides, and to some extent on Cry proteins produced in E. coli or Bacillus species.

There have been comparatively few studies so far that have directly evaluated Bt crops in feeding studies. Those that have been done provide strong evidence that Bt crops are safe for humans and other animals as well as for most non-target invertebrates, with the exception of insect species closely related to target insect pests, and a few other species. There is some evidence based on forced-feeding labouratory studies that non-target organisms such as larvae of the monarch butterfly and lacewings may be adversely affected or killed by Bt proteins produced in Bt crops.

But the limited number of field trials conducted to date have found no evidence that Bt crops adversely affect generalist insect predator populations or parasite populations. In fact, evidence is mounting that Bt crops may enhance biological control and integrated pest management programmes by reducing the amounts of synthetic chemical insecticides used to control insect pests, thereby maintaining the populations of beneficial insects, which are typically decimated by the use of chemical insecticides.

In addition, Bt crops such as Bt maize may actually be safer for consumption by mammals than non-Bt crops because the control of lepidopteran pests reduces the level of certain mycotoxins that result from fungal infections following insect damage. Whereas existing evidence attests to a high degree of safety for Bt crops, these should be more extensively evaluated directly through detailed feeding and toxicological studies, and through more extensive and long-term ecological studies to determine their effects on non-target invertebrate populations under operational field conditions.

Given the public controversy that has arisen over genetically engineered crops, numerous such studies are underway, and will find their way into the refereed scientific literature in the coming years. In designing safety studies for non-target mammals, more emphasis should be placed on feeding studies in which animals are fed Bt crops, rather than relying solely on surrogate data developed from feeding studies using Bt insecticides or Cry proteins produced in E. coli. This should at least be done for a range of Cry proteins used or destined for use in a variety of Bt crops to allay the concerns being raised by public. The costs will be minor compared to the investments being made in Bt crops and the advantages that will accrue long term from having high public confidence in the safety of transgenic crops. Though there are technical hurdles that must be overcome in 'direct' feeding studies due to the low levels of Cry proteins in Bt crops, tests like those developed by Noteborn *et al.* indicate that such studies can be done. The results of less direct methods, such as chemical fingerprinting of metabolic changes that might be detected between Bt crops and their isogenic parents, would appear unnecessary and could easily lead to questionable interpretations.

For example, what if the metabolic pool of a mRNA for a certain enzyme is found to be twice as high in a Bt tomato as opposed to its isogenic parent, but all the direct feeding studies show that the tomato is safe? And what if the mRNA for the same enzyme in another tomato variety is shown to be just as high as the mRNA in the Bt tomato. Do any of these results mean the Bt tomato is not safe? It is important that any new tests developed to evaluate safety be directly relevant to providing data that can serve as a scientific basis for the further assessment of safety.

Of course, all methods used need to be subjected to rigorous validation. Studies that result in data of questionable value or that could be subject to misinterpretation should be avoided. Aside from the scientific rationale for taking this approach, a reason for doing this is that many individuals and environmental groups are opposed to Bt crops simply because they object to the use of genetic engineering technology to modify crops. A good example of this sort of thinking comes from Lord Peter Melchett, a former Labour Minister in the United Kingdom, and a leader in Greenpeace's effort to stop the use of biotechnology in agriculture.

Lord Melchett opposes the use of science to make decisions on food consumption, preferring that selections be made on an emotional basis, stating that 'If its acceptable to choose our car based on emotion and not science, why should it be wrong to choose our food that way?' The public cannot afford to let this sort of thinking dominate or influence views and decisions on what is safe to eat. Therefore, to counter such thinking and avoid the misuse of data, meaningful short- and long-term studies like those carried out by Noteborn *et al.* should be the type used to evaluate Bt crop safety.

Such studies should detect any adverse effects of natural or modified proteins synthesized by crop plants. In addition, the same rationale is relevant to justifying long-term, *i.e.* studies of several years, on the effects of Bt crops on non-target organisms, including soil microorganisms to assess any potential ecological effects, be they positive, neutral or negative. The results of initial studies indicate that Bt crops will be much safer for non-target organisms, especially beneficial insects, than chemical insecticides.

If long-term studies confirm these initial results, dissemination of this knowledge should improve public confidence in this new and important crop protection technology. In summary, Bt crops are one of the first practical examples of new and powerful pest-control technologies emerging from the results of decades of basic research and recombinant DNA technology. An overwhelming amount of evidence based on studies of Bt Cry proteins indicates these crops are safe for consumption by humans and other vertebrates, and are much safer for non-target invertebrates and the environment than synthetic chemical insecticides.

Based on the scientific evidence supporting these conclusions, the US Environmental Protection Agency will continue to register Bt crops for use in the United States. While additional direct testing of Bt crops is warranted to further assess their safety, there is no reason at present to think that these crops present risks greater than those associated with the consumption of non-Bt crops. In fact, Bt crops may be safer for human consumption than conventional crops because they contain lower levels of mycotoxins and residues of chemical insecticides.

IMPORTANCE OF PROTECTED CROPS FOR PLANT PRODUCTION

During the late 50s and early 60s the use of greenhouses spread: initially they were mostly used for vegetable production, with an emerging cut-flower and ornamental plant industry starting, particularly in the UK and in The Netherlands. By 1960, The Netherlands had the most concentrated production of glass-house grown crops, estimated as 5000–6000 ha (75 per cent of which grew tomatoes). At the same time, the UK had 2000 hectares of greenhouses. Hydroponic cultivation started in The Netherlands in the 60s and spread to many countries.

In the USA, hydroponic cultivation became widespread: in the late 60s and early 70s, there were more than 400 ha devoted to hydroponic vegetable production (tomato, followed by cucumber and lettuce), although this surface area has diminished to less than 100 ha today. Moreover, there has been a strong shift from vegetables to ornamentals grown in glasshouses. Nowadays, in the USA, of the total greenhouse production (estimated as 2000 ha), 95 per cent is represented by flowers, potted plants, ornamentals and bedding plants. There has also been a shift in northern Europe, with a delay of about 15 years compared to the USA, from vegetables to added-value ornamental crops.

For example, more than 80 per cent of the greenhouses in The Netherlands were used for vegetables in the 60s, whereas now 60 per cent of the approximate 10,000 ha are used for production of ornamentals. By 1980, there was an estimated 150,000 ha of greenhouses (glass, fibreglass, plastic) world-wide producing high-value crops. In 1995, the surface area had increased to about

280,000 ha. New areas, particularly in Asian and Mediterranean countries, showed a strong increase in protected areas, attracted by cultivation of high-value vegetable crops. The expansion in plasticulture in the Mediterranean area is still going on, again with a gradual transition from the production of vegetables to ornamentals. Spain and Italy have been the leading countries in the 80s and 90s. At present, the North African countries are experiencing a very rapid increase in the area covered with plastic houses, often with very simple structures. This development has been accompanied by a spread in drip irrigation. At the same time, the use of plastic row tunnels, covers and plastic soil mulches has expanded world-wide. These structures will not be discussed further in this book, but it is interesting to know that, for example, in China an area of more than 2.8 million ha of crops was covered with plastic soil mulch in 1995.

Table. Distribution of Protected Cultivation World-Wide

Structure	Ha(×100) in Geographical Area				
	Asia	Mediterranean	North-South America	North Europe	Total
Direct Cover (Floating Types)	5.5	10.3	1.5	27.0	44.3
Low Tunnels (Row Covers)	143.4	90.5	20.0	3.3	257.2
High Tunnels	–	27.6	–	–	27.6
Plastic-Houses	138.2	67.7	15.6	16.7	238.2
Glass-Houses	3.0	7.9	4.0	25.8	40.7

The world greenhouse area is now estimated as 307,000 ha, 41,000 ha of which is covered with glass, 266,000 ha with plastic. The global status of protected cultivation (*sensu lato*) is reported in Table. The distribution and types of crops grown in greenhouses are outlined in Table. Vegetable crops are grown in about 65 per cent of greenhouses, and ornamentals in the remaining 35 per cent.

Cultivation Techniques and Crop Protection

Structures adopted for covering crops vary a lot, from the simple to the sophisticated:

Low Tunnels (Row-Covers)

These are small structures that provide temporary protection to crops. Their height is generally 1 m or less, with no aisle for walking, so that cultural practices must be performed from the outside. Their use enhances early yields and yield volume; they also protect against unfavourable weather. Thermal films of infra-red polyethylene (PE), ethylene vinylacetate (EVA), copolymer, polyvinylchloride (PVC) and conventional PE are used.

High Tunnels (Walk-in Tunnels)

Such structures use the same cover materials as low tunnels and are high enough to perform cropping practices inside. Moderately tall crops are grown. Statistics concerning high tunnels are often included in the same category as low cost plastic houses since the materials used are similar.

Greenhouses

These differ from other protection structures in that they are sufficiently high and large to permit a person to conveniently stand upright and work within. Greenhouses appeared when glass became available for covering. Later, the introduction of plastic films permitted world-wide expansion of the greenhouse industry.

Greenhouses protect crops against cold, rain, hail and wind, providing plants with improved environmental conditions compared to the open field. In greenhouses, crops can be produced out-of-season year-round with yields and qualities higher than those produced in the open field. Greenhouses have also allowed the introduction of new crops, normally foreign to the region. There are two basic types of greenhouse.

The first type seeks maximum control in an environment to optimize productivity. In Europe, optimal conditions for year-round production are provided in the glasshouses of The Netherlands, Belgium, the UK and Scandinavia. The other type of greenhouse, which is very common throughout the Mediterranean area, provides minimal climatic control, enabling the plants grown inside to adapt to suboptimal conditions, survive and produce an economic yield. The choice of greenhouse depends on location, crop and financial resources.

There is a strong relationship between local conditions, greenhouse design, cladding materials and insulation needs. The structure of a greenhouse depends on the climate and the cladding used. There are various roof, space and height geometries with single-span materials such as bamboo, used in low cost structures, particularly in China and in semi-tropical and tropical areas. Cladding materials were limited to glass until the middle of the 20th century. From 1950, plastic films, because of their low cost, light weight and adaptability to different frame designs, became available, permitting world-wide development of the greenhouse industry, particularly in the semi-tropical areas. But plastic covers are not acceptable in northern Europe because of low light transmission compared to glass.

A full range of conventional and modified plastic films is now available: all coverings can perform well, depending on the desired use and location. Single plastic films prevail in warm climates; inflated double plastic film or rigid single plastic panels are more common in cool areas. A combination of high and low technology may be seen in countries such as Korea and Israel. Nets are used in tropical areas or during hot weather in temperate zones: they may reduce pest damage and the extremes of temperature and air humidity. Moreover, nets have a windbreak effect and reduce the damage from heavy rain and hail. The greenhouse design (particularly its height, shape, opening systems and cladding material) strongly influences climatic conditions inside, thus having a profound impact on pest and disease development. Plastic houses almost always have a more humid climate, large diurnal temperature variation and are more difficult to ventilate. Typically, they result in more problems with high humidity-dependent diseases, such as grey mould, downy mildews and rusts.

Regulating the atmosphere throughout the day and night is important for disease control and for reducing the total amount of chemicals sprayed. This has been demonstrated in the case of grey mould in tomato and cucumber, and of downy mildew (*Bremia lactucae* Regel) in lettuce. With respect to the cladding material used, in some cases a possible effect on diseases has been reported, mostly through the direct influence of radiation on sporulation. Certain UV-absorbing plastic coverings for greenhouses that absorb light at 340 nm have been exploited to inhibit the sporulation of *Sclerotinia sclerotiorum* (Lib.) de Bary, and species of *Alternaria* and *Botrytis squamosa* J.C. Walker. Reuveni observed a reduction in the number of infection sites of *B. cinerea* on tomato and cucumber when a UV-absorbing material was added to polyethylene film to increase the ratio of blue light to transmitted UV light.

Recently, blue photoselective polyethylene sheets have been suggested for their ability to reduce grey mould on tomato and downy mildew on cucumber. Green-pigmented polyethylene reduced the conidial load and grey mould in commercial tomato and cucumber greenhouses by 35–75 per cent. *Sclerotinia sclerotiorum* on cucumber, *Fulvia fulva* (Cooke) on tomato and cucumber powdery mildew were also reduced.

The technologies for environmental control in the most sophisticated greenhouses have been characterized by many new developments over the past three decades. The variables of light, temperature, air and soil humidity, and CO_2 content of the atmosphere are computer-programmed 24 h a day to achieve maximum crop yield.

Further refinements and improvements for adjusting the greenhouse climate to optimal crop productivity can be expected. In the less sophisticated structures of the sub-tropical and tropical regions, it is much more difficult to manipulate the greenhouse climate. In tropical and subtropical areas greenhouses often simply have an umbrella effect, using just roofs, with sides left open. The influence of greenhouse structures and covers on greenhouse climatic regimes may have strong consequences for pests and their natural enemies, as they have for diseases. A typical case of climate influence on pests and natural enemies concerns the spider mite and its predator *Phytoseiulus persimilis* Athias-Henriot: low humidity regimes may constrain effective use of *P. persimilis*. In high-tech greenhouses, regulation of temperature and water pressure deficit enables the creation of conditions less favourable to pathogens and, in some cases, more favourable to biocontrol agents. The use of heating to limit development of a number of pathogens is well known: however, heating is not economically feasible in all greenhouse systems. Recently, with the development of soilless systems, the effect of managing the temperature of the circulating solution has been studied, and has proven to be effective against certain pathogens.

The use of high root temperatures in wintergrown tomatoes in rock wool offers a non-chemical method of controlling root rot caused by *Phytophthora cryptogea* Pethybr. and Lafferty. The high temperature was shown to enhance root growth while simultaneously suppressing inoculum potential and infection, and, consequently, reducing or preventing aerial symptoms. Careful control of the temperature also proved important in the case of hydroponically grown spinach and lettuce, in which it prevented or reduced attack by both *Pythium dissotocum* Drechs. and *Pythium aphanidermatum* (Edson) Fitzp. Recently, attacks of *P. aphanidermatum* on nutrient film technique (NFT) grown lettuce in Italy were related to the high temperature (>29 °C) of the nutrient solution.

Root rot was inhibited by reducing the temperature below 24 °C. Much less exploited are the effects of temperature and water pressure deficit on biocontrol agents, although the first models, resulting in advice for optimal climate control for insect natural enemies, are now becoming available. In the case of biological control of plant pathogens, most of the studies carried out are related to the effect of environmental conditions on *Trichoderma harzianum* Rifai, used as biocontrol agent of *B. cinerea* and of several hyperparasites of *Sphaerotheca fusca* (Fr.) Blumer. [= *Sphaerotheca fuliginea* (Schlechtend.:Fr.) Pollacci]. In the case of *T. harzianum*, populations of the antagonist are promoted by low vapour pressure deficit; in commercial greenhouses significant control of grey mould of cucumber has been correlated with low water pressure deficit but not with conditions of air saturation and dew deposition.

In the case of *Ampelomyces quisqualis* Cesati:Schltdl., hyperparasite of *S. fusca*, a period of 24 h with low vapour pressure deficit is necessary. Low vapour pressure deficit also favours the activity of *Sporothrix flocculosa* Traquair, Shaw and Jarvis. More studies in this field are necessary, both in order to keep conditions close to the optimum for biocontrol agents within the greenhouse and for selecting biocontrol agents more adapted to the greenhouse environment. Greenhouses were initially built in areas with long, cold seasons to produce out-ofseason vegetables, flowers and ornamental plants. Northern Europe is the paradigm of pioneering areas of greenhouse cultivation.

The development of international exchanges of agricultural products and the availability of a variety of cheap plastic materials for covering simple structures has led to a spectacular increase in the area of protected crops in wanner regions like the Mediterranean basin and East and Southeast Asia. These new regions are commonly characterized by low or irregular annual precipitation and poor vegetation development. The insertion of greenhouse patches leads to drastic changes in the structure and ecology of the landscape. In early stages of greenhouse cultivation in a new area, greenhouses are isolated spots, like oases, where some phytophagous insects find good seasonal conditions for rapid increases in density. But optimal weather and host-plant conditions rarely last throughout the year and for a few months – usually the hottest – the increase in the herbivore population is interrupted. When greenhouses become more common in the area, the mosaic pattern may evolve to a large area of protected crops, with a succession of crops throughout most of the year and with polyphagous pests. These pests are able to feed on many agricultural plants and migrate between greenhouses. Additionally, field crops may be excellent refuges for pests in hot seasons, when the temperature is too high for greenhouse cultivation.

This has several consequences, as the immigration of pests into the greenhouse causes sudden and largely unpredictable pest density increases. Exotic pests quickly become established, especially if ornamental plants are cultivated. Polyphagous pests (like whiteflies, spider mites, thrips, leafminers, several aphids species, especially *Aphis gossypii* Glover, leaf-eating caterpillars and soilworms), which may exploit several crops successively, become prevalent. As pest densities increase, crops are increasingly sprayed with insecticides, native natural enemies becomc very rare, and natural control loses effectiveness. Unexpected and high pest pressure from the outside makes biological control very difficult. Under such conditions, a more holistic approach would consider the fields outside the greenhouse and the crop inside the greenhouse as a single entity for applying integrated strategies against pests and diseases. Programmes for conserving native or introduced natural enemies in the area should both lower pest pressure on greenhouse crops and incorporate beneficial fauna into the outside-inside greenhouse cycle of the pest-natural enemy complex

Pest and Disease Development

Well-grown and productive crops are generally less susceptible to diseases, but in many cases compromises have to be made between optimum conditions for economic productivity and conditions for disease and pest prevention. Well-fertilized and irrigated crops are, however, often more sensitive to pests, like aphids, whiteflies and leafminers. Groundbed crops are rarely rotated, so soilborne pathogens and pests pupating in the soil accumulate if the soil is not disinfested. Soil disinfestation, although effective, creates a "biological vacuum". Major changes in cultural techniques include the use of hydroponic and soilless cultures and artificial substrates controlled by computerized systems. Although these changes are ultimately intended to reduce production costs and maximize profits, precise environmental and nutritional control that pushes plants to new limits of growth and productivity can generate chronic stress conditions, which are difficult to measure, but are apparently conducive to diseases caused by pathogens such as *Penicillium* spp. or *Pythium* spp. Some soil substitutes and soilless systems do not always provide sufficient competition for pathogens, due to their limited microflora. High host plant densities and the resulting microclimate are favourable to disease spread. Air exchange with the outside is restricted, so water vapour transpired by the plants and evaporated from warm soil tends to accumulate, creating a low vapour pressure deficit (high humidity). Therefore, the environment is generally warm, humid and wind-free inside the greenhouse.

Such an environment promotes the fast growth of most crops, but it is also ideal for the development of bacterial and fungal diseases, of insects vectoring viruses and of herbivorous insects. For bacteria and many fungi (causal agents of rusts, downy mildews, anthracnose, grey mould, etc.) infection is usually accomplished in a film or drop of water on the plant surface. Unless temperature, humidity and ventilation are well regulated, this surface water can persist in the greenhouse until infection becomes assured. Many of the energy saving procedures adopted during the past three decades are favourable to disease development, since they favour increases in relative humidity, but they may lead to pest suppression as temperatures are generally somewhat lower.

Most greenhouse crops are labour-intensive, and for long periods require daily routine operations (such as tying, pruning, harvesting). The risks of spreading pathogens through workers and machinery are increased by the risks deriving from accidental wounds and from the exposure of large areas of tissues by pruning. Greenhouses are designed to protect crops from many adverse conditions, but most pathogens and several pests are impossible to exclude. Windblown spores and aerosols containing bacteria enter doorways and ventilators; soilborne pathogens enter in windblown dust, and adhere to footwear and machinery. Aquatic fungi can be present in irrigation water; insects that enter the greenhouse can transmit viruses and can carry bacteria and fungi as well. Once inside a greenhouse, pathogens and pests are difficult to eradicate.

Cultural Techniques used in Protected Cultivation

In most greenhouses of northern Europe continuous cropping is practised, without a fallow crop-free interval. This has profound implications for diseases and pests. In the case of plant pathogens, it leads to the build-up of soilborne pathogens and an increased importance of foliar pathogens with a broad host-spectrum (*i.e. B. cinerea*). The same can be said for insects that pupate in the soil such as leafminers and thrips. Greenhouse crops are grown in various soils and soilless media whose physical and chemical properties are adjusted to obtain maximum productivity. These properties, such as heat conservation, water-holding capacity, fertilizer levels and pH can also be manipulated to reduce the amount of inoculum of pests and pathogens and the probability of infection. Systems for growing crops in the greenhouse vary widely in terms of complexity. The most common rooting media are soil and various soil mixtures, incorporating peat, vermiculite, perlite and several other materials which are added to the soil in order to modify its structure.

In the 60s, bench cultivation was adopted for high value crops (*i.e.* carnations), permitting better results in soil disinfestation. In the 80s and 90s, soilless substrates gained more and more importance, particularly in the northern European countries, because they eliminate or reduce the need for soil disinfestation. Among soilless substrates, rock wool has been widely used in northern Europe, while in the tropics and sub-tropics cheaper substrates have been exploited. The nutrient film technique, originally devised to improve precision in crop nutrition, reduces soilborne diseases and removes the cost of soil disinfestation. In fact, it confers relative freedom from diseases, although severe epidemics can still occur. During the past two decades, various composted organic wastes and sewage sludges have partially replaced peat in container media used for production of ornamentals.

Recycling of these wastes has been adopted for economic and production reasons. The cost of these composts can be lower than peat. Production costs may also be decreased because some of the compost-amended media, particularly those amended with composted bark, suppress major soilborne plant pathogens, thus reducing plant losses. The low pH of sphagnum peat, pine bark and composts could theoretically have beneficial side effects for some plants. For example, Phytophthora root rot of rhododendron (*Phytophthora cinnamomi* Rands) is suppressed at pH < 4.0, because the low pH reduces sporangium formation, zoospore release and motility. This may be important during propagation of rhododendron cuttings under mist. Moreover, chemical inhibitors of *Phytophthora* spp. have been identified in composted hardwood bark. These inhibitors do not affect *Rhizoctonia solani* Kühn. Soilless cultivation can affect pests that need the soil/substrate to complete their development, as in the case of leafminers or thrips. The thermal and gas exchange properties of rooting media affect the growth of roots as well as the activities of pathogens. Peat, a common rooting medium, used either alone or in mixture, often suppresses pathogen activity, depending on its origin.

However, pathogens, including species of pathogenic *Pythium* and *Fusarium* have been isolated from commercial peat compost. The design of benches is important due to the effect on the ventilation of seedling trays and potted plants. Correct spacing prevents the establishment of a microclimate conducive to foliar diseases and the rapid spread of pathogens and pests from plant to plant in crops grown in groundbeds. Altered greenhouse and bench design can improve air movement, thus reducing the risk of diseases. Bottom heating of benches, a traditional means of avoiding Phytophthora, Pythium and Rhizoctonia root rots, is enhanced in cutting and seedling trays with upward air movement between the young plants. Through-the-bench air movement is perhaps the most neglected and simplest means of reducing seedling rots in tangled plant masses.

Every crop species and cultivar requires a special fertilizer regime in order to obtain maximum productivity and to prevent stress on the plant. Fertilizer requirements change as the crop ages from seeding to harvest. In general, excessive nitrogen leads to excessive foliage that is intrinsically more succulent and susceptible to damage and necrotrophic pathogens, such as *B. cinerea*, and also stimulates development of pests such as aphids and leafminers. Nitrogen generally has to be balanced with potassium; for many diseases, susceptibility decreases as the potassium-nitrogen ratio increases.

Calcium generally enhances resistance, due to its role in the integrity of the cell wall. No general practical recommendations can be made for controlling diseases by adjusting the fertilizer levels supplied to plants: each host-pathogen combination reacts differently. However, optimal, instead of maximal fertilization, results in lower pest and disease pressure. General recommendations can be given concerning irrigation. First of all, the factors that determine irrigation demand in greenhouse crops can all be closely regulated.

From a general point of view, overhead irrigation must be carried out early in the day and should be limited late in the afternoon in order to avoid long periods of leaf wetness, which favour diseases such as downy mildews, rusts, grey mould, leaf spots, etc. When it is necessary to wet foliage for any reason (including pesticide spraying), it is always essential to maintain environmental conditions under which the foliage can dry out within a very short period of time. Also, it is important to avoid excess water in the soil: this creates conditions that are very favourable for the development of root rots. The effects of irrigation on pests are mainly through the relative humidity of the environment or through the water-status of the plants. For instance, plants under stress are more easily colonized by thrips and spider mites.

7

The Impact of Diseases and Pests on Crop Yield

Knowledge of the impact of diseases and pests on crop yield is needed for decisionmaking in disease/pest control. Decision-making is based on the damage relationship, that is, the level of yield loss associated with different amounts of disease or levels of pest attack. Yield loss can be either loss of quality or loss of quantity or both. Usually, damage relationships for a certain pathosystem or pest/plant interaction are expressed in quantity of yield, either in relative or absolute units.

Most frequently, damage relationships are assessed empirically and analysed with regression analysis to estimate yield or yield loss from observations on the amount of infestation. The main drawback of this approach is that the resulting models are descriptive and do not take into account the physiological processes underlying the yield. Despite this, the descriptive models can be very useful in integrated control.

Following Campbell and Madden, five descriptive models to calculate the amount of disease and yield loss relationship may be distinguished:

1. *The single-point models or critical-point models:* These models estimate yield loss by determining the amount of disease on one given moment, usually determined by the physiological status of the crop, for example the onset of flowering. Less frequently, these models have been developed with time variables, for example the number of disease-free days or the time until a certain level of disease is reached. Single-point models have been developed mainly for diseases in cereals. Their use is limited in crops in which yield accumulates over a longer period of time or harvesting takes place more than once, as for example in greenhouse vegetables.
2. *Multiple-point models:* These models use several disease assessments to estimate yield loss. This type of model is most useful in situations where disease progress can be highly variable, depending on the host plant or the environment.

3. *Integral models*: These models use the summed disease pressure over a specific period of crop growth which is relevant to yield. This is determined by calculating the area under the disease progress curve (AUDPC). These models can not distinguish between an early moderate epidemic and a more severe epidemic which starts later with the same AUDPC. This can be overcome by assigning weighting factors to the disease assessments made on different times or by incorporating another factor, for example the number of disease-free days, into the model.
4. *Response surface models*: These models predict yield loss by using two different types of variables, for example disease severity and crop growth stage.
5. *Synoptic models*: They are multivariate models that estimate yield loss by incorporating all independent variables in one equation.

Much of the conceptual framework to estimate the relationship between amount of disease and yield loss may be applied as well to damage relationship concerning arthropod pests.

For decision-making purposes, a linear function of the yield response to insect infestation is generally assumed. In case the crop is able to compensate limited injury, there is a level of tolerance associated with low pest density. Crop tolerance to pest attack may be relatively high when pests injure the leaves of fruit vegetables like tomato, pepper, cucurbits or egg plants; often even 30–40 per cent of leaf injury does not result in yield reduction.

Sigmoid yield responses to pest infestation are more difficult to fit, but logarithmic or power transformations may linearize the damage relationship. The consideration of more than one pest or disease and crop variables render complex polynomial relationships which are difficult to interpret and to use in decision-making. If a linear yield response may be assumed or derived, damage relationship can be "easily" found with field data as it has been mentioned above in single- and multiple-point models.

When pests are multivoltine and their numbers are quite variable along the season, the use of insect*days instead of seasonal mean insect densities may be more meaningful as noted in the above mentioned integral models. Methods and techniques for this kind of studies may be found in Teng. A different approach to determining crop loss is the use of dynamic simulation models. Generally, a model for the development of the pest or disease is combined with a model for crop growth and production.

This approach generates explanatory models, which are expected to have a greater predictive value than descriptive models. However, the development of simulation models requires a lot more basic information on the physiological processes and on the effect of environmental parameters on epidemics, and are therefore more difficult to develop than models based on regression analysis.

DAMAGE AND ACTION THRESHOLDS

In combination with the damage relationship for a given pathosystem or pest/crop interaction, it is important to assess the damage and action thresholds. The damage threshold is the maximum level of disease or pest attack below which yield losses do not occur. The action threshold is the level of disease or pest attack at which control action should be taken to prevent the epidemic or pest to reach the damage threshold. Because there are often no fully effective techniques to control a disease or a pest immediately, the action threshold is lower than the damage threshold.

Generally speaking, action thresholds for diseases lie below the logistic increase in the DPC. Damage and action thresholds are an important tool in integrated control when several control alternatives are available. Whereas the damage threshold essentially depends on the disease/pest level and yield loss relationship, the action threshold may greatly vary according to the efficacy of each of the control alternatives and how long they take to be effective. Action threshold to control a disease – or a pest – will be probably higher if we choose a quick acting pesticide than with biological control, where natural enemies need time to react. For example, the action threshold for greenhouse whitefly control may be up several adults per leaf if we rely on insecticides, whereas it is around one adult per leaf when the parasitoid *Encarsia formosa* Gahan must be used in seasonal inoculative releases for the biological control of the whitefly. The knowledge and application of action thresholds generally reduce the amount of control inputs compared to general practice.

Determination of the Thresholds is Not Always Easy

Yield loss can be defined as loss in weight of the harvested product or the loss in economic value. This potential economic loss can be compared to the cost of a control measure. The translation of yield loss in weight to economic yield loss depends on the expected price of the harvested product and is therefore difficult to perform. In greenhouse crops that are harvested continuously, this is further complicated by the fluctuating prices within one growing season. For example, one kilo yield loss in cucumber or tomato in The Netherlands will be much more costly for the grower at the beginning (early spring) and end (late fall) of the season when prices are higher than in summer.

Further complexity in the determination of the action thresholds leads to some other considerations:

- Long term consequences of the current decisions for the disease or pest levels (instead of considering just one generation or infection cycle);
- Influence of control actions on crop revenue (it may be different if the decision is made at farmer or regional level);

- The risk attitude of the grower. Regarding the latter, stochastic models (in which an occurrence probability is associated to each decision) are more useful thresholds for quantifying risks than the deterministic damage and action model.

Thresholds in Greenhouse Crops

Despite the importance of knowing damage relationships and damage and action thresholds for integrated control, very limited specific information is available for most greenhouse diseases and pests. This can be explained in flower and pot plant crops for the extremely low tolerance of their aesthetic value to the most common diseases and pests. When known, damage thresholds in ornamental crops are near zero, as in the case of powdery mildew in roses, where the damage threshold is only 5 pustules/m^2. The same pest may have quite different damage thresholds if vegetable or ornamental crops are considered. For example, tomato may tolerate relatively high leafminer infestation (*e.g.* several dozens of mines per plant) with no yield loss, whereas 1–2 mines on chrysanthemum leads to cosmetic damage. Some thresholds are available for greenhouse vegetable diseases and pests. Currently, the damage relationship for powdery mildew fungi in greenhouse vegetables is determined in The Netherlands. For cucumber, the best fitting model to describe the damage relationship for powdery mildew caused by *Sphaerotheca fusca* (Fr.) Blumer was an integral model using AUDPC. The slope of the regression line between yield and AUDPC was similar in several seasons and for different cultivars.

In this case, early disease was also compared to late, more severe disease. Similar AUDPC values and similar yield losses were found for these situations, so the inability of integral models to distinguish between early and late epidemics does not seem to be very important here. An additional problem for the determination of damage relationships is the fact that some pests can inflict more than one type of damage concurrently. This is the case of greenhouse whitefly, that feeds on phloem sap, with the consequent debilitating effect on the plant, but it also damages leaves and fruits by producing honeydew on which sooty mould develops, hampering photosynthesis and respiration and rendering fruits unmarketable.

Damage thresholds may be quite different depending on which type of damage is considered to first occur as whitefly populations increase and this is decisively influenced by RH. In very humid greenhouse environments – particularly common in northern Europe – damage by sooty mould development occur at whitefly densities lower than those needed for damage derived from whitefly feeding activity.

A density of 2500 greenhouse whitefly nymphs per leaf has been reported to cause yield reduction on tomatoes as direct consequence of phloem extraction, whereas a much lower density of 60 nymphs/leaf has been observed to produce sufficient honeydew to induce sooty mould development on tomatoes if RH reaches at least 80 per cent for eight hours. For *Bemisia* species, a third type of damage is known which is that related to their ability to transmit tomato yellow leaf curl virus (TYLCV).

This adds even more complexity to the pest density and yield loss relationship. The damage threshold is highly dependent on the amount of virus inoculum present in or near the greenhouse. Examples of damage thresholds for thrips pests in vegetable crops may illustrate the variability of values that can be found in the literature. For *Thrips palmi* Karny on aubergine the damage threshold is 0.08 individuals/leaf, whereas the same author gives a value of 4.4 thrips/leaf for cucumber. For sweet pepper, Kawai gives, for the same pest, a threshold of 0.11 thrips/flower. There is unlikely to be a single damage threshold for a given pest on a given crop, but many, depending on the market and climatic conditions.

Even more variable are the action thresholds of greenhouse pests in various IPM programmes implemented in the world. If natural control must be considered in biological control of greenhouse pests, abundance of naturally occurring predators and parasitoids is an additional element to monitor and consider in decision-making. This is the case of action thresholds to release *Diglyphus isaea* (Walker) in Mediterranean greenhouses for the control of leafminers; the standard action threshold can be lowered if the parasitoid, that occurs naturally in the area, comes into greenhouses and establishes early in the season.

INTERACTION OF DISEASES AND PESTS

The pest tetrahedron is used to envisage the interaction of diseases and pests with their environment. The tetrahedron consists of four components, which can all influence each other and together determine the level of the disease or pest.

The four components are the plant pathogen/pest, the host plant, the environment, and human activity.

Generally, it can be said that chemical control is only aimed at influencing the pest or pathogen directly, whereas integrated control may reduce the level of disease or pest by influencing any or several of the four components of the tetrahedron.

A thorough knowledge on the influence of different factors on pests and diseases offers the basis for integrated control.

INFECTION CYCLE OF PLANT PATHOGENS

The infection cycle of fungal plant pathogens consists of the following phases:

- Infection (germination of spores, penetration of plant tissue and colonization of plant tissue);

- Sporulation (production of spores and maturation of spores);
- Dissemination (spore liberation, spore dispersal and spore deposition).

Bacterial pathogens and viruses generally have a similar but simpler infection cycle. Some pathogens only complete one infection cycle per season, but most pathogens complete several or sometimes many cycles per season. The amount of disease that develops is the sum of the successes of the different phases in the cycle. All phases are subject to influences from the environment, the host plant and human activity.

LIFE CYCLE OF PEST ORGANISMS

Although pest organisms belong to various taxonomic groups and consequently their life cycle varies accordingly, a general pattern for phytophagous arthropods may be described. Dispersal usually occurs at adult stage; adults look for an exploitable habitat and then a plant to feed and oviposit on. Once a plant is recognized as suitable for feeding and ovipositing, an adult female lays eggs (ovipary) or deposits nymphs or larvae (vivipary). Progeny commonly feed and develop on the plant where oviposition took place or on neighbouring host plants until reaching the mature stage.

The adults may feed and oviposit on the same plant or move to younger and more suitable host plants. Many arthropods spend all their life on a plant but others, particularly among holometabolous insects, either have soil-inhabiting stages or the adults feed on a different host plant than the immatures did.

Most greenhouse pests are multivoltine (several generations within a year), and univoltinism (one generation per year) is rather rare. Research on disease epidemiology and pest population biology, can study the effect of different factors on the phases in the infection/pest cycle or can determine the effect on the resulting amount of disease or pest density directly. Studying the effect on each of the processes described – for instance germination and spore production in the infection cycle, or host plant selection and exploitation in a pest cycle – often yields a good understanding of the effect of a factor on a disease or amount of damage caused by a pest. This kind of research is best done under controlled conditions, which allows the fluctuation of only one or two factors. Research on epidemiology of a disease or pest population biology in a whole crop is usually based on monitoring many factors and analysing their respective effects through regression analysis. It is important that in such research all the relevant factors are monitored.

IMPORTANT FACTOR OF HOST PLANT

Traditionally, the host plant has been considered as a more important factor for the development of a disease than for determining the damage caused by an arthropod pest. This may be one of the reasons why plant resistance has been used more frequently to control diseases than to control pests. In fact, the tetrahedron scheme is common in plant pathology texts but rare in applied entomology books. However, this situation has changed recently as entomologists have intensified studies on insect-plant relationships. Different cultivars may vary in their susceptibility to certain diseases or pests, even if none of them is completely resistant.

This partial resistance may influence disease development by decreasing the number of successful infections by the pathogen, by increasing the latency period, by reducing the rate of lesion expansion or sporulation, or by any combination of these processes. The result will be a slower development of the epidemic.

Similarly, different cultivars may influence the host plant selection by a pest and/or its oviposition, development and survival, and thus the rate of population increase. Many partially resistant cultivars express the same resistance during their entire life, but some resistance may depend on the physiological age of the plant. Agerelated resistance can be either adult plant resistance or young plant resistance. On another scale, certain parts of the plant may be more or less susceptible to disease because of their age. The physiological status of the plant is affected by temperature, humidity and nutrition. Nutrient deficiencies or certain climatic conditions may predispose plants to the development of diseases and pests, but unbalanced fertilization may also increase this kind of problems. For example, excess of nitrogen amendments renders the plant more susceptible to *Botrytis cinerea* Pers.:Fr. and enhances the population increase of homopteran pests like aphids and whiteflies. A thorough knowledge of the factors affecting the incidence of diseases and pests through their host plant may help to prevent outbreaks by applying correct crop management practices. Because the host plant is such an important factor in disease epidemics and pest population dynamics, it is important to carefully monitor the host plant in any epidemiological research.

This means not only that cultivar and planting date should be recorded, but also plant spacing, nutrition, development stage of the plant and plant growth. Recording plant growth is also important for another reason: whether the disease is assessed as a percentage of total plant area – for example leaf area covered – or as a pest density – for example, the number of insects per leaf area – it is important to know the size of the host plant. Also, a count of the number of lesions for example may yield the same number for different crops, but if the number of leaves or other aspects of the size of the host plant are different, the impact of this number of lesions will be different and results difficult to compare between crops.

COMPONENTS OF THE ENVIRONMENT

Many components of the environment can indirectly influence the severity of disease and pest injury through the host plant or by a direct effect on the pathogen and pest. Here, only the direct impact of the environment on the pathogen and pest population is dealt with. The main influences on pathogens stem from temperature, relative humidity (RH), radiation and wind. Pathogens are often affected by climatic conditions in most of the phases of the infection cycle. Germination of spores and superficial germ tube growth often show an optimum curve for temperature. For most pathogens, germination only occurs above a certain (high) level of RH or in the presence of wetness on the plant surface.

Lesion expansion is often influenced by air temperature, since this also affects the temperature of the plant tissue, but less by RH. Sporulation is affected by temperature and RH, whereas spore dispersal is often mostly influenced by air humidity and movement.

It is important to realise that all these environmental factors will be different at different crop heights. It is therefore necessary to measure microclimatic conditions at a height where the pathogen is expected to attack the host plant. Manipulation of the environment by, for example, changing radiation with different covers, and/or changing temperature, RH and ventilation by using different heating and/or ventilation regimes, may influence diseases and pests.

Concerning pests, their density is directly affected by many biotic and abiotic factors of the environment other than the host plant, such as competing herbivores, natural enemies and climatic factors. Among the latter, temperature, RH, light (quantity, quality and periodicity) and air movement are usually the most decisive in determining behaviour, phenology, survival, fecundity and dispersal of pests.

All these abiotic factors may also affect natural enemy populations and, therefore, pests in an indirect, but important, way. Understanding all these complex interactions is crucial for detecting the most decisive factors that govern pest population dynamics and managing the environment for pest control accordingly.

CONCLUSION

Pest control is the study of the development and spread of plant diseases and arthropod pests and the factors affecting these processes. The level of disease or pest infestation is the result of many interacting factors and this level determines the yield loss that the grower suffers from the pathogen or pest.

In many respects, the methodology of research and the underlying concepts are very similar for bacterial and fungal plant pathogens, insects, mites, viruses and nematodes.

8

Biological Control and IPM in Protected Crops: Control of Pests

Biological control of pests in agriculture is a method of controlling pests (including insects, mites, weeds and plant diseases) that relies on predation, parasitism, herbivory, or other natural mechanisms. It can be an important component of integrated pest management (IPM) programmes.

Biological control is defined as the reduction of pest populations by natural enemies and typically involves an active human role. Natural enemies of insect pests, also known as biological log control agents; and include predators, parasitoids, and pathogens. Biological control agents of plant diseases are most often referred to as antagonists.

Biological control agents of weeds include herbivores and plant pathogens. Predators, such as birds, lady beetles and lacewings, are mainly free-living species that consume a large number of prey during their whole lifetime. Parasitoids are species whose immature develops on or within a single insect host, ultimately killing or fatally infecting the host. Most have a very narrow host range. Many species of wasps and some flies are parasitoids. Pathogens are disease-causing organisms including bacteria, fungi, and viruses. They kill or debilitate their own host and are relatively specific. There are three basic types of biological control strategies; conservation, classical biological control, and augmentation.

The greenhouse environment in Japan is characterized by:

- Cultivation in small plastic greenhouses;
- High temperatures from spring to autumn and relatively low temperatures in winter;
- High humidity all the year round;
- Immigration of pests from the surroundings because of poor isolation from the outside environment.

These conditions cause several problems in pest and disease control, *i.e.*, the need for frequent application of pesticides and fungicides, rapid population growth of pests, and unfavourable conditions for some of the natural enemies. Chemical control of pests is not easy and is always accompanied by the risk of development of resistance to pesticides. Also, the application of biological control is more difficult than in Western Europe. Most of the greenhouse pests in Japan are cosmopolitan. They are small in size and suck fluids from plants.

They belong to several groups of insects or mites, *i.e.*, aphids, whiteflies, thrips, mites, leaf miners and nematodes. Noctuid moth species such as *Spodoptera litura* (Fabricius) (Lepidoptera:Noctuidae) sometimes outbreak in the southwestern region. Another important fact is that several exotic pests have become serious in protected cultivation. After the rapid development of greenhouse cultivation in the 1960s, *Trialeurodes vaporariorum* (Westwood) (Homoptera: Aleyrodidae) and *Thrips palmi* Karny (Thysanoptera: Thripidae) invaded in 1974 and around 1980, respectively.

More recently, *Bemisia argentifolii* Bellows and Perring (Homoptera:Aleyrodidae), *Liriomyza trifolii* (Burgess) (Diptera:Agromyzidae) and *Frankliniella occidentalis* (Pergande) (Thysanoptera:Thripidae) were reported as new pests around 1990.

These species have established as major pests in protected cultivation in Japan. Another important fact is the high activity of indigenous natural enemies, which sometimes completely suppress the pests in greenhouses after immigrating into greenhouses from surrounding environment.

Table. Major Greenhouse Insect Pests in Japan

Crop	Pest
Tomato	Trialeurodes Vaporariorum, Bemisia Argentifolii, Liriomyza Trifolii, Myzus Persicae, Aphis Gossypii, Aculops Lycopersici, Meloidogyne Incognita
Eggplant	Thrips Palmi, T. Vaporariorum, B. Argentifolii, M. Persicae, A. Gossypii, Tetranychus Urticae, Tetranychus Kanzawai, Polyphagotarsonumus Latus, Spodoptera Litura
Sweet Pepper	Frankliniella Occidentalis, Frankliniella Intonsa, T. Palmi, M. Persicae, A. Gossypii, P.latus, S. Litura
Cucumber	T. Palmi, T. Vaporariorum, B. Argentifolii, M. Persicae, A. Gossypii, Rhopalosiphum Rufiabdominalis, M. Incognita
Watermelon	T. Palmi, M. Persicae, A. Gossypii, T. Urticae, T. Kanzawai, M. Incognita
Strawberry	F. Occidentalis, A. Gossypii, Chaetosiphon Minor, Aphis Forbesi, T. Urticae, T. Kanzawai, Nothotylenchus Acris, Pratylenchus Vulnus, S. Litura, Anomala Cuprea

BIOLOGICAL CONTROL IN PROTECTED CROPS IN JAPAN

Natural enemies released in greenhouses are produced in a large scale. In many countries, natural enemies are produced as commercialized products for pest control.

Table. Comparison of Aspects Related to the Application of Chemical and Augmentative Biological Control

	Chemical Control	*Biological Control*
Risk of Resistance	Large	Small
Specificity	Low	High
Enviromental Safety	Depend on Chemicals	Safe in Most Cases
Dependence of Effectiveness on Environment	Low	High
Cost	Depend on Chemicals	Often Expensive

Although biological control has the advantage with respect to risk of resistance development of pests against pesticides and environmental safety, it has the disadvantage with respect to specificity and dependence of effectiveness on environment.

Table. Registered Arthropod Natural Enemies in Japan, 2002

Natural Enemy	*Pest*	*Crop*
Phytoseiulus	Spider Mites	Strawberry, Eggplant, Persimilis Cucumber, Watermelon, Sweet Pepper, Three Other Crops
Encarsia Formosa	Trialeurodes Vaporariorum, Bermisia Argentifolii	Tomato, Cucumber, Melon, Eggplant
Diglyphus Isaea	Liriomyza Trifolii	Tomato, Eggplant
Dacnusa Sibirica	L. Trifolii	Tomato, Eggplant
Amblyseius Cucumeris	Thrips Palmi, Frankliniella	Cucumber, Strawberry, Watermelon, Melon, Eggplant, Sweet Pepper
Aphidius	Aphids	Cucumber, Strawberry, Colemani Watermelon, Melon, Eggplant, Sweet Pepper
Aphidoletes	Aphids	Cucumber, Melon, Aphidimyza Strawberry
Orius Sauteri	T. Palmi,	Sweet Pepper, Cucumber, F. Occidentalis Eggplant
Orius Strigicollis	T.Palmi, F. Intonsa	Sweet Pepper, Eggplant F. Occidentalis
Chrysoperla	Aphids	Strawberry, Eggplant, Sweet Carnea Pepper

In Japan, registration as biopesticides is necessary if natural enemies are sold as commercial products. Ten arthropod natural enemies have been registered as biopesticides until 2002. After the first registration of *Phytoseiulus persimilis* Athias-Henriot (Acarina:Phytoseiidae) and *Encarsia formosa* Gahan (Hymenoptera:Aphelinidae) in 1995, eight imported species have been registered.

Two *Orius* species and indigenous and some other promising indigenous species will be registered soon.

P. persimilis has a long history of study in Japan, where it was introduced in 1966. After extensive fundamental and applied studies on its use, several production facilities for the predator were constructed and the predators produced were distributed in several prefectures.

However, its use by farmers was not developed until recently because of limitation of the supply and the lack of integrated control systems for greenhouse crops. *P. persimilis* was first registered as a biotic pesticide and commercialized in 1995 for controlling spider mites on strawberries. Since then, *P. persimilis* has been registered to control spider mites on eight greenhouse crops and the area of release is the largest among commercialized natural enemies in Japan.

E. formosa was imported from the U.K. in 1975, the year after the first recognition of the occurrence of *T. vaporariorum* in Japan. The effects of temperature and release methods on the control capability of *E. formosa* were evaluated in greenhouse trials and by developing a simulation model of the host-parasite population interaction. Use of the imported bumblebee, *Bombus terrestris* (L.) (Hymenoptera:Apidae), for pollination is increasing among tomato growers in Japan. This is helpful for the production of good quality fruits, and the application of insecticides must be regulated after introduction of the bumblebee in greenhouses. *E. formosa* is expected to control whiteflies in this case.

After its registration as a biotic pesticide in 1995, the area in which the parasitoid is being released in tomato greenhouses is increasing. The use of *P. persimilis* and *E. formosa* is not sufficient to develop a practical IPM system for each greenhouse crop. Aphids are very difficult to control as indigenous pests. Imported pests after 1980 also became serious on many crops. Imported natural enemies, which have been commercialized in Europe and in North America, are tested in many agricultural experimental stations for possible registration as biotic pesticides for commercial use and another six species have been registered to control pests on many kinds of crops.

Trials have not always been successful with some species of natural enemies. This was partly due to the lack of sufficient knowledge and experience about such exotic natural enemies on the part of the researchers involved. Although most serious pests are exotic species from abroad, indigenous natural enemies can be used to control them if they are released artificially.

This is also preferable to the use of imported species from the point of view of the possible effect of imported species on indigenous species in agro-ecosystems. Promising results were obtained in the use of *Orius* spp. for biological control of *T. palmi*. *Orius sauteri* (Poppius) (Hemiptera:Anthocoridae) suppressed the increase of *T. palmi* on eggplants in greenhouse trials. Rearing techniques of *Orius* spp. in the laboratory have been developed using eggs of *Ephestia kuehniella* Zeller(Lepidoptera:Pyralidae). With respect to commercial application, *Orius strigicollis* (Poppius) (Hemiptera: Anthocoridae) is preferred to *O. sauteri* because of ease of mass production and weaker diapause response. *O. strigicollis* and *Amblyseius cucumeris* (Oudemans) (Acarina:Phytoseiidae) are widely used to control thrips on eggplants and sweet peppers in Kochi and Miyazaki prefectures.

Factors Limiting Use of Natural Enemies

Technically, extreme temperatures in greenhouses and the effect of chemicals are two major factors affecting control capability of *P. persimilis* in commercial greenhouses. Monitoring of the early infestation by spider mites to determine the time of introduction of *P. persimilis* is also crucial for the commercial use of the predatory mite because spider mite outbreaks occur in spring more readily than in Western Europe. Delay of release in greenhouse and low temperature condition in winter are two major factors of failure of use of *E. formosa*. Monitoring of the early infestation of whiteflies is of paramount importance, because one of the reasons for failure of the use of *E. formosa* is delay in its release. Trapping of adult whiteflies by yellow sticky traps and check of occurrence of adults on each plant based on binomial sampling were developed as monitoring methods. Most of the insecticides are harmful to *E. formosa* except for some insect growth regulators (IGRs), while many acaricides and fungicides are less harmful. The most important technical factors for commercializing natural enemies is integration of use of natural enemies with other control measures for controlling other pests on a crop. If the control measures for other pests are incompatible with use of natural enemies, practical use of natural enemies is almost impossible.

Social limiting factors in biological control are:

- Low tolerance of pest damage;
- Supply of natural enemies;
- Poor advisory service for farmers;
- Need of registration.

Low tolerance of pest damage is the main reason for difficulty in biological control in flower production. Lack of the system for supplying enough natural enemies of good quality was the limiting factor in Japan, because private companies had been not interested in biological control in Japan until recently. However, 10 species arthropod natural enemies have been registered and sufficient number of imported or indigenous natural enemies is supplied by private companies now. Advisory service for farmers is very important to obtain good performance of natural enemies for promoting biological control.

IPM in Protected Crops

Since the number of available natural enemies for biological control is limited and the technical development of the use of other natural enemies is still preliminary, integrated use of natural enemies and other control measures is essential for commercial application.

Table. A Scheme of IPM for Tomatoes in Greenhouses in Japan

Period	Control Measures	Pests
Middle of July	Application of Granules in Soil	Thrips, Aphids
Late August	Sulfur Application	Rust Mites
September-October	E. Formosa, D. Isaea	Whiteflies, Leaf Miners
Early October	IGR Application	Noctuid Moths
Middle of January	IGR Application	Thrips
April-May	E. Formosa, D. Isaea	Whiteflies, Leaf Miners

Suitable crops for integrated control using natural enemies can be assumed to be crops attacked by a small number of serious pests, crops for which a low level of pest damage is tolerable, crops requiring insect pollinators. Tomato, strawberry and sweet pepper are three good crops for developing integrated control in Japan. Table shows an integrated control scheme for tomatoes combining use of natural enemies and selective pesticides.

Noctuid moths sometimes cause severe damage in southwestern Japan, but they can be controlled by *Bacillus thuringiensis* Berliner or IGR. Soil application of systemic pesticides and spot treatment with non-selective pesticides are also possible options in integrated control. In Western Europe, integration of the use of natural enemies and selective pesticides is emphasized. However, the greenhouse environment in Japan is less favourable for biological control because of the extreme temperature conditions and incomplete isolation of the greenhouse environment from the surroundings. Integration with other control techniques is needed for more satisfactory biological control. For example, cultural control, including prevention of immigration of pests, removal of host weeds within and outside greenhouses and avoidance of cultivation of host plants near greenhouses, are fundamental for effective control. After infestation, pests increase almost exponentially in the high temperatures from spring to autumn. In such cases, delay of release of natural enemies often results in failure of biological control. Monitoring systems of early infestation by pests in greenhouses are crucial for successful biological control. Use of resistant varieties of crops should also be studied for integrated control. Partial resistance is adequate for integrated control including the use of natural enemies. In conclusion, integration of all possible techniques is needed for the successful and reliable use of natural enemies in Japan.

Other Control Measures for IPM

Although cultural and physical control measures do not have a curative effect after heavy infestations of greenhouse pests, their preventive effect on the rapid increase of pests is of great importance in integrated control. Their effectiveness is related to regulation of life histories or regulation of dispersal. *T. palmi* is a species of subtropical or tropical origin and is considered to be unable to survive in the field in Japan in winter. Suppression of *T. palmi* in greenhouses before the winter should decrease the density in the next season. *B. argentifolii* is also difficult to survive in winter. Most of the greenhouse pests are polyphagous and can infest many kinds of crops and weeds as host plants. If suitable host crops are cultivated, or host weeds grow, near greenhouses, the pests can migrate from the greenhouse to surrounding fields and *vice versa*. Host weeds growing in greenhouses are also important for the survival of pests after harvesting. Removal of host weeds inside and near greenhouses and avoidance of cultivation of host crops are fundamental conditions for preventive cultural control.

Helgesen and Tauber pointed out three possible routes of introduction of T. vaporariorum into greenhouses:

1. Artificial introduction of the infested seedlings;
2. Infestation from weeds or other crops in the greenhouse;
3. Infestation from host plants outside but near the greenhouse.

This suggests several possibilities for control by prevention of artificial introduction or dispersal of pests. Cultivation of uninfested seedlings in nurseries, removal of host weeds in greenhouses and the use of nets to cover the openings of greenhouses to prevent immigration of pests from outside. These techniques are also applicable to other pests. Plastic films that absorb ultraviolet rays (UVA) are commonly used as covering materials for plastic greenhouses in Japan.

One of its important and interesting effects is the prevention of immigration of pests from outside. It is effective against aphids, whiteflies and thrips. Another important physical control measure is the use of coloured traps. Winged aphids and whitefly adults are attracted to yellow and while thrip adults prefer white or blue. Many types of colored sticky traps have been developed and used for mass trapping or monitoring of these pests. Probably, their usefulness consists in the use in supervised control in chemical or biological control. Yellow sticky traps can be used to determine the time of release of *E. formosa* and to evaluate the effectiveness of control.

INTEGRATED PEST MANAGEMENT IN THE FIELD

Some studies have been carried out to develop IPM systems for eggplants in the field. Potential of *O. sauteri* to control *T. palmi* on eggplants was first evaluated by insecticidal check methods in Okayama prefecture. Potted eggplants were introduced into a screen cage, and both *T. palmi* and *O. sauteri* were released. After spraying with fenthion, which controls *O. sauteri* but does not affect *T. palmi*, was sprayed, *T. palmi* increased rapidly. Without application of fenthion, *T. palmi* was kept at a very low level. Similar experiments were carried out in open fields of eggplants. When the eggplants were treated with fenthion, the peak density of *T. palmi* was four times as large as that on the untreated plants. Further experiments were also conducted using the selective pesticide, pyriproxyfen, which has a control effect on the pupal stage of *T. palmi* but is harmless to *O. sauteri. T. palmi* on eggplants decreased very rapidly after application of this pesticide, due to a combined effect of the pesticide and *O. sauteri* on the *T. palmi* population.

On the other hand, when eggplants were treated with a non-selective pesticide, carbaryl, which is very harmful to *O. sauteri* and harmless to *T. palmi, O. sauteri* disappeared immediately, which caused the increase of *T. palmi*. Effectiveness of the IPM programme with a combination of naturally occurring *Orius* spp. and selective pesticides was evaluated in eggplant fields in Fukuoka prefecture. In most of the surveyed fields, the number of insecticide applications and the damage on the fruit by *T. palmi* were remarkably reduced in comparison with conventional fields. In these successful IPM fields, indigenous thrips species were abundant when *Orius* spp. colonized the eggplant fields. With regard to vegetation around eggplant fields, *O. sauteri* and *Orius nagaii* Yasunaga were predominant on white clovers and on paddy rice, respectively. It is likely that surrounding vegetation serves as a main reservoir for *Orius* populations on eggplants. Indigenous natural enemies were well conserved in this IPM system and effectively suppressed other pests such as cotton aphids, spider mites, and broad mites.

BIOLOGICAL CONTROL WITH MICRO-ORGANISMS

Various microbial insect diseases occur naturally, but may also be used as biological pesticides. When naturally occurring, these outbreaks are density-dependent in that they generally only occur as insect populations become denser. .

COMBINED USE OF PARASITOIDS AND PATHOGENS

In cases of massive and severe infection of invasive pests, techniques of pest control are often used in combination. An example being, that of the emerald ash borer (*Agrilus planipennis* Fairmaire, family Buprestidae), an invasive beetle from China, which has destroyed tens of millions of ash trees in its introduced range in North America. s part of the campaign against the emerald ash borer (EAB), American scientists in conjunction with the Chinese Academy of Forestry searched since 2003 for its natural enemies in the wild leading to the discovery of several parasitoid wasps, namely *Tetrastichus planipennisi*, a gregarious larval endoparasitoid, *Oobius agrili*, a solitary, parthenogenic egg parasitoid, and *Spathius agrili*, a gregarious larval ectoparasitoid.

These have been been introduced and released into the United States of America as a possible biological control of the emerald ash borer. Initial results have shown promise with *Tetrastichus planipennisi* and it is now being released along with *Beauveria bassiana*, a fungal pathogen with known insecticidal properties.

PLANTS TO REGULATE INSECT PESTS

Choosing a diverse range of plants for the garden can help to regulate pests in a variety of ways, including:

- Masking the crop plants from pests, depending on the proximity of the companion or intercrop.
- Producing olfactory inhibitors, odours that confuse and deter pests.
- Acting as trap plants by providing an alluring food that entices pests away from crops.
- Serving as nursery plants, providing breeding grounds for beneficial insects.
- Providing an alternative habitat, usually in a form of a shelterbelt, hedgerow, or beetle bank where beneficial insects can live and reproduce. Nectar-rich plants that bloom for long periods are especially good, as many beneficials are nectivorous during the adult stage, but parasitic or predatory as larvae. A good example of this is the soldier beetle which is frequently found on flowers as an adult, but whose larvae eat aphids, caterpillars, grasshopper eggs, and other beetles.
- Some plants have chemical defences in order to regulate pests. The geranium has developed such a defence against Japanese beetles, one of the most damaging and expensive pests to control when it comes to ornamental and turf plants. The geranium's petals contain a chemical compound that paralyzes the beetle within 30 minutes of ingestion. The beetle will remain paralyzed for several hours and will typically regain movement within 24 hours. However, while paralyzed the beetle is very vulnerable to its predators and is usually hunted before the paralysis subsides. Agricultural Research Service (ARS) scientists are working to isolate the chemical compound in geraniums that causes the paralysis in the beetles. Scientists hope to one day use this natural pesticide to control the population of beetles. In addition to this research ARS scientists are studying ways to help geranium leaves better hold on to pesticide chemicals that are sprayed on them, that way less pesticides will have to be applied to the leaves.

BACTERIA AND BIOLOGICAL CONTROL

Bacteria used for biological control infect insects via their digestive tracts, so insects with sucking mouth parts like aphids and scale insects are difficult to control with bacterial biological control. *Bacillus thuringiensis* is the most widely applied species of bacteria used for biological control, with at least four sub-species used to control Lepidopteran (moth, butterfly), Coleopteran (beetle) and Dipteran (true flies) insect pests.

FUNGI AND BIOLOGICAL CONTROL

Fungi that cause disease in insects are known as entomopathogenic fungi, including at least fourteen species of entomophthoraceous fungi attack aphids. Species in the genus *Trichoderma* are used to manage some soilborne plant pathogens. *Beauveria bassiana* is used to manage different types of pest such whiteflies, thrips, aphids and weevils.

Examples of entomopathogenic fungi:

- *Beauveria bassiana* (against white flies, thrips, aphids and weevils)
- *Paecilomyces fumosoroseus* (against white flies, thrips and aphids)
- *Metarhizium* spp. (against beetles, locusts, Hemiptera, spider mites and other pests)
- *Lecanicillium lecanii* (against white flies, thrips and aphids)
- *Cordyceps* species (sometines teleomorphs of the above: that infect a wide spectrum of arthropods)

CONSERVATION OF NATURAL ENEMIES

The conservation of natural enemies is probably the most important and readily available biological control practice available to homeowners and gardeners. Natural enemies occur in all areas, from the private garden to the open field. They are adapted to the habitat and to the target pest, and their conservation is generally simple and cost-effective. Lacewings, lady beetles, hover fly larvae, and parasitized aphid mummies are almost always present in aphid colonies. Fungus-infected adult flies are often common following periods of high humidity. These naturally occurring biological controls are often susceptible to the same pesticides used to target their pest hosts. Preventing the accidental eradication of natural enemies is termed "simple conservation."

To conserve and encourage pest insect eating birds, including native plants and ornamental plants that supply berries, acorns, nuts, seeds, nectar, and other vegetative foods, and also bird nest building materials, encourages their presence, health, and new generations. These qualities can also increase the visible population to enjoy in a garden. Using companion planting and the birds' insect cuisine habits is a traditional method for biological control agent pest control in an organic garden and any landscape, and in organic farming and sustainable agriculture. Installing specified nest boxes for mosquito-eating bats reduces a pest and increases endangered species conservation.

EFFECTS ON INVASIVE SPECIES

Invasive species are closely associated with biological controls because the environment in which they are invasive most likely does not contain their natural enemies. If invasive species are not controlled, biodiversity may be at great threat in the affected area. An example of an invasive species is the alligator weed. This plant was introduced to the United States from South America. This aquatic weed spreads rapidly and causes many problems in lakes and rivers. The weed takes root in shallow water causing major problems for navigation, irrigation, and flood control. The alligator weed flea beetle and two other biological controls were released in Florida. Because of their success, Florida banned the use of herbicides to control alligator weed three years after the controls were introduced. Similarly, *Galerucella calmariensis*, a leaf beetle, has been introduced in North America as a control agent for purple loosestrife (*Lythrum salicaria*).

Biological controls for invasive species also can have a negative impact on biodiversity. The cane toad, as mentioned previously, is an example of trying to control an invasive species. The cane toad was introduced to eradicate an invasive species. It became invasive, thus altering the biodiversity. The introduction of the cane toad could have caused more of a disturbance in biodiversity than the targeted species did.

EFFECTS OF BIOLOGICAL CONTROL

Biological control can potentially have positive and negative effects on biodiversity. Usually a biological control is introduced to an area to protect a native species from an invasive or exotic species that has moved into its area. The control is introduced to lessen the competition between native and introduced species. However, the introduced control does not always target only the intended species. It can also target native species.

When introducing a high biological control to a new area, a primary concern is the host- or prey-specificity of the control agent. Generalist feeders (control agents that are not restricted to a single species or a small range of species) often make poor biological control agents, and may become invasive species themselves.

For this reason, potential biological control agents should be subject to extensive testing and quarantine before release into any new environment. If a species is introduced and attacks a native species, the biodiversity in that area can change dramatically. When one native species is removed from an area, it may have filled an essential ecological niche. When this niche is absent it may directly affect the entire ecosystem.

Because they tend to be generalist feeders, vertebrate animals seldom make good biological control agents, and many of the classic cases of "biocontrol gone awry" involve vertebrates. For example, the cane toad, *Bufo marinus*, was introduced as a biological control and had significant negative impact on biodiversity. The cane toad was intentionally introduced to Australia to control the introduced French's Cane Beetle and the Greyback Cane Beetle. When introduced, the cane toad thrived and did not only feed on cane beetles but other insects as well. The cane toad soon spread very rapidly, thus taking over native amphibian habitat. The introduction of the cane toad also brought foreign disease to native reptiles. This drastically reduced the population of native toads and frogs. "The cane toad also exudes and can squirt poison from the parotid glands on their shoulders when threatened or handled. This toxin contains a cocktail of chemicals that can kill animals that eat it. Freshwater crocodiles, goannas, tiger snakes, dingos and northern quolls have all died after eating cane toads, as have pet dogs. This example shows how small mis-introduced organisms can alter the native biodiversity in large ecosystems in a rapid manner. A pyramid effect can take place if native species are reduced or eradicated. The domino effect keeps on going and can potentially exude on other bordering ecosystems until an equilibrium is reached.

A second example of a biological control agent that subsequently crossed over to native species is the *Rhinocyllus conicus*. The seed feeding weevil was introduced to North America to control exotic thistles. However, the weevil did not target only the exotic thistles, it also targeted native thistles that are essential to various native insects. The native insects rely solely on native thistles and do not adapt to other plant species. Therefore, they cannot survive. Biological controls do not always have negative impacts on biodiversity.

Successful biological control reduces the population density of the target species over several years, thus providing the potential for native species to re-establish. In addition, regeneration and reestablishment programmes can aid to the recovery of native species. Native species can be affected in a positive way as well.

To develop or find a biological control that exerts control only on the targeted species is a very lengthy process of research and experiments. In the late 19th century, the citrus industry was in great fear when the cottony cushion scale was discovered. This organism could cause a great deal of economic loss to the industry. However, a biological control was introduced. The vedalia beetle and a parasitoid fly were introduced to control the pest. Within a few years time, the cottony cushion scale was controlled by the natural enemies and the citrus industry suffered little financial loss. Many exotic or invasive species can suppress the development of native species. The introduction of an effective biological control that reduces the population of the invasive species allows the rejuvenation of the native species. Biological controls can reduce competition for biotic and abiotic factors which can result in the re-establishment of the once over ran native species.

CLASSICAL BIOLOGICAL CONTROL OF NATURAL ENEMIES

Classical biological control is the introduction of natural enemies to a new locale where they did not originate or do not occur naturally. This is usually done by government authorities. In many instances the complex of natural enemies associated with an insect pest may be inadequate. This is especially evident when an insect pest is accidentally introduced into a new geographic area without its associated natural enemies. These introduced pests are referred to as exotic pests and comprise about 40 per cent of the insect pests in the United States. Examples of introduced vegetable pests include the European corn borer (*Ostrinia nubilalis*), one of the most destructive insects in North America.

To obtain the needed natural enemies, scientists turned to classical biological control. This is the practice of importing, and releasing for establishment, natural enemies to control an introduced (exotic) pest, although it is also practiced against native insect pests.

The first step in the process is to determine the origin of the introduced pest and then collect appropriate natural enemies associated with the pest or closely related species. The natural enemy is then passed through a rigorous quarantine process, to ensure that no unwanted organisms (such as hyperparasitoids) are introduced, then they are mass produced, and released. Follow-up studies are conducted to determine if the natural enemy becomes successfully established at the site of release, and to assess the long-term benefit of its presence.

There are many examples of successful classical biological control programmes:

One of the earliest successes in the west was in controlling *Icerya purchasi*, the cottony cushion scale, a pest that was devastating the California citrus industry in the late 19th century. A predatory insect *Rodolia cardinalis* (the Vedalia Beetle), and a parasitoid fly were introduced from Australia by Charles Valentine Riley. Within a few years the cottony cushion scale was completely controlled by these introduced natural enemies.

Damage from *Hypera postica* Gyllenhal, the alfalfa weevil, a serious introduced pest of forage, was substantially reduced by the introduction of several natural enemies. 20 years after their introduction the population of weevils in the alfalfa area treated for alfalfa weevil in the Northeastern United States was reduced by 75 per cent. A small wasp, *Trichogramma ostriniae*, was introduced from China to help control the European corn borer making it a recent example of a long history of classical biological control efforts for this major pest. Many classical biological control programmes for insect pests and weeds are under way across the United States and Canada. The population of *Levuana irridescens* (the Levuana moth), a serious coconut pest in Fiji, was brought under control by a classical biological control programme in the 1920s.

Classical biological control is long lasting and inexpensive. Other than the initial costs of collection, importation, and rearing, little expense is incurred. When a natural enemy is successfully established it rarely requires additional input and it continues to kill the pest with no direct help from humans and at no cost. Classical biological control does not always work. It is usually most effective against exotic pests and less so against native insect pests. The reasons for failure are not often known but may include the release of too few individuals, poor adaptation of the natural enemy to environmental conditions at the release location, and lack of synchrony between the life cycle of the natural enemy and host pest.

EXAMPLES OF PREDATORS

Ladybugs, and in particular their larvae which are active between May and July in the northern hemisphere, are voracious predators of aphids such as greenfly and blackfly, and will also consume mites, scale insects and small caterpillars.

The ladybug is a very familiar beetle with various colored markings, whilst its larvae are initially small and spidery, growing up to 17 mm long. The larvae have a tapering segmented grey/black body with orange/yellow markings and ferocious mouthparts. They can be encouraged by cultivating a patch of nettles in the garden and by leaving hollow stems and some plant debris over winter so that they can hibernate.

Hoverflies resemble slightly darker bees or wasps and they have characteristic hovering, darting flight patterns. There are over 100 species of hoverfly whose larvae principally feed upon greenfly, one larva devouring up to fifty a day, or 1000 in its lifetime.

They also eat fruit tree spider mites and small caterpillars. Adults feed on nectar and pollen, which they require for egg production. Eggs are minute (1 mm), pale yellow white and laid singly near greenfly colonies. Larvae are 8–17 mm long, disguised to resemble bird droppings, they are legless and have no distinct head. Semi-transparent in a range of colours from green, white, brown and black.

Hoverflies can be encouraged by growing attractant flowers such as the poached egg plant *(Limnanthes douglasii)*, marigolds or phacelia throughout the growing season.

Dragonflies are important predators of mosquitoes, both in the water, where the dragonfly naiads eat mosquito larvae, and in the air, where adult dragonflies capture and eat adult mosquitoes. Community-wide mosquito control programmes that spray adult mosquitoes also kill dragonflies, thus removing an important biocontrol agent, and can actually increase mosquito populations in the long term.

Other useful garden predators include lacewings, pirate bugs, rove and ground beetles, aphid midge, centipedes, spiders, predatory mites, as well as larger fauna such as frogs, toads, lizards, hedgehogs, slow-worms and birds. Cats and rat terriers kill field mice, rats, June bugs, and birds. Dogs chase away many types of pest animals. Dachshunds are bred specifically to fit inside tunnels underground to kill badgers.

More examples:

- *Phytoseiulus persimilis* (against spider mites)
- *Amblyseius californicus* (against spider mites)
- *Amblyseius cucumeris* (against spider mites)
- *Typhlodromips swirskii* (against spider mites, thrips, and white flies)
- *Feltiella acarisuga* (against spider mites)
- *Stethorus punctillum* (against spider mites)
- *Macrolophus caluginosus* (against spider mites)

COMPRISE OF PARASITOID INSECTS

Most insect parasitoids are wasps or flies. Parasitoids comprise a diverse range of insects that lay their egg on or in the body of an insect host, which is then used as a food for developing larvae.

Parasitic wasps take much longer than predators to consume their victims, for if the larvae were to eat too fast they would run out of food before they became adults. Such parasites are very useful in the organic garden, for they are very efficient hunters, always at work searching for pest invaders. As adults they require high energy fuel as they fly from place to place, and feed upon nectar, pollen and sap, thereby pollinating plenty of flowering plants, particularly buckwheat, umbellifers, and composites will encourage their presence.

Four of the most important groups are:

1. *Ichneumonid wasps*: (5–10 mm). Prey mainly on caterpillars of butterflies and moths.
2. *Braconid wasps*: Tiny wasps (up to 5 mm) attack caterpillars and a wide range of other insects including greenfly. A common parasite of

the cabbage white caterpillar- seen as clusters of sulphur yellow cocoons bursting from collapsed caterpillar skin.

3. *Chalcid wasps*: Among the smallest of insects (<3 mm). Parasitize eggs/larvae of greenfly, whitefly, cabbage caterpillars, scale insects and Strawberry Tortrix Moth (*Acleris comariana*).
4. *Tachinid flies:* Parasitize a wide range of insects including caterpillars, adult and larval beetles, true bugs, and others.

Examples of parasitoids: wasp

- *Encarsia formosa* (against white flies)
- *Eretmocerus* spp. (against white flies)
- *Aphidius colemani* (against aphids)

BIOLOGICAL CONTROL INVOLVES IN AUGMENTATION

This third type of biological control involves the supplemental release of natural enemies. Relatively few natural enemies may be released at a critical time of the season (inoculative release) or literally millions may be released (inundative release). Additionally, the cropping system may be modified to favour or augment the natural enemies. This latter practice is frequently referred to as habitat manipulation. An example of inoculative release occurs in production of several crops. Periodic releases of the parasitoid, *Encarsia formosa*, are used to control whitefly, and the predaceous mite, *Phytoseiulus persimilis*, is used for control of the two-spotted spider mite.

Lady beetles, lacewings, or parasitoids such as those from the genus *Trichogramma* are frequently released in large numbers (inundative release). Recommended release rates for Trichogramma in vegetable or field crops range from 5,000 to 200,000 per acre (1 to 50 per square metre) per week depending on level of pest infestation. Similarly, entomopathogenic nematodes are released at rates of millions and even billions per acre for control of certain soil-dwelling insect pests.

The spraying of octopamine analogues (such as 3-FMC) has been suggested as a way to boost the effectiveness of augmentation. Octopamine, regarded as the invertebrate counterpart of dopamine plays a role in activating the insects' flight-or-fight response. The idea behind using octopamine analogues to augment biological control is that natural enemies will be more effective in their eradication of the pest, since the pest will be behaving in an unnatural way because its flight-or-fight mechanism has been activated. Octopamine analogues are purported to have two desirable characteristics for this type of application: (1) they affect insects at very low dosages (2) they do not have a physiological effect in humans (or other vertebrates).

Habitat or environmental manipulation is another form of augmentation. This tactic involves altering the cropping system to augment or enhance the effectiveness of a natural enemy. Many adult parasitoids and predators benefit from sources of nectar and the protection provided by refuges such as hedgerows, cover crops, and weedy borders. Also, the provisioning of natural shelters in the form of wooden caskets, boxes or (turnaround) flowerpots is a form of this. For example, the stimulation of the natural predator *Dermaptera* is done in gardens by hanging up turnaround flowerpots with straw or wood wool.

Mixed plantings and the provision of flowering borders can increase the diversity of habitats and provide shelter and alternative food sources. They are easily incorporated into home gardens and even small-scale commercial plantings, but are more difficult to accommodate in large-scale crop production. There may also be some conflict with pest control for the large producer because of the difficulty of targeting the pest species and the use of refuges by the pest insects as well as natural enemies.

Examples of habitat manipulation include growing flowering plants (pollen and nectar sources) such as Buckwheat near crops to attract and maintain populations of natural enemies. For example, hover fly adults can be attracted to umbelliferous plants in bloom.

Biological control experts in California have demonstrated that planting prune trees in grape vineyards provides an improved overwintering habitat or refuge for a key grape pest parasitoid. The prune trees harbour an alternate host for the parasitoid, which could previously overwinter only at great distances from most vineyards. Caution should be used with this tactic because some plants attractive to natural enemies may also be hosts for certain plant diseases, especially plant viruses that could be vectored by insect pests to the crop. Although the tactic appears to hold much promise, only a few examples have been adequately researched and developed.

BOTANICAL AND CHEMICAL INSECTICIDES

Botanical insecticides have long been touted as attractive alternatives to synthetic chemical insecticides for pest management because botanicals reputedly pose little threat to the environment or to human health. The body of scientific literature documenting bioactivity of plant derivatives to arthropod pests continues to expand, yet only a handful of botanicals are currently used in agriculture in the industrialized world, and there are few prospects for commercial development of new botanical products. Pyrethrum and neem are well established commercially, pesticides based on plant essential oils have recently entered the marketplace, and the use of rotenone appears to be waning.

A number of plant substances have been considered for use as insect antifeedants or repellents, but apart from some natural mosquito repellents, little commercial success has ensued for plant substances that modify arthropod behaviour. Several factors appear to limit the success of botanicals, most notably regulatory barriers and the availability of competing products (newer synthetics, fermentation products, microbials) that are cost-effective and relatively safe compared with their predecessors. In the context of agricultural pest management, botanical insecticides are best suited for use in organic food production in industrialized countries but can play a much greater role in the production and postharvest protection of food in developing countries.

The practice of using plant derivatives, or botanical insecticides as we now know them, in agriculture dates back at least two millennia in ancient China, Egypt, Greece, and India. Even in Europe and North America, the documented use of botanicals extends back more than 150 years, dramatically predating discoveries of the major classes of synthetic chemical insecticides (*e.g.*, organochlorines, organophosphates, carbamates, and pyrethroids) in the mid-1930s to 1950s. What is clear from recent history is that synthetic insecticides effectively relegated botanicals from an important role in agriculture to an essentially trivial position in the marketplace among crop protectants.

However, history also shows that overzealous use of synthetic insecticides led to numerous problems unforeseen at the time of their introduction: acute and chronic poisoning of applicators, farmworkers, and even consumers; destruction of fish, birds, and other wildlife; disruption of natural biological control and pollination; extensive groundwater contamination, potentially threatening human and environmental health; and the evolution of resistance to pesticides in pest populations. Governments responded to these problems with regulatory action, banning or severely restricting the most damaging products and creating policies to replace chemicals of concern with those demonstrated to pose fewer or lesser risks to human health and the environment.

In the United States, these policies are reflected by the definition of "reduced risk" pesticides by the Environmental Protection Agency in the early 1990s with their favoured regulatory status, and by the Food Quality Protection Act (1996), which, in reappraising safe levels of pesticide residues in foods, is having the net effect of removing most synthetic insecticides developed before 1980 from use in agriculture. These changes in the regulatory "environment" appeared to heighten the impetus for the discovery and development of alternative pest management products—those with reduced health and environmental impacts—including insecticides derived from plants.

Indeed, the scientific literature of the past 25 years describes hundreds of isolated plant secondary metabolites that show feeding deterrent or toxic effects to insects in laboratory bioassays, and botanical insecticides have been the subject of several recent volumes. Yet in spite of the scale of this research enterprise, only a handful of botanical insecticides are in commercial use on vegetable and fruit crops today, with significant commercial development of only two new sources of botanicals in the past 20 years. In this chapter I review current botanicals and their trends in use, discuss the few botanical materials with potential for future commercialization, suggest why so few botanicals reach the marketplace, and finally suggest in what contexts botanicals could prove effective in the years to come.

TYPES OF BOTANICAL PRODUCTS

At present there are four major types of botanical products used for insect control (pyrethrum, rotenone, neem, and essential oils), along with three others in limited use (ryania, nicotine, and sabadilla). Additional plant extracts and oils (*e.g.*, garlic oil, *Capsicum* oleoresin) see limited (lowvolume) regional use in various countries, but these are not considered here. In discussing the extent to which each of the more important botanical insecticides is used, I often refer to data published annually by the State of California's Department of Pesticide Regulation.

Although not necessarily representative of uses in other jurisdictions, total pesticide use (in terms of active ingredient applied) in California totaled more than 175 million pounds, or 80,000 tonnes, in 2003. This amount represents approximately 6 per cent of global pesticide use, of which 91 per cent was used for agriculture. Pesticide use data in California are reported by active ingredient and by crop or other use and as such are perhaps the world's most accurate and detailed records of pesticide use available to the general public.

Neem

Two types of botanical insecticides can be obtained from seeds of the Indian neem tree, *Azadirachta indica* (Meliaceae). Neem oil, obtained by cold-pressing seeds, can be effective against soft-bodied insects and mites but is also useful in the management of phytopathogens.

Apart from the physical effects of neem oil on pests and fungi, disulfides in the oil likely contribute to the bioactivity of this material. More highly valued than neem oil are medium-polarity extracts of the seed residue after removal of the oil, as these extracts contain the complex triterpene azadirachtin. Neem seeds actually contain more than a dozen azadirachtin analogs, but the major form is azadirachtin and the remaining minor analogs likely contribute little to overall efficacy of the extract.

Seed extracts include considerable quantities of other triterpenoids, notably salannin, nimbin, and derivatives thereof. The role of these other natural substances has been controversial, but most evidence points to azadirachtin as the most important active principle. Neem seeds typically contain 0.2 per cent to 0.6 per cent azadirachtin by weight, so solvent partitions or other chemical processes are required to concentrate this active ingredient to the level of 10 per cent to 50 per cent seen in the technical grade material used to produce commercial products. Azadirachtin has two profound effects on insects. At the physiological level, azadirachtin blocks the synthesis and release of molting hormones (ecdysteroids) from the prothoracic gland, leading to incomplete ecdysis in immature insects.

In adult female insects, a similar mechanism of action leads to sterility. In addition, azadirachtin is a potent antifeedant to many insects. The discovery of neem by western science is attributed to Heinrich Schmutterer, who observed that swarming desert locusts in Sudan defoliated almost all local flora except for some introduced neem trees. Indeed, azadirachtin was first isolated based on its exceptional antifeedant activity in the desert locust, and this substance remains the most potent locust antifeedant discovered to date. Unlike pyrethrins, azadirachtin has defied total synthesis to this point.

Promoted in the United States by Robert Larson (with assistance from the U.S. Department of Agriculture), neem rapidly became the modern paradigm for development of botanical insecticides. Enthusiasm for neem was fostered by several international conferences in the 1980s and 1990s, and several volumes dedicated to neem and neem insecticides have been published. Unfortunately, neem's commercial success has fallen well short of the initial hype fueled by the explosive scientific literature surrounding it.

In part this is due to the relatively high cost of the refined product and the relatively slow action on pest insects. Nonetheless, several azadirachtinbased insecticides are sold in the United States and at least two such products in the European Union.

In California, azadirachtin-based insecticides constituted about one third of the botanicals used in agriculture in 2003 (~600 kg). In practice, reliable efficacy is linked to the physiological action of azadirachtin as an insect growth regulator; the antifeedant effect, which is spectacular in the desert locust, is highly variable among pest species, and even those species initially deterred are often capable of rapid desensitization to azadirachtin. What is clear is that azadirachtin is considered non-toxic to mammals (rat oral acute LD_{50} is >5000 mg kg^{-1}), fish, and pollinators. The influence of azadirachtin on natural enemies is highly variable. Like the pyrethrins, azadirachtin is rapidly degraded by sunlight. For example, on olives growing in Italy, azadirachtin has a half-life of approximately 20 h. On the other hand, azadirachtin has systemic action in certain crop plants, greatly enhancing its efficacy and field persistence.

Plant Essential Oils

Steam distillation of aromatic plants yields essential oils, long used as fragrances and flavorings in the perfume and food industries, respectively, and more recently for aromatherapy and as herbal medicines. Plant essential oils are produced commercially from several botanical sources, many of which are members of the mint family (Lamiaceae). The oils are generally composed of complex mixtures of monoterpenes, biogenetically related phenols, and sesquiterpenes.

Examples include 1,8-cineole, the major constituent of oils from rosemary (*Rosmarinus officinale*) and eucalyptus (*Eucalyptus globus*); eugenol from clove oil (*Syzygium aromaticum*); thymol from garden thyme (*Thymus vulgaris*); and menthol from various species of mint (*Mentha* species). A number of the source plants have been traditionally used for protection of stored commodities, especially in the Mediterranean region and in southern Asia, but interest in the oils was renewed with emerging demonstration of their fumigant and contact insecticidal activities to a wide range of pests in the 1990s.

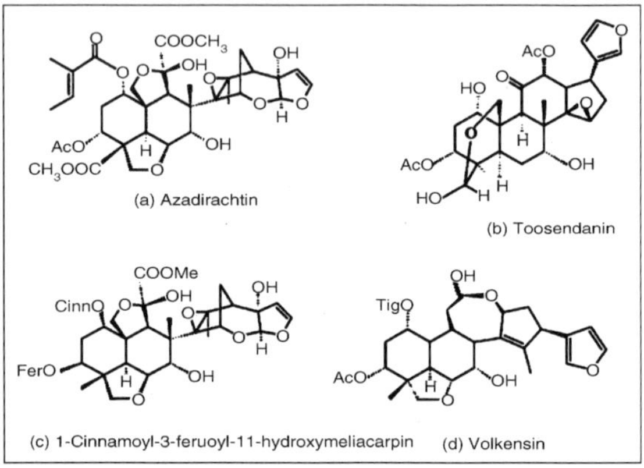

(a) Azadirachtin (b) Toosendanin (c) 1-Cinnamoyl-3-feruoyl-11-hydroxymeliacarpin (d) Volkensin

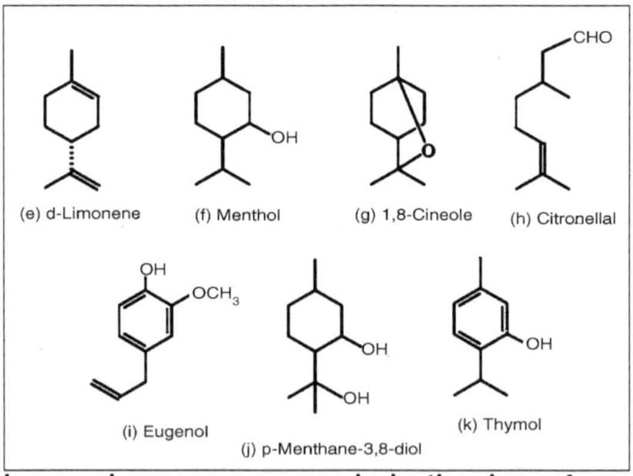

(e) d-Limonene (f) Menthol (g) 1,8-Cineole (h) Citronellal (i) Eugenol (j) p-Menthane-3,8-diol (k) Thymol

The rapid action against some pests is indicative of a neurotoxic mode of action, and there is evidence for interference with the neuromodulator octopamine by some oils and with GABA-gated chloride channels by others. Some of the purified terpenoid constituents of essential oils are moderately toxic to mammals, but, with few exceptions, the oils themselves or products based on oils are mostly non-toxic to mammals, birds, and fish. However, as broad-spectrum insecticides, both pollinators and natural enemies are vulnerable to poisoning by products based on essential oils.

Owing to their volatility, essential oils have limited persistence under field conditions; therefore, although natural enemies are susceptible via direct contact, predators and parasitoids reinvading a treated crop one or more days after treatment are unlikely to be poisoned by residue contact as often occurs with conventional insecticides. In the United States, commercial development of insecticides based on plant essential oils has been greatly facilitated by exemption from registration for certain oils commonly used in processed foods and beverages.

This opportunity has spurred the development of essential oil-based insecticides, fungicides, and herbicides for agricultural and industrial applications and for the consumer market, using rosemary oil, clove oil, and thyme oil as active ingredients. Interest in these products has been considerable, particularly for control of greenhouse pests and diseases and for control of domestic and veterinary pests, with several private companies (*e.g.*, EcoSMART Technologies, Inc., United States) moving towards or into the marketplace. Another factor favouring development of botanical insecticides based on plant essential oils is the relatively low cost of the active ingredients, a result of their extensive worldwide use as fragrances and flavorings. In contrast, pyrethrum and neem are used primarily for insecticide production.

Pyrethrum

Pyrethrum refers to the oleoresin extracted from the dried flowers of the pyrethrum daisy, *Tanacetum cinerariaefolium* (Asteraceae). The flowers are ground to a powder and then extracted with hexane or a similar non-polar solvent; removal of the solvent yields an orange-colored liquid that contains the active principles. These are three esters of chrysanthemic acid and three esters of pyrethric acid. Among the six esters, those incorporating the alcohol pyrethrolone, namely pyrethrins I and II, are the most abundant and account for most of the insecticidal activity.

Technical grade pyrethrum, the resin used in formulating commercial insecticides, typically contains from 20 per cent to 25 per cent pyrethrins. The insecticidal action of the pyrethrins is characterized by a rapid knockdown effect, particularly in flying insects, and hyperactivity and convulsions in most insects.

These symptoms are a result of the neurotoxic action of the pyrethrins, which block voltage-gated sodium channels in nerve axons. As such, the mechanism of action of pyrethrins is qualitatively similar to that of DDT and many synthetic organochlorine insecticides.

In purity, pyrethrins are moderately toxic to mammals (rat oral acute LD_{50} values range from 350 to 500 mg kg^{-1}), but technical grade pyrethrum is considerably less toxic (~1500 mg kg^{-1}). Pyrethrins are especially labile in the presence of the UV component of sunlight, a fact that has greatly limited their use outdoors. A recent study indicated that the half-lives of pyrethrins on field-grown tomato and bell pepper fruits were 2 hours or less.

This problem created the impetus for the development of synthetic derivatives ("pyrethroids") that are more stable in sunlight. The modern pyrethroids, developed in the 1970s and 1980s, have been highly successful and represent one of the rare examples of synthetic pesticide chemistry based on a natural product model. However, note that the modern pyrethroids bear little structural resemblance to the natural pyrethrins, and their molecular mechanism of action differs as well.

Pyrethrum use data from California in 2003 clearly demonstrate the dominance of this material among botanicals: Pyrethrum accounted for 74 per cent of all botanicals used that year, but only 27 per cent of that amount was used in agriculture (~800 kg). Major uses of pyrethrum in California are for structural pest control, in public health, and for treatment of animal premises. Pyrethrum is the predominant botanical in use, perhaps accounting for 80 per cent of the global botanical insecticide market.

For many years world production of pyrethrum was led by Kenya, with lesser quantities produced in Tanzania and Ecuador. In the past five years, Botanical Resources Australia, with plantings in Tasmania, has become the second largest producer in theworld (~30 per cent of world production at present). Pyrethrum produced in Tasmania is qualitatively similar to that produced in East Africa and elsewhere, but the market share achieved by the Australian producer may not increase owing to World Bank grants and government subsidies to producers in Kenya and China.

Rotenone and Other Extant Botanicals

As an insecticide, rotenone has been in use for more than 150 years, but its use as a fish poison dates back even further. Rotenone is one of several isoflavonoids produced in the roots or rhizomes of the tropical legumes *Derris*, *Lonchocarpus*, and *Tephrosia*. Most rotenone used at present comes from *Lonchocarpus* grown in Venezuela and Peru and is often called cub´e root. Extraction of the root with organic solvents yields resins containing as much as 45 per cent total rotenoids; studies indicate that the major constituents are rotenone (44 per cent) and deguelin (22 per cent). Rotenone is commonly sold as a dust containing 1 per cent to 5 per cent active ingredients for home and garden use, but liquid formulations used in organic agriculture can contain as much as 8 per cent rotenone and 15 per cent total rotenoids. Rotenone is a mitochondrial poison, which blocks the electron transport chain and prevents energy production. As an insecticide it is considered a stomach poison because it must be ingested to be effective. Pure rotenone is comparable to DDT and other synthetic insecticides in terms of its acute toxicity to mammals (rat oral LD^{50} is 132 mg kg^{-1}), although it is much less toxic at the levels seen in formulated products.

Safety of rotenone has recently been called into question because of:

- Controversial reports that acute exposure in rats produces brain lesions consistent with those observed in humans and animals with Parkinson's disease,
- The persistence of rotenone on food crops after treatment.

A study of rotenone residues on olives conducted in Italy determined that the halflife of rotenone is 4 days, and at harvest residue levels were above the tolerance limit. Moreover, residues were concentrated in oil obtained from the olives. As an agricultural insecticide, use of rotenone is limited to organic food production. In California, about 200 kg are used annually, mostly on lettuce and tomato crops. Sabadilla is a botanical insecticide obtained from the seeds of the South American lily *Schoenocaulon officinale*. In purity, the active principles, cevadine-type alkaloids, are extremely toxic to mammals (rat oralLD_{50} is ~13 mg kg^{-1}), but commercial preparations typically contain less than 1 per cent active ingredient, providing a margin of safety. The mode of action of these alkaloids is remarkably similar to that of the pyrethrins, despite their lack of structural similarity. Sabadilla is used primarily by organic growers; in California about 100 kg is used annually, primarily on citrus crops and avocado.

Another botanical in declining use is ryania, obtained by grinding the wood of the Caribbean shrub *Ryania speciosa* (Flacourtiaceae). The powdered wood contains <1 per cent ryanodine, an alkaloid that interferes with calcium release in muscle tissue. It is used to a limited extent by organic apple growers for control of the codling moth, *Cydia pomonella*. More information on sabadilla and ryania can be found in a recent review.

Like pyrethrum and rotenone, nicotine, an alkaloid obtained from the foliage of tobacco plants (*Nicotiana tabacum*) and related species, has a long history as an insecticide. Nicotine and two closely related alkaloids, nornicotine and anabasine, are synaptic poisons that mimic the neurotransmitter acetylcholine. As such, they cause symptoms of poisoning similar to those seen with organophosphate and carbamate insecticides. Owing to the extreme toxicity of pure nicotine to mammals (rat oral LD_{50} is 50 mg kg^{-1}) and its rapid dermal absorption in humans, nicotine has seen declining use, primarily as a fumigant in greenhouses against soft-bodied pests. However, there remains some interest in preparing stable nicotine fatty acid soaps, presumably with reduced bioavailability and toxicity to humans.

INSECT ANTIFEEDANTS AND REPELLENTS

Antifeedants

The possibility of using nontoxic deterrents and repellents as crop protectants is intuitively attractive. The concept of using insect antifeedants (=feeding deterrents) gained strength in the 1970s and 1980s with the demonstration of the potent feeding deterrent effect of azadirachtin and neem seed extracts to a large number of pest species. Indeed, considerable literature, scientific and otherwise, touts neem as a successful demonstration of the antifeedant concept. In reality, it is the physiological actions of azadirachtin that appear most reliably linked to field efficacy of neem insecticides; although purely behavioural effects cannot be ruled out, there is hardly any irrefutable evidence or documentation of field efficacy based on the antifeedant effects of neem alone. As an academic exercise, the discovery and demonstration of plant natural products as insect antifeedants has been unquestionably successful.

In addition to the neem triterpenoids, extensive work has been performed on clerodane diterpenes from the Lamiaceae and sesquiterpene lactones from the Asteraceae.

On the other hand, not a single crop protection product based unequivocally on feeding or oviposition deterrence has been commercialized. Two main problems face the use of antifeedants in agriculture. The first is interspecific variation in response—even closely related species can differ dramatically in behavioural responses to a substance—limiting the range of pests affected by a particular antifeedant.

Repellents

For many chemists, an effective alternative to DEET (N,N-diethyl-*m*-toluamide) for personal protection against mosquitoes and biting flies is the holy grail. In spite of five decades of research, no chemical has been found that provides the degree of protection against biting mosquitoes or persistence on human skin afforded by DEET. Concerns with the safety of DEET, especially to children, have resulted in the introduction of several plant oils as natural alternatives. Some personal repellents in the U.S. marketplace contain oils of citronella, eucalyptus, or cedarwood as active ingredients; 2-phenethylpropionate, a constituent of peanut oil, and *p*-menthane-3,8-diol (obtained from a particular species of mint) are also used in consumer products. All of these materials can provide some protection, but the duration of their effect can be limited (often <1 h). In tropical areas where mosquito-borne disease is a threat (*e.g.*, yellow fever, dengue, malaria), DEET probably remains the only reliable repellent. Oil of citronella or the constituent citronellal is also used in mosquito coils to repel mosquitoes from outdoor areas. Several veterinary products for flea and tick control on domestic pets contain *d*-limonene (from citrus peels) as the active ingredient.

Other uses for repellents under investigation include perimeter treatments of buildings to exclude termites and the use of essential oils to repel cockroaches from kitchens and flies from dairy barns. Another important use of plant essential oil constituents is in fumigation of beehives to manage economically important honey bee parasites, the Varroa mite (*Varroa jacobsoni*) and the tracheal mite (*Acarapis woodi*). In North America, menthol (from peppermint) is widely used for this purpose), and in Europe thymol (from garden thyme) is most often used.

Deter Feeding

Some substances that deter feeding by one pest can even serve as attractants or stimulants for other pests. The second is the behavioural plasticity in insects—pests can rapidly habituate to feeding deterrents, rendering them ineffective in a matter of hours. This has been recently demonstrated not only for pure substances like azadirachtin, but also for complex mixtures (plant extracts). Whereas a highly mobile (flying) insect may leave a plant upon first encountering an antifeedant, a less mobile one (larva) may remain on the plant long enough for the deterrent response to wane. Such behavioural changes are important in light of the observation that some plant substances are initially feeding deterrents but lack toxicity if ingested. Azadirachtin is clearly an exception to this rule, as ingestion leads to deleterious physiological consequences, but many other compounds or extracts with demonstrated antifeedant effects lack toxicity when administered topically or via injection.

POTENTIAL NEWBOTANICALS

Annonaceous Acetogenins

Botanical insecticides have been traditionally prepared from the seeds of tropical *Annona* species, members of the custard apple family (Annonaceae). These include the sweetsop (*A. squamosa*) and soursop (*A. muricata*), important sources of fruit juices in Southeast Asia. Detailed investigations in the 1980s led to the isolation of a number of long-chain fatty acid derivatives, termed acetogenins, responsible for the insecticidal bioactivity. The major acetogenin obtained from seeds of *A. squamosa* is annonin I, or squamocin, and a similar compound, asimicin, was isolated from the bark of the American pawpaw tree, *Asimina triloba*.

McLaughlin and colleagues (66) hold a U.S. patent on insecticides based on acetogenins from *A. triloba*; Bayer AG (Germany) holds a similar patent based on *Annona* acetogenins. These compounds are slow acting stomach poisons, particularly effective against chewing insects such as lepidopterans and the Colorado potato beetle (*Leptinotarsa decemlineata*). Further investigations revealed that the acetogenins have a mode of action identical to that of rotenone, *i.e.*, they block energy production in mitochondria in both insects and mammals. In purity certain acetogenins are toxic to mammals (LD^{50} is <20 mg kg^{-1}), an impediment to regulatory approval, even though standardized extracts from *Annona* seeds and *Asimina* bark are much less toxic.

McLaughlin and associates have isolated hundreds of acetogenins from the Annonaceae, and for many their potential as anticancer agents exceeds their value as insecticides. In spite of the patents based on the insecticidal activities of these materials, no commercial development has proceeded with the exception of a head lice shampoo that contains a standardized pawpaw extract among its active ingredients (Nature's Sunshine Products, Inc., United States). *Annona* seed extracts may prove more useful in tropical countries where the fruits are commonly consumed or used to produce fruit juice, in which case the seeds are a waste product. For example, Leatemia and Isman recently demonstrated that crude ethanolic extracts or even aqueous extracts of seeds from *A. squamosa* collected at several sites in eastern Indonesia are effective against the diamondback moth (*Plutella xylostella*).

Melia Extracts

The remarkable bioactivity of azadirachtin from the Indian neem tree (*Azadirachta indica*) led to the search for natural insecticides in the most closely related genus, *Melia*.

Seeds from the chinaberry tree, *M. azedarach*, contain a number of triterpenoids, the meliacarpins, that are similar but not identical to the azadirachtins, and these too have insect growth regulating bioactivities. But in spite of the abundance of chinaberry trees in Asia and other tropical and subtropical areas to which they were introduced, development of commercial insecticides has not paralleled that of the neem insecticides. The main reason is the presence, in chinaberry seeds, of additional triterpenoids, the meliatoxins, that have demonstrated toxicity to mammals.

However the chemistry of chinaberry varies considerably across its natural and introduced range, and seeds of *M. azedarach* growing in Argentina lack meliatoxins but produce triterpenoids (most notably meliartenin) that are strong feeding deterrents to pest insects and could prove useful for pest management. Similar results have been obtained from South Africa using aqueous extracts of chinaberry leaves, presumably lacking meliatoxins but efficacious against the diamondback moth.

In the early 1990s a botanical insecticide produced in China was based on an extract of bark of *Melia toosendan*, a tree considered by most taxonomists to be synonymous with *M. azedarach*. The extract contains a number of triterpenoids based on toosendanin, a substance reported to be a stomach poison for chewing insects. Later studies suggest that this substance acts primarily as a feeding deterrent but can also serve as a synergist for conventional insecticides. Although relatively nontoxic to mammals, it is unclear whether this material remains in production or whether it is sufficiently efficacious as a stand-alone crop protectant.

When *M. toosendan* came under scientific scrutiny, investigation of the east African *M. volkensii* demonstrated bioactivity in insects from seed extracts of this species. The active principles in *M. volkensii* include the triterpenoid salannin, also a major constituent of neem seed extracts, and some novel triterpenoids such as volkensin. Collectively these function as feeding deterrents and stomach poisons with moderate efficacy against chewing insects and as a mosquito larvicide. Although a standardized seed extract has been made in quantities sufficient for research, commercial production appears unlikely owing to a lack of infrastructure for harvesting seeds in addition to regulatory impediments.

Sucrose Esters

In the early 1990s scientists at the U.S. Department of Agriculture discovered that sugar esters naturally occurring in the foliage of wild tobacco (*Nicotiana gossei*) were insecticidal to certain soft-bodied insects and mites. Although patented, extraction of these substances on a commercial scale from plant biomass proved impractical, leading to the development of sucrose esters manufactured from sugar and fatty acids obtained from vegetable oils. AVA Chemical Ventures (United States) has patented and registered an insecticide/miticide based on C_8 and C_{10} fatty acid mono-, di-, and triesters of sucrose octanoate and sucrose dioctanoate.

The product, first registered in 2002, contains 40 per cent active ingredient. Functionally, this product appears to differ little from the insecticidal soaps based on fatty acid salts developed in the 1980s, particularly potassium oleate. Both products are contact insecticides that kill small insects and mites through suffocation (by blocking the spiracles) or disruption of cuticular waxes and membranes in the integument, leading to desiccation. Although useful in home and garden products and in greenhouse production, the utility of these materials for agriculture remains to be seen.

CURRENT TRENDS IN THE USE OF BOTANICALS

Trends and Changes in Registration

Given the ongoing negative perception of pesticides by the general public, government response to that perception, and increasing documentation of environmental contamination, it is hard to imagine pesticide regulators easing toxicological requirements for new pesticides, with the possible exception of certain plant oils and extracts widely used in human foods. Globalization of agricultural commodities will serve only to tighten restrictions on pesticide use in developing countries where fresh produce for export to wealthier countries is an important source of revenue. All produce imported into the European Union, United States, and Japan (for example) must comply with pesticide regulations in the respective importing country, meeting the same standards as their own domestic produce.

As a result, pesticide regulations set in the wealthiest countries have global reach—they affect growers directly in developing countries who are forced to comply.

In short, a lack of confidence in the safety of a specific botanical insecticide by the European Union could make that product unfavourable in a tropical country, even where it makes sense for poorer growers providing agricultural produce for their domestic markets, and perhaps where the botanical source material grows and could be inexpensively prepared for crop protection. If nothing else, this reviewshould highlight the fact that few new botanical insecticides are likely to see commercialization on a meaningful scale in the near future, in spite of the continual discovery of plant natural products with bioactivity against insect pests.

Africa

Data on regulated insecticides are not readily available for most African countries. Among botanicals, only pyrethrum is approved for use in South Africa. As in Latin America, numerous crude plant extracts and oils are likely in local use in the poorer countries.

Europe

Although considered by many in the agrochemical industry to be especially restrictive with respect to pesticide registrations, the European Union permits the use of pyrethrum, neem, rotenone, and nicotine, along with "components of etheric oils of plant origin". Individual countries in western Europe show considerable variation in the botanicals they permit. For example, Hungary permits pyrethrum and nicotine, although the latter is severely restricted. Denmark permits only pyrethrum and rotenone, Germany pyrethrum and neem. The Netherlands permits pyrethrum alone. In spite of years of research on azadirachtin and neem in the United Kingdom, this botanical has never achieved registration there, leaving pyrethrum, rotenone, and nicotine as the only approved botanicals

Pacific/Asia

India appears to embrace botanicals more than many other countries in the region, permitting all of the materials (save sabadilla) mentioned herein and allowing new products provisional registration while toxicological and environmental data in support of full registration are acquired. New Zealand has registrations for pyrethrum, rotenone, and neem, whereas Australia has yet to approve neem in spite of almost two decades of research and development in that country. Likewise, neem has yet to be approved for use in the Philippines, where pyrethrum is the only approved botanical insecticide.

North America

At present, the United States allows the broadest range of botanical insecticides among industrialized countries, with registrations for pyrethrum, neem, rotenone, several essential oils, sabadilla, ryania, and nicotine. Several azadirachtinbased (neem) insecticides are sold in the United States, and a number of plant essential oils are exempt from registration altogether. Canada has been more conservative with respect to pesticide registrations, allowing pyrethrum, rotenone, and nicotine, but only a handful of essential oils. Neem has yet to achieve full registration in Canada, to the disappointment of many organic growers. Mexico allows the use of most products sold in the United States, although there is no specific exemption for plant oils.

Latin America

In Brazil, each state has autonomous regulatory authority. Botanicals registered in most states include pyrethrum, rotenone, neem, and garlic, although nicotine and extracts of native plants are used to a small extent. Throughout Latin America plant oils and extracts are produced by cottage industries on a small scale and used outside of any regulatory system on a regional basis.

DRAWBACKS AND BARRIERS TO COMMERCIALIZATION

In reviewing this subject previously, three main barriers are identified to commercialization for botanical insecticides: sustainability of the botanical resource, standardization of chemically complex extracts, and regulatory approval. For each of these there are also important cost considerations. Other drawbacks or limitations are the slow action of many botanicals—growers must gain confidence in insecticides that do not produce an immediate "knockdown" effect—and the lack of residual action for most botanicals.

Sustainability

To produce a botanical insecticide on a commercial scale, the source plant biomass must be obtainable on an agricultural scale and preferably not on a seasonal basis. Unless the plant in question is extremely abundant in nature, or already grown for another purpose (*e.g.*, sweetsop, *Annona squamosa*, grown for its edible fruit; rosemary, *Rosmarinus officinale*, as a flavoring), it must be amenable to cultivation. Pyrethrum and neem meet this criterion; the latter has been extensively introduced into Africa, Australia, and Latin America, more so as a shade tree, windbreak, or source of firewood than for its yield of natural medicines or insecticides. Research aimed at producing azadirachtin from neem tissue culture provided proof of concept, but economic feasibility has yet to be attained. In the not-too-distant future it may be possible to produce botanical insecticides by "phytopharming," *i.e.*, through genetic engineering of an existing field crop to produce high-value natural products originally isolated from a different botanical source.

But as progress in plant biotechnology continues at a rapid pace, it may prove just as easy to modify the plants we wish to protect from pests directly, such that they produce the natural product protectant constitutively, alleviating the need to obtain the desired botanical product through extraction, formulate it, and then apply it to the crop we wish to protect.

These sorts of technological advancements seem far more likely now than they did even a decade ago; however, the cost of these technologies will dictate that the traditional means of obtaining botanical insecticides, and indeed their minor uses (on small acreage specialty crops) or uses in developing countries on lesser value crops will continue for many years to come. For example, neem seed oil had a long history of use in India for the production of soaps and low grade industrial oil. When extraction companies began purchasing neem seeds in bulk to produce insecticides, the price of seeds increased 10-fold. In contrast, certain plant essential oils have numerous uses as fragrances and flavorings, and the massive volumes required to satisfy these industries maintain low prices that make their use as insecticides attractive.

Regulatory Approval

Regulatory approval remains the most formidable barrier to the commercialization of new botanical insecticides. In many jurisdictions, no distinction is made between synthetic pesticides and biopesticides, including botanicals. Simply put, the market for botanicals in industrialized countries—based mostly on uses in greenhouse production and organic agriculture—is too small to generate sufficient profits to offset multimillion dollar regulatory costs. Unfortunately, this situation may prevent many "green" pesticides from reaching the marketplace in countries where the demand is greatest. I am not making the case that botanicals should be exempt from all regulatory scrutiny; as discussed above, nicotine is as toxic and hazardous as many synthetic insecticides, and strychnine, still used for rodent and insect control in some regions, is responsible for some human poisonings.

Natural products can pose risks, and safety cannot be assumed. But most of the botanicals discussed in this review are characterized by low mammalian toxicity, reduced effects on non-target organisms, and minimal environmental persistence. As noted, several plant essential oils and their constituents are exempt from registration in the United States, attributed to their long use history as food and beverage flavorings or as culinary spices. This exemption has facilitated the rapid development and commercialization of insecticides based on these materials as active ingredients. Although other jurisdictions have yet to follow the lead of the United States in this regard, there are proposals in some Asian countries to exempt some types of pesticides from registration for specific uses in public health, for example, in head lice preparations or for cockroach and fly suppression. It seems that regulatory agencies continue to focus their efforts on protecting the general public from miniscule traces of pesticides in the food supply rather than focusing on the safety of applicators and farmworkers, for whom, arguably, the more demonstrable hazards occur.

Standardization of Botanical Extracts

An often cited drawback to the adoption of botanical insecticides by growers is the variation in performance of a particular product, even when prepared by the same process. Natural variation in the chemistry of a plant-based commodity should come as no surprise to anyone who enjoys coffee, tea, wine, or chocolate.

Recent investigation of seed extracts from sweetsop (*A. squamosa*) collected in Indonesia demonstrated both geographical and annual variation in their insecticidal potency.

For a botanical insecticide to provide a reliable level of efficacy to the user, there must be some degree of chemical standardization, presumably based on the putative active ingredient(s). This has certainly been achieved with more refined products based on pyrethrum, neem, and rotenone, but crude preparations often contain low concentrations of active ingredients without adequate quantitation.

To achieve standardization, the producer must have an analytical method and the equipment necessary for analysis and may need to mix or blend extracts from different sources, which requires storage facilities and is partially dependent on the inherent stability of the active principles in the source plant material or extracts thereof held in storage.

ROLE OF BOTANICALS IN THE FUTURE

Botanical Insecticides in Ccrop Protection

What role can botanical insecticides play in crop protection and for other uses in the near future? In industrialized countries it is hard to imagine botanicals playing a greater role than at present, except in organic food production. Organic production is estimated to be growing by 8 per cent to 15 per cent per annum in Europe and in North America, and it is in those marketplaces that botanicals face the fewest competitors. Even there, however, microbial insecticides and spinosad have proven efficacious and cost-effective. Rather than considered as stand-alone products, botanicals might be better placed as products in crop protectant rotations, especially in light of the documented resistance of the diamondback moth to *Bacillus thuringiensis* and spinosad due to overuse. In conventional agriculture, botanicals face tremendous competition from the newest generation of "reduced risk" synthetic insecticides such as the neonicotinoids. Between 1998 and 2003, use of reduced risk pesticides in California increased more than threefold (from 138 to 483 tons), whereas biopesticide use declined (from 652 to 472 tons). Botanicals, constituting less than 1 per cent of biopesticide use in California, are also in decline.

Benefits of Botanical Insecticides

Overall, it is hard not to conclude that the best role for botanicals in the wealthier countries is in public health (mosquito, cockroach abatement) and for consumer (home and garden) use. The real benefits of botanical insecticides can be best realised in developing countries, where farmers may not be able to afford synthetic insecticides and the traditional use of plants and plant derivatives for protection of stored products is long established.

Even where synthetic insecticides are affordable to growers (*e.g.*, through government subsidies), limited literacy and a lack of protective equipment result in thousands of accidental poisonings annually. Recent attention has been paid to traditional plants used in West Africa for postharvest protection against insects. Some of the more efficacious plants used have well-known active principles (*e.g.*, rotenoids from *Tephrosia*, nicotine from *Nicotiana*, methyl salicylate from *Securidaca*, and eugenol from *Ocimum*); some of these are volatile and act as natural fumigants that kill adult pests and their progeny.

At least one study indicates that these materials are relatively safe in the forms in which they are used. Some plant products could even be useful in industrialized countries for the protection of grain from storage pests. Many of the plants discussed in this review are tropical in distribution and theoretically available to growers in developing countries. However, efficacy against pests is only one factor in the adoption of botanicals—logistics of production, preparation, or use of botanicals can mitigate against their use. Perhaps it is time to refocus the attention of the research community towards the development and application of known botanicals rather than screen more plants and isolate further novel bioactive substances that satisfy our curiosity but are unlikely to be of much utility.

PLANTS TO REGULATE PLANTS

The legume vine *Mucuna pruriens* is used in the countries of Benin and Vietnam as a biological control for problematic *Imperata cylindrica* grass. *Mucuna pruriens* is said not to be invasive outside its cultivated area. *Desmodium uncinatum* can be used in push-pull farming to stop the parasitic plant, *Striga*.

NEGATIVE CONSEQUENCES OF BIOLOGICAL PEST CONTROL

In some cases, biological pest control could have unforeseen averse consequences that outweigh all benefits, often by becoming an invasive species.

For example:

- When the mongoose was introduced to Hawaii in order to control the rat population, it preyed on the endemic birds of Hawaii, especially their eggs, more often than it ate the rats. (Note, however, that the introduction of the mongoose was not undertaken based on scientific—or perhaps any—understanding of the consequences of such an action. The introduction of a generalist mammal for biocontrol of anything would be unthinkable by any reasonable standards today.)
- Cane toads (*Bufo marinus*) were introduced to Australia in the 1930s in a failed attempt to control the cane beetle, a pest of sugar cane

crops. 102 toads were obtained from Hawaii and bred in captivity to increase their numbers until they were released into the sugar cane fields of the tropic north in 1935. It was later discovered that the toads could not jump very high and so they could not eat the cane beetles which stayed up on the upper stalks of the cane plants. The toad population grew dramatically and eventually out-competed native species. Not only were these toads very harmful to the Australian environment, they were also very toxic to would-be predators such as the native snakes.

- 5 cats brought to the subantarctic Marion Islands to catch mice in 1949 multiplied to 3,400 in about two decades and started to threaten local extinction of birds. They had to be infected with feline distemper virus. The rest were shot and completely eliminated by the 1990s.
- The sturdy and prolific mosquito fish was introduced from around the Gulf of Mexico to around the world in the 1930s and 40s to combat malaria; however, it was found to cause the decline of local fish and frogs through competition of other food source as well as eating their eggs.

Living organisms, through the process of evolution, may achieve increased resistance to biological, chemical, and physical methods of control over time. In the event the target pest population is not completely exterminated or is still capable of reproduction (were the pest control means a form of sterilization), the surviving population could acquire a tolerance to the applied pressures - this can result in an evolutionary arms race with the control method.

DIRECTLY INTRODUCING BIOLOGICAL CONTROLS

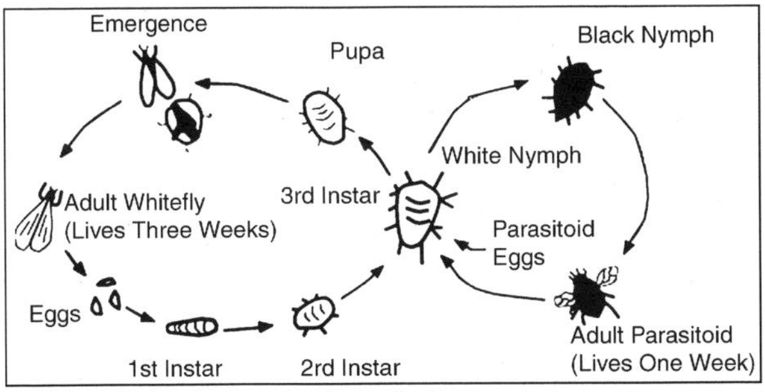

Most of the biological controls listed above depend on providing incentives in order to 'naturally' attract beneficial insects to the garden. However there are occasions when biological controls can be directly introduced. Common biocontrol agents include parasitoids, predators, pathogens or weed feeders. This is particularly appropriate in situations such as a largely artificial environment and are usually purchased by mail order.

Some biocontrol agents that can be introduced include:

- *Encarsia formosa*: This is a small predatory chalcid wasp which is parasitical on whitefly, a sap-feeding insect which can cause wilting and black sooty moulds. It is most effective when dealing with low level infestations, giving protection over a long period of time. The wasp lays its eggs in young whitefly 'scales', turning them black as the parasite larvae pupates. It should be introduced as soon as possible after the first adult whitefly are seen. Should be used in conjunction with insecticidal soap.

- *Red spider mite:* Another pest can be controlled with the predatory mite *Phytoseilus persimilis*. This is slightly larger than its prey and has an orange body. It develops from egg to adult twice as fast as the red spider mite and once established quickly overcomes infestation.
- A fairly recent development in the control of slugs is the introduction of 'Nemaslug', a microscopic nematode (*Phasmarhabditis hermaphrodita*) which will seek out and parasitize slugs, reproducing inside them and killing them. The nematode is applied by watering onto moist soil, and gives protection for up to six weeks in optimum conditions, Nemaslug nematodes are mainly effective with small and young slugs under the soil surface.
- A bacterial biological control which can be introduced in order to control butterfly caterpillars is *Bacillus thuringiensis*. This available in sachets of dried spores which are mixed with water and sprayed onto vulnerable plants such as brassicas and fruit trees. The bacterial disease will kill the caterpillars, but leave other insects unharmed. There are strains of *Bt* that are effective against other insect larvae. *Bt israelensis* is effective against mosquito larvae and some midges.
- A viral biological control which can be introduced in order to control the overpopulation of European rabbit in Australia is the rabbit haemorrhagic disease virus that causes the rabbit haemorrhagic disease.
- A biological control being developed for use in the treatment of plant disease is the fungus *Trichoderma viride*. This has been used against Dutch Elm disease, and to treat the spread of fungal and bacterial growth on tree wounds. It may also have potential as a means of combating silver leaf disease.
- Several species of dung beetle were introduced to Australia from South Africa and Europe during the Australian Dung Beetle Project (1965–1985) led by Dr. George Bornemissza of the Commonwealth Scientific and Industrial Research Organisation in order to biologically control the population of pestilent bush flies and parasitic worms.
- The parasitoid *Gonatocerus ashmeadi* (Hymenoptera: Mymaridae) has been introduced to control the glassy-winged sharpshooter *Homalodisca vitripennis* (Hemipterae: Cicadellidae) in French Polynesia and has successfully controlled ~95 per cent of the pest density.

9

Process of Selecting and Implementing: Decision Support Systems

Greenhouse pest and disease problems are often the result of complex interactions among many variables such as greenhouse environment, nutrition, production practices, growing media, other pest and disease outbreaks, economics and environmental and social concerns. As a result, managing or preventing pest and disease outbreaks requires an interdisciplinary approach, which will vary according to the problem.

Greenhouse industry is a very technologically-advanced agro-food industry with computerized climate control and fertigation systems in widespread commercial use. These systems offer precise and versatile tools for controlling and manipulating the greenhouse and plant environment, but also affect pest and disease outbreak dynamics. Biological control agents are commercially-available for most of the major insect and mite pests and cultural control measures are also viable management strategies to chemical control, especially for disease prevention.

With all these management strategies and other variables that can impact upon IPM, the grower can use as much assistance as possible to collect, collate, understand and integrate, where necessary, the information needed to choose the most viable solution for the problem at that point in time. The purpose of this chapter is to provide an overview of the decisionmaking process and decision tools as they apply to IPM of greenhouse crops.

Decision-making is the process of selecting and implementing an action with the intention of producing a favourable outcome. The quality of decisions can be enhanced by using a structured, analytic methodology to decision-making. Analytic decisionmaking is based on logic and considers all available data and alternatives. The structured decision-making process consists of five basic steps: problem recognition and definition, alternative generation, alternative evaluation, alternative selection and decision implementation. These steps do not necessarily follow one another sequentially without deviation, but often decisionmakers must backtrack and repeat some steps.

Problem recognition and definition begins with recognition of a deviation between actual conditions and established standards or desired conditions. A clear, concise problem statement, defining what the variance is and is not, when and where the variance occurs, etc., should be developed. The problem statement must go beyond the symptoms and identify the true cause of the problem. For example, if *Botrytis cinerea* Pers.:Fr. infects your crop and you only apply fungicides, you are treating the symptom (*Botrytis*) and are doing nothing to alleviate the cause of the infection. The cause of the infection may be poor sanitation, inadequate climate control or excessive plant stress.

Alternative generation is a creative process whereby alternative solutions are identified. Brainstorming at this time can result in some very novel ideas and also some non-feasible suggestions. Not all suggestions may be used, but discussion of them may help improve upon the more feasible solutions. Alternative evaluation involves setting goals to be achieved by solving the problem and quantifying each alternative in terms of its value, cost, risk and other decision criteria.

Establishing specific and measurable goals assists the decision-maker in quantifying a problem. Most greenhouse growers have many goals, including maximizing profit. Other goals may include increased productivity, increased product quality and employee safety. Decision criteria are attributes of a solution that can be measured or estimated. These attributes are used to evaluate the different alternatives that are generated in step two. Decision criteria for selecting a pesticide may include cost, efficacy, compatibility with biological control agents, safety and days to harvest. Decision aids or tools, such as decision matrix, decision tree, linear programming, simulation models, expert systems and decision support systems, can be used to more fully understand the scope of the problem, the differences among alternatives, and the relative worth of each.

Based upon the evaluation, the alternative that best satisfies the goal(s) is selected. Numerous methods or decision rules have been suggested for selecting among alternatives such as the dominance rule (choose A1 over A2 if A1 is better than A2 on at least one attribute and not worse than A2 on all other attributes), lexicographic rule (choose A1 over A2 if it is better than A2 on the most important attribute; if this requirement is not met, base the choice on the next important attribute) and addition of utilities rule (choose the alternative with the greatest sum of weighted values across all attributes). Further analysis of the selected alternative may be conducted to verify the decision and identify possible adverse consequences. Sometimes the most challenging phase of decision-making is trying to implement the selected alternative. An implementation plan that specifies the barriers and obstacles to acceptance of the decision, and ways that these can be overcome, is as important as the decision itself.

When making IPM decisions, it is vital that the decision-maker search for information that will help solve the problem. The search for information can help in all steps of the decision process. It may reveal facts about the situation that will result in redefinition of the problem. Valuable insight into the different alternatives and data by which they can be evaluated can be provided. The information search can also reveal how the selected alternative may be implemented. For greenhouse growers, we only found one survey that was related to sources of information for decision-making. This survey listed growers' journals and study groups as the most important sources for Dutch growers. Surveys of other types of agricultural producers found that significant sources of information include a grower's own experience and records, extension publications and bulletins, extension specialists, grower magazines, universities, colleges and research institutions, other growers, private industry salesmen (chemical, equipment, etc.) and independent consultants.

Greenhouse growers can and do also obtain their decisionmaking information from similar sources. A grower's own experience and records can be one of the most important sources of decision-making information. If a pest problem reoccurs, a grower can use their records to see how well previously implemented alternatives performed. Records can also be used to obtain evaluation data such as cost and effectiveness of chemicals and biologicals. Extension publications can provide general recommendations for IPM in greenhouse crops, while detailed information on specific pests and diseases can be obtained from books or other publications. In addition, every grower is advised to own a good pest and disease identification reference and a nutritional disorder identification reference.

These references can assist growers in quickly identifying crop disorders. Commercially produced grower magazines are widely used by growers and often report on new ideas and techniques for IPM. Government extension advisors have traditionally been the main source of pest and disease management information for growers in many countries. Recently, government cutbacks in several countries have severely reduced the number and availability of the extension advisors. As a result, there has been an increase in the number of private consultants in the greenhouse industry. Sales representatives can also be a valuable source of information, providing advice on the use of their products. Other greenhouse growers, especially study groups, are also an important source of information. Association with other growers allows one the opportunity to obtain, discuss and compare information on new IPM practices and innovations.

Technically, any aid that assists a decision-maker could be defined as a decision support system (DSS). However, for this discussion we will consider DSS as computer programmes that help decision-makers solve problems through direct interaction with data and models. DSS usually use a combination of decision tools including ES, database systems, simulation and other computer models. DSS have been used to solve pest management problems for field crops such as cotton and apples. The cotton DSS uses a cotton crop simulation model and expert advice on pests, diseases and weeds to provide recommendations on the timing of irrigation, fertilizer and pesticides.

In the greenhouse, Jones *et al.* described a DSS where crop models were combined with an ES to choose optimal environmental setpoints for greenhouse tomato. The ES contained a knowledge base for variables that have not been well modelled, such as the length of time that humidity may remain high without a disease outbreak.

BOTMAN makes decisions concerning whether to spray the biological control agent Trichodex (developed from an isolate of *Trichoderma harzianum* Rifai T39) or fungicides for integrated biological and chemical control of *B. cinerea* in non-heated greenhouse vegetable production. BOTMAN uses weather forecasts, past weather and a *B. cinerea* risk index to predict the severity of outbreaks of grey mould. Based upon the expected severity, application of the biological control agent or a fungicide is recommended.

Results show that BOTMAN controlled *B. cinerea* as well as weekly fungicide applications while significantly reducing the number of fungicide applications. Compared to a strategy of weekly Trichodex applications, BOTMAN was also significantly better. Another system, GREENMAN, was developed to deal with other greenhouse diseases. It is based on criteria that are similar to BOTMAN and, likewise, controls diseases such as leaf mould and white mould [*Sclerotinia sclerotiorum* (Lib.) de Bary]. A DSS for integrated management of greenhouse vegetables has been developed at the Greenhouse and Processing Crops Research Centre, Harrow, Ontario, Canada.

The HGM contains modules for the following:

- Expert diagnosis of insect and mite pests of the crop;
- Expert diagnosis of crop diseases and physiological disorders;
- IPM recommendations for pest, disease and physiological disorder control, and identification of conflicting recommendations;
- Cost allocation, including pest control expenses, to crops;
- Record-keeping capabilities including crop production, labour, insect counts, disease occurrence and control measures that were implemented;
- Tools to determine tank mixes for fertigation systems;
- Analysis section to analyse relationships between any recorded entity (such as insect counts and crop yield);
- Climate data retrieval from climate control systems that are BACnet compatible.

Currently, HGM contains the knowledge for greenhouse cucumber and tomato crops. The approach in the development of the HGM was to provide a framework for integrated crop management (ICM). ICM is a multidisciplinary approach that integrates pest and disease protection strategies with routine cultural practices and environmental and fertigation regimes into a common decision-making process.

It is not acceptable to manage one component of the greenhouse in isolation since the component can potentially affect all other aspects of the greenhouse crop. DSS have a lot of potential in greenhouse pest management, particularly if greenhouse climate is integrated with control strategies. To utilize climate in controlling diseases and insects, DSS will need to control the microclimate at the leaf surface. DSS will need to integrate models that predict the microclimate at the plant surface from spatially averaged climate data with crop and pest simulation models. Computerized DSS will be required at the grower level to enable growers to interpret all the information necessary for ICM.

Decision-making is a very important part of greenhouse pest management. However, it is becoming more complicated and demanding in an industry that rapidly changes yearly.

The grower can no longer rely on the old tools for information gathering and decision-making. Greenhouse operations are often 2–10 ha in size and must be operated more like a business corporation rather than a family-owned operation. The Internet can provide a readily accessible and up-to-date source of information that can link a grower to current technology that is being used throughout the world.

In the future, databases, DSS and other decision tools are going to become the main method for assisting the grower in making decisions as these tools can handle large data sets in an organized fashion and quickly form conclusions or solve problems. Interpreting and managing technical information for decision-making without computers will be beyond the means of individual growers. First, the quantity of data required is large and the management of this data is impossible without computer systems. Computerized climate-control systems can quickly generate megabytes of data. Add crop production data, fertigation records, pest count records, etc. and the data quickly becomes unmanageable without computer technology. Second, the relationships between the various crop production factors are complex. For example, greenhouse climate can affect the effectiveness of entomopathogens, but, at the same time, provide conditions that are conducive to a disease outbreak, such as *B. cinerea*. In this case, climate directly affects the crop, a biological control agent and the epidemiology of a plant pathogen.

Expert knowledge or simulation models can help improve our understanding of how all these factors interact. Although the systems developed to date are useful for providing solutions to greenhouse crop management problems, the technology is still a long way from a DSS that provides true ICM strategies.

To meet this goal, an improved understanding of the response of crops, pests, pathogens and biological controls to climate is required. Also those responses need to be mathematically or heuristically modelled and incorporated into DSS. Finally, DSS have to be validated in commercial greenhouses under actual production conditions.

AH of the issues relating to database systems, simulation models and expert systems apply to DSS. In addition, we have found that the user interface and data entry procedures play a much bigger role with DSS. With pest control, future control actions are dependent upon previously implemented controls. For example, using *Encarsia* for whitefly control may limit the chemicals available for control of grey mould. However, the HGM does not know *Encarsia* is in the greenhouse unless the grower has recorded it in the database. Therefore, user interface and data entry procedures must be structured to streamline the time that it takes for the grower to enter the required data. Growers are showing a strong interest in databases and DSS and are beginning to incorporate them into their daily operational decision-making processes.

PREPARATION OF BIO-INSECTICIDES AND PESTICIDES

Today the rapid increase in population and demand of food materials has initiated the large use of insecticides and pesticides. These toxic chemical insecticides and pesticides are resulting in harmful effects and biomagnifications due to which our environment is continuously polluting and fertile lands are acquiring infertility. No doubt they are providing hopeful results in eradication of insects, pests and diseases but they are also killing useful organism present in soil due to which the fertility of the soil is rapidly declining. The conventional farming practices which uses chemical methods to kill both useful and harmful life forms indiscriminately, resulting in the malfunctioning of food chain and food web.

Bio-control is the best method to cope with the losses done by the chemicals. In these method insects, pests and pathogens are removed using biological methods without harming the environment and other organism. This is based on natural predation rather than introduced chemicals. The use of bio-insecticides and pesticides also comes under this category.

Today due to awareness about the harmful effects of the chemical insecticides and pesticides, most of the farmers are diverting towards the organic farming. In our local area many such plants, waste matter etc. are available from which these bio-insecticides and pesticides can be prepared by using natural means only. Conventional pesticides are generally synthetic materials that directly kill or inactivate the pest. Being single chemical entity, chemical pesticides have resulted in increased resistance in pests. Bio pesticides also known as Biological Pesticides are pesticides derived from natural materials as animals, plants, bacteria, and certain minerals. Bio pesticides are less toxic and also reduce the pollution problems caused by conventional pesticides. The use of bio-insecticides and bio-pesticides also fall under this category only. Organic agriculture is a unique production management system which promotes and enhances agro-ecosystem health, including biodiversity, biological cycles and soil biological activity, and this is accomplished by using on-farm agronomic, biological and mechanical methods in exclusion of all synthetic off-farm inputs. Organic farming generally produces somewhat lower yields but sustains better yields during drought years, allowing it to reap higher yields in some cases.

Studies thus far have shown that organic farming requires less water, uses few and always natural pesticides, prevents soil erosion, leaches dramatically fewer nitrates, and has been shown to have improved nutrient qualities including as much as double the flavonoids, an important antioxidant. This project is mainly based on the preparation, usage, effects and advantages of bio-insecticides and bio-pesticides.

Different types of bio-insecticides and bio-pesticide:

- Cow Urine
- Fermented curd water
- Dashparni Extract
- Neem -cow urine extract
- Mixed leaves extract
- Chili –garlic extract
- Broad spectrum formulation 1
- Broad spectrum formulation 2

Preparation and Application

Collect 5 liter of pure and concentrated cow urine and dilute it with 40 liters of water and spray it using the sprayer in one hectare of field at the evening time.

Effects and Advantages

- Due to high content of urea in it which is toxic to most of the organisms, the pests and insects etc. will not attack the leaves and buds of the crop plants.
- Due to pungent and bad smell of the extract most of the pests and insects which are attracted due to nectar and fragrance get repelled, preventing the plant.

Cow Urine Extract

In India, as farming goes hand in hand with cattle rearing, therefore the cow urine is available easily in the rural areas. It has acquired an important place in the hindu religion since ancient times. Any religious ceremony and rituals can't be performed without cow urine which is commonly known as gomutra. We will here use the cow urine for the control of pests and as a growth promoter for the growing crops.

Preparation and Application

Take the distasteful butter milk and add equal amount of water in it. Keep it for two days in a semi-shaded place.

Now take it and add 40 liter of water in 10 liter of the extract to form 50 liter of the solution. Spray this solution in 1 hectare of field in such a way that all plants get bath in the fogging spray in the early morning time.

Effects and advantages:

- The fermented butter milk contains billions of bacteria that suppress the growth of other bacteria, fungi and protozoa. The inhibitors released by these bacteria works against the unwanted pathogens.
- Moreover, fermented butter milk also contains the useful nutrients for the growth and maturation of the crop plant.

DASHPARNI EXTRACT

Materials Required

Neem leaves 5 kg, vitex negundo leaves 2 kg, Aristolochia leaves 2 kg, papaya (carica papaya) 2 kg, Tinospora cordifolia leaves 2 kg, Annona squamosa (Custard apple) leaves 2 kg, Pongamia pinnata (Karanja) leaves 2 kg, Ricinus communis (Castor) leaves 2 kg, Nerium indicum 2 kg, Calotropis procera leaves 2 kg, Green chili paste 2 kg, Garlic paste 250 gm, Cow dung 3 kg and Cow Urine 5 lit.

Method of Preparation

Crush all of the above material and mix it in 200 liters of water to ferment for one month. Shake this mixture regularly three times a day. Filter it after one month. This extract can be stored up to 6 months.

Application: Take this extract in fogging machine and apply it on the plants. The above prepared extract is sufficient for one acre crop.

Effects and Advantages

- The extract prepared is very useful for the control of wide variety of pests such as Thrips, Leaf folder, Leafhopper etc.
- It contains neem which has ovi-position deterrent effect, cow dung and cow urine which has bio-fertilizer role also helps in enhancing useful microbial activity in soil.
- It can be prepared easily and can be stored for a large time interval for further use.

Fermented Curd Water

In many states of rural India milk production is the main occupation associated with the farming also known as dairy management. The dairy products can also prove a major alternative in controlling the insects and pests in farmlands.

One of the dairy product known as butter milk or Chaach can be used for this purpose. It is made by the centrifugation of curd in which byproduct ghee (fats) is also obtained. There is much greater demand of butter milk in the summer season and it gets distasteful if it is kept for more than two days. This distasteful butter milk can be used to make the extract.

Neem-cow Urine Extract

Materials Required: 5 kg of neem leaves, 5 liter of cow urine, 2 kg of cow dung, 100 liter of water

Method of Preparation

Crush the above substances and ferment for 24 hours with intermittent stirring, filter and squeeze the extract and dilute to 100 liters of water.

Application

Use this extract to fill in the spray machine and spray it over one acre of the crop.

Effects and Advantages

- Neem is a broad spectrum pesticide and even compatible with most chemical pesticides. Neem works by intervening at several stages of the life of an insect. It may not kill the pests instantaneously but incapacitate it in several other ways.
- Neem acts in various ways like as anti-feedant, as repellent, as growth inhibitor and also has ovi-position deterrent effect.
- Cow urine has high content of urea in it which is toxic to most of the organisms, the pests and insects etc.

Mixed Leaves Extract

- Materials Required: 3 kg of neem leaves, 10 liters of cow-urine, custard apple leaves 2 kg, 2 kg papaya leaves, 2 kg pomegranate leaves, 2 kg guava leaves.
- Method of Preparation: Crush all of the above recommended material and add 5 liters of water in it. Boil the above mixture 5 times after some intervals of time till the mixture becomes half of the initial. Keep it for 24 hours, filter and squeeze the extract. This can be stored in bottles for 6 months.
- Application: Dilute the above prepared extract 2-2.5 liters in 50 liters of water. Fill the diluted solution in fogging machine and spray it in one acre area.

Effects and Advantages

- Guava leaves has achieved a special place in ayurveda due to its medical use for curation of some diseases.
- Pomegranate leaves have resistive nature against many insects and pests due to presence of some special compounds in it.

Chili-garlic Extract

- *Materials Required:* Ipomea leaves 1 kg, 500 gm hot chili, 500 gm garlic, 5 kg neem leaves, 10 liter cow urine
- Method of Preparation: Crush the above recommended material and boil the suspension till it becomes half of the initial. Filter and squeeze the above extract and store it in glass or plastic bottles.
- Application: Take 2-3 liters of the extract and dilute it with 50 liters of water. Now mix it thoroughly and use it as a foliar spray for one acre of crop.
- *Effects and advantages:*
  - Garlic contains sulphur which is an antibacterial.
  - Chili has the property to avoid fungal and bacterial infection due to its preservative property.

Broad Spectrum Formulation 1

- Materials Required: 3 kg fresh crushed neem leaves, 1 kg neem seed kernel powder, 10 lit of cow urine, 500 gm green chilies and 250gm of garlic
- Method of Preparation: Take a copper container and mix 3 kg neem leaves, neem seed kernel powder with 10 liters of cow urine. Seal the container and allow the suspension to be ferment for 10 days. After 10 days boil the suspension, till the volume becomes half of the initial. Now ground 500 gm of green chilies in 1 liter of water and keep overnight. In another container crush 250 gm of garlic, add 1 liter of water and keep it overnight. Next day mix the above boiled extract, chili extract and garlic extract. Mix it vigorously and filter it. Store the concentrate in glass or plastic containers.
- Application: Dilute 250 ml of concentrate with 10 liters of water and mix it thoroughly. Now use it as a foliar spray on the crop.

Effects and Advantages

- Due to the formation of complex compounds of Cu in this formulation which are poisonous for many bacteria, fungi, protozoa etc., it is very useful for microbial control.
- The compounds produced by the fermentation of the above extract are poisonous to many micro-organisms and insects.

Broad Spectrum Formulation 2

- Materials Required: 5 kg neem seed kernel powder, 1kg Karanj seed powder, 5 kg chopped leaves of besharam (Ipomeas sp.), 5kg chopped neem leaves, 10-12 lit of cow urine
- Method of Preparation: Suspend 5 kg neem seed kernel powder, 1kg Karanj seed powder, 5 kg chopped leaves of besharam (Ipomeas sp.) and 5kg chopped neem leaves in a 20 liter drum. Add 10-12 lit of cow urine and fill the drum with water to make 150 lit. Seal the drum and allow it to ferment for 8-10 days. After 8 days mix the contents and distil in a distiller.
- Application: Distillate obtained from 150 litre liquid will be sufficient for one acre. Dilute in appropriate proportion and use as foliar spray.

Effects and Advantages

- Neem effectively controls common pests like Thrips, Whitefly, Leaf folder, Bollworms, Aphids, Jassids, Pod borer, Fruit borer, Stem borer, Leafhopper, Caterpillars, Diamond back moth.
- Karanj contains poisonous chemicals in it and can be used to check microbial growth.

IPM INFORMATION IS THE INTERNET

A new source of IPM information is the Internet. The Internet has many features that can be used to assist in the management of greenhouse crops. Provided one knows the address, one can send and receive messages from anyone connected to the Internet including other growers, extension advisors and researchers. Another useful tool is the browser. A browser is an application that knows how to interpret and display documents that it finds on the Internet.

Most browsers can access other Internet services including Anonymous FTP (File Transfer Protocol for downloading files), e-mail and news groups. One way for growers to use the Internet is to find information relating to pest and disease management. Many useful sites can be found on the Internet that are related to horticulture and greenhouse management. Extension/research sites provide many extension documents and information on current research projects for growers. In addition to product information and pricing, commercial sites also provide lots of related information. All sites contain good links to other related sites.

When using information from the Internet, however, one should exercise caution. Anyone can put up a web site and publish anything on the site. Therefore, be aware of the source and quality of the information. Unlike books and journal articles, web documents are not peer reviewed so there is no guarantee that the information is accurate. As well, the main purpose of commercial sites is to advertise their products. Another consideration when using the Internet for decision support is finding the relevant information that one requires. Searching the Internet using a search engine can generate thousands of matches.

For example, a search for IPM generated 3714 matching sites. Determining which sites are truly helpful can take a considerable amount of time. This problem can be alleviated somewhat by carefully choosing keywords to search. Searching for IPM and greenhouse reduced the number of matching sites to 81. Another option is to find and search topic specific databases.

APPLICATION OF DECISION TOOLS FOR IPM

Decision tools are techniques for modelling actual systems and are thus simplifications of actual conditions. They have become widely accepted for several reasons. Decision tools or models are less expensive and disruptive than experimenting with the actual systems and can force the decision-maker to analyse the problem in a logical and systematic manner. Decision tools allow managers to ask "what if" questions and evaluate different scenarios. They can also reduce the time needed to make a decision. On the other hand, models can be expensive and time consuming to develop. The results may be misused and misunderstood because of the complexity of models and because models may use assumptions that oversimplify actual systems. As stated earlier, many tools are available to assist growers in making IPM decisions in the greenhouse. Practical applications of many decision tools in IPM are reviewed in Norton and Mumford. Although none of the many examples presented are specific to greenhouse IPM, the techniques presented can be applied to many greenhouse IPM problems.

Simulation Models

A model is a description of a system. Models may be scaled physical objects, mathematical equations and relations, or graphical representations of actual systems. For purposes of this discussion, a simulation model is a mathematical-logical representation of a system which can be exercised in an experimental fashion on a digital computer. In terms of decision-making, simulation models allow us to examine different alternatives and how these alternatives perform under different conditions.

Simulation models have been used to study various field pests. Pest population growth, fungal passive dispersal, insect active dispersal and predator-prey interactions have been simulated. Pest systems have also been modelled in the greenhouse environment. Nachman simulated the dispersal of two-spotted spider mite (*Tetranychus urticae* Koch) and its predator *Phytoseiulus persimilis* Athias-Henriot in a greenhouse cucumber crop. Spider mite oviposition rate, death rate and emigration rate to other plants are dependent on the health of plants.

The birth, death and emigration rates of the predator are linked to the predation rate. The simulation model was used to study fluctuations in overall population densities within the greenhouse. Biological control of the leafminer species *Liriomyza trifolii* (Burgess) in greenhouse-grown chrysanthemums and *Liriomyza bryoniae* (Kaltenbach) in greenhouse tomato by the parasitoid *Diglyphus isaea* (Walker) have been simulated. Heinz assumed a constant greenhouse temperature of 27 °C and that the population dynamics of the leafminer are independent of the quality of host plant. Simulation results indicated that successful biological control was unlikely when parasitoid releases are initiated later than 14 d after planting regardless of the release rate. sing a different approach, Boot used ambient temperature and tomato leaf nitrogen content to determine the population dynamics of *L bryoniae.* The timing and growth of leafminer generations were simulated and the results validated with greenhouse experiments, although no practical strategies for parasitoid were given.

These two models could be used to explore different strategies for biological control of leafminers. Disease infection and progression and arthropod pest populations are dependant upon plant nutrition. Crop growth simulation models, such as those developed for greenhouse tomato and cucumber could supply input parameters to pest population models. A feedback loop could potentially predict crop yield reductions due to the pest. Greenhouse microclimate also affects disease development and insect population dynamics. Microclimate models that simulate the climate within the crop canopy could be combined with disease and pest models to investigate climate control strategies for pest and disease management. Currently, pest management simulation models are being used at the research level to better understand interactions between pests and their control agents.

An example is the simulation model developed by van Roermund to evaluate release strategies for the parasitoid, *Encarsia formosa* Gahan, for control of greenhouse whitefly, *Trialeurodes vaporariorum* (Westwood) on greenhouse crops. As personal computers become more powerful, we envision simulation models being used by growers to evaluate pest management strategies.

However, before this happens more research needs to be done to develop models for other pests and diseases. As well, these models must be validated and in a format that non-modellers can understand how to operate the models and interpret the outputs. Model validation can be difficult due to inadequate experimental data, inappropriate assumptions and lack of knowledge regarding some of the physical processes being modelled. Despite these problems, simulation models certainly have a lot of potential for analysing various pest management strategies and understanding interactions between the pest, its biological control agents, the crop and the crop microclimate.

Decision Tables and Trees

Decision tables and trees are simple yet powerful tools to assist in the decision-making process. These tools can be used to logically and systematically select among alternatives and the structure provided by these tools can give a valuable framework for further investigations.

Decision Matrix

Criteria	Weight	Chemical 1	Chemical 2	Chemical 3
Cost	40	10 / 400	10 / 400	7 / 280
Effectiveness	25	10 / 250	9 / 225	10 / 250
Compatibility with Biologicals	20	5 / 100	9 / 180	8 / 160
Days to Harvest	15	5 / 75	8 / 120	5 / 75
Total	100	825	925	765

Rating Scale, R
Excellent =9-10
Good =7-8
Fair =5-6
Poor =3-4
Unsatisfactory =0-2

Rating: 5 / 100 (R x W)

The decision matrix is used to select among alternatives using the addition of utilities rule. Consider a hypothetical situation where one must select a fungicide from three different alternatives. In this example, four evaluation attributes or criteria are established and weights of importance are assigned to each.

These weights reflect the beliefs, concerns and experiences of the decision-maker. Each alternative is evaluated and graded on a 0 to 10 scale on how well it satisfies the criterion. The grade is multiplied by the weight and recorded. The alternative with the greatest sum of weighted values across all criteria (chemical 2) is selected. If the lexicographic rule had been used, then chemical 1 would be selected.

Database Systems

Database systems (DBS) consist of a collection of interrelated data and a set of application programmes to access and manipulate the data. The different data items are stored in related files or tables. The application programmes usually provide functions to enter, edit, browse, query and analysis the data. DBS can help solve pest and disease management problems in several ways. First, the development of a DBS can help to better organize and understand the problem. One of the first steps in developing a DBS is to develop a data model. The data model identifies the paths of information flow, specific data items and relationships among data items. DBS can also assist decision-making by storing detailed records of past pest and disease management strategies, along with the outcomes and costs of these strategies. These records can help the decision-maker select among different strategies based upon past results.

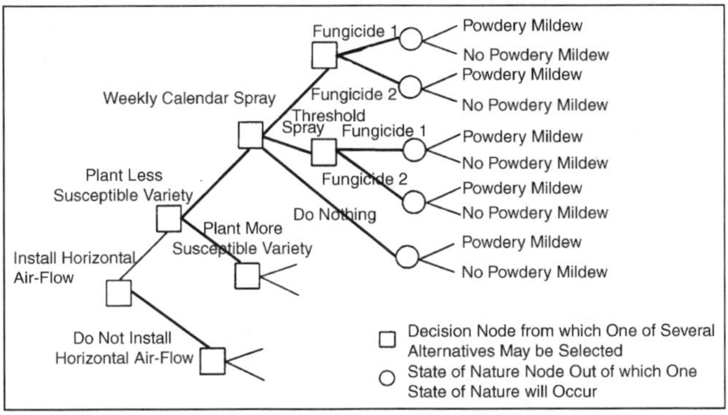

There are many commercial greenhouse cost accounting and financial management DBS available. These packages usually provide facilities to record costs (including pest control) and sales throughout the cropping season. These DBS can assist in tracking pest management costs in the greenhouse. A DBS for greenhouse pest surveillance, Emerald ICM, is commercially available. Pest survey data is collected with a hand-held computer in the greenhouse and uploaded to a personal computer.

The data is used to generate colour maps of pests and their severity over time. Applications of pesticides, fungicides and biologicals are also recorded. Emerald ICM allows the grower to monitor the progress of pest and predator movement and analyse the effectiveness of different control strategies.

Pay-off Matrix

A pay-off matrix helps the decision-maker economically evaluate alternatives. Pay-off matrices can be used both for decision-making under risk (where the decision-maker knows the probability of occurrence of the outcomes for each alternative) and decisionmaking under uncertainty (whether probabilities are unknown).

A possible pay-off matrix is presented for thrips control on sweet pepper in Ontario greenhouses under uncertainty. The pay-off for each combination of alternative and state of nature (an occurrence or situation over which the decisionmaker has little or no control) is included in the matrix. In this example, the states of nature are low, medium and high levels of thrips attack. The do nothing alternative shows the cost of damage caused by the three levels of thrips attack. The other alternatives include both the cost of the control strategy and its ability to reduce thrips levels and damage. The chemical alternative also includes an estimate of yield reduction resulting from crop injury.

Table. Pay-off Matrix for Decision-Making under Uncertainty

Strategy	**Level of Thrips Infestation**			**Medium Outcome in Row**	**Minimum Outcome**	**Row Average**
Low	**Medium**	**High**	**in Row**	**in Row**		
Do Nothing	-6,000	-18,000	-60,000	-6,000[2]	-60,000	-28,000
Chemical Control	-32,500	-36,200	-49,500	-32,500	-49,500	-39,400
Biological Control	-7,000	-16,000	-32,000	-7,000	-32,000	-18,333

Maximax↑ Maximin↑ Equally

Likely↑

With decision-making under uncertainty, the decision-maker can use three different rules for selecting among the strategies. The maximax (optimistic) rule selects the alternative (do nothing) that maximizes the maximum outcome for every alternative.

The maximin (pessimistic) rule selects the alternative (biological control) that maximizes the minimum outcome for every alternative. The equally likely rule finds the alternative (biological control) with the highest average outcome and assumes that each state of nature is equally likely to occur.

With a situation where a grower has kept detailed records of thrips levels in the greenhouse, the probabilities of thrips attacks can be calculated. A pay-off matrix can be developed for this situation where the decision is being made under risk.

The expected monetary value (EMV) for alternative i is:

$$\mathrm{EMV}(i) = \sum_{j=1}^{n} \$if * pj$$

where *n is* the total number of outcomes, \$ij is the payoff of alternative *i* for outcome *j,* and p_j is the probability of outcome *j.* A risk-neutral grower, who is unconcerned with year to year variations in outcomes, would choose biological control, which has the highest EMV (in our example, the lowest crop loss). Most growers are more likely to be risk-adverse and choose a strategy that gives acceptable outcomes at high pest levels.

In this case, an extremely riskadverse grower would also choose biological control, which has the best outcome under the worst conditions. Similarly, a risk taker may choose to do nothing, which has the best outcome under the best conditions.

Although pay-off matrices help to economically select among alternatives, they do not allow for non-economical criteria to be considered (such as compatibility with *Bombus* spp. pollinators). If a cost can be determined for these criteria, then they should be included in the analysis.

Table. Pay-off Matrix for Decision-Making under Risk

Strategy	Probability of Thrips Level			EMV
	Low (0.2)	Medium (0.5)	High (0.3)	
Do Nothing	–6,000[1]	–18,000	–60,000	–28,200
Chemical Control	–32,500	–36,200	–49,500	–39,450
Biological Control	–7,000	–16,000	–32,000	–19,000[2]

Decision Trees

Many problems consist of sequential decisions over time. When more than one set of decisions are necessary, a decision tree is appropriate. A decision tree is a graphic display of the decision process which indicates decision alternatives, states of nature and their respective probabilities, and pay-offs for each combination of alternative and state of nature.

A decision tree is shown for powdery mildew management in figure (state of nature probabilities and pay-offs are not included). Note that as the crop season progresses, the number of options decreases.

If the probabilities of powdery mildew occurrence and the pay-offs are added to the tree, the EMV can be calculated for each branch and the best decision determined. Even if probabilities and pay-offs are not known, decision trees are still very useful by laying out all possible options and providing a framework for deciding which options and strategies need further investigation.

Expert Systems

Expert systems (ES) are computer programmes that emulate the decision-making ability of a human expert. ES contain knowledge in one specific problem area or domain as opposed to knowledge about general problem-solving techniques. ES usually consist of a set of rules that were obtained from an expert to solve a particular problem, and an inference engine that decides which rules to execute. In terms of decision-making, ES can be used as tools for summarizing information and knowledge, selecting among alternative solutions, exploring and evaluating alternative scenarios, assessing risks, diagnosing problems, outlining approaches to problem solving and teaching nonexperts the problem-solving approaches of experts. ES technology has been applied to greenhouse pest management. The most common type of application is the diagnosis of crop diseases and pests. Several ES have been developed to identify greenhouse tomato disorders and recommend possible control actions for the identified disorder.

With these ES, the user is prompted to enter information about the symptoms displayed by the tomato plant. The ES then uses expert rules to match the observed symptoms with a disorder. Boulard developed an ES to determine the climatic setpoints to control the climate for greenhouse tomato. In addition to information on the outside weather conditions and the current greenhouse climate, the ES also incorporated expert rules on climate and tomato diseases and physiological aspects.

This system attempted to optimize conditions for energy use, crop growth and disease prevention and control. In a separate research effort and using a different approach, Manera developed an ES with similar objectives for greenhouse production and pest management under Mediterranean conditions. Although not specific to greenhouses, other pest management-related ES have been published.

Logan developed an expert system to automatically assemble a model describing insect population phenology. The programme offers time savings and compares well with a human expert. Messing describe NERISK, an expert system that assesses the impact of pesticides on beneficial arthropod predators and parasitoids in agricultural systems.

Similar ES could be developed and be useful for greenhouse pest management. Although ES are very good at assisting in the decision-making process, certain issues should be considered before undertaking the development of one. First, it takes considerable time and resources to complete an ES.

The time and commitment of an expert(s) is required as well as a knowledge engineer (an individual who extracts, organizes and programmes the knowledge of an expert). The intended user of the ES must be consulted and involved during the development. Ongoing maintenance is required as the knowledge contained in the ES may become outdated. These and other issues relating to ES development are reviewed by Clarke.

BIO-RATIONAL PESTICIDES

While, in principle, any pesticide use is discouraged in organic systems, a rather wide range of bio-rational pesticides is permitted. The frequency of pesticide use varies considerably with crop and location. For example, there is virtually no use of pesticides on organic row crops in the Cornbelt. By contrast, organic tree fruits in the Midsouth routinely receive heavy applications of several fungicides and insecticides allowable in organic production. The pesticides permitted in organic farming fall predominantly into several classes.

Minerals: These include sulphur, copper, diatomaceous earth, and clay-based materials like Surround.

Botanicals: Botanicals include common commercial materials such as rotenone, neem, and pyrethrum. Less common botanicals include quassia, equisetum, and ryania. Tobacco products like Black-Leaf 40® and strichnine are also botanicals but are prohibited in organic production due to their high toxicity.

Soaps: A number of commercial soap-based products are effective as insecticides, herbicides, fungicides, and algicides. *Detergent*-based products are *not* allowed for crop use in organic production.

Pheromones: Pheromones can be used as a means to confuse and disrupt pests during their mating cycles, or to draw them into traps.

Biologicals: One of the fastest-growing areas in pesticide development, biopesticides present some of the greatest hope for organic control of highly destructive pests. Among the most well-known biopesticides are the *Bacillus thuringiensis* (Bt) formulations for control of lepidopterous pests and Colorado potato beetle.

FOLIAR FERTILIZATION

Foliar fertilization or feeding entails the application—via spraying—of nutrients to plant leaves and stems and their absorption at those sites. It is not specifically an organic practice, though it is commonly used by many organic growers. The fertilizer materials used are typically soluble fish-and seaweed-based products, naturally chelated nutrients, humic acid extracts, and teas made from plants, dried blood, manure, guano, or compost. At first glance, the use of foliar feeding appears contradictory to the organic notion that one feeds the soil to feed the plant. Organic growers rationalize the use of this approach on two points.

1. Foliar feeding is strictly supplemental fertilization; it is not used as a substitute for traditional soil building practices.
2. Foliar fertilization is understood to increase the production of root exudates, which stimulates biological activity in the rhizosphere (soil area adjacent to plant roots). The soil bio-life gets considerable benefit in this indirect way from foliar feeding.

BUFFERS AND BARRIERS

Field buffers are strips of sod or other permanent vegetation at the edge of or surrounding crop fields. There is considerable interest in both conventional and alternative agriculture in these buffers, since they assist in reducing soil erosion and improving water quality. If managed properly, some buffers also serve as beneficial insect habitat. In certified organic production, however, field buffers have an added purpose in reducing crop contamination from chemicals used on adjacent land. Most agencies require a minimum 25-foot buffer along "uncontrolled" borders where there is a hazard of chemical use. Wider borders may be required where hazards are great—for instance, where adjacent farms have synthetic pesticides applied by aircraft.

Organic Farming: Niche Market or Viable Alternative?

The pioneers of organic farming considered organics the preferred direction for the whole of agriculture to take. It is likely that most contemporary proponents still hold that view. While recent growth in the organic industry is definitely encouraging, much of the impetus is tied to its growth as a niche market, not as a serious shift in the direction of mainstream agriculture. Unfortunately, a likely reason for the newfound "tolerance" of organic agriculture in many land grant universities and other formerly hostile environs is its perception as a niche market opportunity only. As such, it is not a serious threat to the status quo. Is it practical and responsible to promote organic agriculture as the dominant approach to farming in the future? While some are quick to answer, "Yes!", most are buffaloed by the nagging question, "Can organic agriculture feed the world?" Questions about the productivity and the prospects of widespread starvation have long been an effective tactic for stonewalling serious discussion of organics. While it is not the purpose of this publication to settle such a compelling issue, there are a few points that can and should be made regarding the "starving billions" scenario.

- Critics warn that, were organic farming to be adopted on a wider scale, per-acre agricultural productivity would decline sharply; that meeting food needs would necessitate the plowing of even more erodible hillsides and the draining of more wetlands. Such scenarios are often based on pre-chemical era yield data that ignore the advancements of modern organic farming. Contemporary research on organic systems—as limited as it is— indicates that the per-acre productivity of organic and conventional systems is not vastly different. Other scenarios are sometimes based on the perceived need for vast quantities of grain—particularly corn. Since overproduction of grain has led to prices below costs of production for several years, this is a questionable basis for argument. It also ignores the fact that most of the grain produced is fed to livestock and that many livestock species can eat forage instead. Ruminant livestock—cattle and sheep especially—are designed by nature to thrive on forage and can even be finished for market on pasture.
- The paucity of good research on the productivity of organic systems also spotlights the lack of practical research for organic producers. That organic farms are as productive as they appear to be is remarkable in light of minimal research and extension support for many decades. It has often been argued that, had comparable resources been put into organic research, its widespread feasibility would be unquestionable. This assertion is well-supported by the development, in recent years, of a number of alternative, organically acceptable pesticides, that are certain to expand organic production of many crops in regions where it had previously been near impossible on a commercial basis.
- Increasing agricultural production alone does not alleviate hunger.

The amount of grain produced in the world in 1999 could, by itself, sustain 8 billion people – 2 billion more than our current population. Total food production is estimated to provide each human being with at least 3500 calories per day. The issue of hunger, it appears, is not so much a lack of food, but lack of *entitlement* to food. People are shocked to learn that while many Bangladeshees starved during the 1974 floods, roughly 4 million tons of rice produced in that country were stacked in warehouses for want of buyers. The people were simply too poor to buy it. They are surprised to learn that Ireland exported grain during the Irish potato famine while 1 million of its citizens died and even more emigrated; that India regularly exports food and animal feed despite an estimated 200 million in starvation. Those lacking the ready cash to buy food or the resources to produce it themselves seem destined for hunger no matter what miracles agricultural technology provides. The world's nations will need to deal with issues of equity and democracy first, if hunger is ever to be effectively addressed.

- No agriculture can continue to feed a growing population if it depletes or fouls its resource base. The path undertaken by conventional agriculture is ultimately a dead end in this regard, though there is an almost mystical faith that genetic engineering and other complex technologies will always triumph. Agriculture needs to be sustainable. Therefore, those who promote organic agriculture as a true alternative are well advised to do their part in ensuring that certification and regulation does not create a "compliance agriculture" in which sustainability becomes little more than an afterthought.

ESOTERIC PRACTICES

There are a number of farming and gardening practices based on belief in a non-physical world closely aligned with our physical reality. Those who use these practices believe that conditions in this unseen realm influence, or even dictate, what happens on a material level. They take pains to understand these influences and adjust their farm activities. In some instances, they act to influence events. Obviously, these practices are rooted more deeply in metaphysics than in the conventional sciences.

The most well-known of these esoteric practices is the scheduling of planting or other field operations according to lunar or other astrological signs. Dowsing and other forms of *divining* may also be used to guide scheduling, fertilizer selection, and other facets of farming. The use of potentized *preparations*, as done in Biodynamic farming, is also a metaphysical practice. Radionics, energy balancing towers, "medicine" wheels, prayer—these are further examples of practices used by some farmers to bring a tangible spiritual element into their farm operation. Esoteric practices like these (there are many others) are not inherently part of organic agriculture. In all likelihood, only a small-to-modest minority practice them or believe in their efficacy. Still, esoteric practices are sometimes associated with organic agriculture, and it is almost a given that practitioners of "energetic agriculture" are either certified organic or actively pursue a sustainable form of agriculture.

The rationale for this association of organic with the esoteric is not difficult to understand. Most metaphysical practices are founded on a world-view similar to that of ecology, in which everything is related to everything else. The difference is that ecology considers this interrelationship only on a hard physical level; metaphysics perceives it also on a non-physical level. This belief tends to cast humankind as part of nature, not merely as an inhabitant. This perspective would tend to make one more concerned about the environment and, therefore, more readily drawn to organics as a philosophy and pragmatic farming approach. To put it in a nutshell, it is natural for the metaphysically-inclined to choose organics; it is far less likely, however, that the average organic farmer is drawn to metaphysical practices.

A further factor plays into the association between esoteric farming practices and organics. It relates to the fact that organic agriculture has, for many years, been marginalized and treated as pseudoscience by the mainstream—an experience shared by practitioners of metaphysical arts. It is not surprising for underdogs to seek common ground.

SEEDS, SEEDLINGS, AND GMOS

One of the challenges faced by organic growers is getting suitable planting stock. Certified production requires that seed not be treated with pesticides and be organically grown.

Transplants must be purchased from a certified organic source or otherwise be grown using organic methods. This presents some difficulty, as most commercial potting mixes contain prohibited fertilizers and wetting agents, requiring special ordering or the added effort of making a homemade mix. Generally, planting stock for perennial crops like tree fruits and berries is available from conventional sources, even if treated with pesticides. However, no production can be sold as organic for a minimum of 12 months following transplanting to an organic field.

In all instances, organic growers are not permitted to use varieties of crops that have been developed through genetic engineering (GE). At first glance, this might seem perplexing. GE crops promise further means of non-chemical pest control and the possibility of nutritionally enhanced foods. However, the organic community is concerned about environmental, economic, and social impacts from this new technology which, they feel, have not been adequately studied.

This same concern applies to GE foods. Natural foods consumers—a large segment of the organic market—do not want them. This was made abundantly clear to the USDA when its first draft of the proposed rule, which suggested permitting genetically modified organisms (GMOs), was released in 1997. The backlash was so strong that not only GE crop varieties, but all inputs, such as GMO-derived biopesticides, are prohibited from organic production.

Unfortunately the GE issue does not resolve itself so simply for organic farmers. The proliferation of GMOs in the marketplace and across the landscape has created a host of new challenges that have particular implications for organic agriculture:

- *Genetic Drift onto Crops:* Pollen drifting from adjacent fields can contaminate organic crops. Some clients in the lucrative European Union market accept 0 per cent GMO contamination; others accept only very low levels. Pollen from cross-pollinating crops like corn easily travels hundreds of yards and may be carried for miles under the right conditions. No reasonable amount of non-crop buffer can prevent contamination except in highly isolated locations.
- *Genetic Drift onto Soils:* Concerns have been expressed over the persistence of GE pollen in soils. Should contaminated soils be decertified?
- *Non-labelled Products:* Should soybean, cottonseed, and corn gluten meals be prohibited as fertilizers, since these may come from GE varieties? Should pesticidal use of vegetable oils be prohibited?
- *Pest Control Problems:* The rapid proliferation of corn, cotton, and other crops featuring the Bt gene from *Bacillus thuringiensis* is likely to accelerate the development of resistance to this natural pesticide. Organic growers may lose one of their most useful pest management tools.
- *Domination of the Food System:* In the late 1990s, the public was made aware of two specific developments of agricultural biotechnology dubbed the *terminator* and *traitor* technologies. Terminator technology is the genetic altering of seed so that the subsequent generation will not germinate. This prevents farmers from saving a portion of the harvested crop to replant the next year — a traditional practice among farmers worldwide. Traitor or *suicide* technology is the genetic altering of seed so that it will fail to germinate or mature unless a proprietary chemical is applied. Such technologies are authoritarian and advance the centralization and industrialization of the food system. Organic farmers perceive additional challenges from genetic drift impacts and difficulty in finding non-GE seed in the marketplace.

CONTROL PESTS OF PESTICIDES

Pesticides are used extensively to control pests, which can harm crops and other plant species. There are many different types of pesticides. They are mostly synthetic chemicals designed to affect a specific target. Concern over pesticide use is a major issue among many organizations. This chapter describes pesticides, why they are used, and why they generate so much concern.

There are many plants and animals which co-exist with crops in a farmers field. Less than one per cent of these plants and animals are pests. Pests are organisms which attack crops and harm them in some way. It is critical that these pests are controlled for a farm to earn the best income. The use of synthetic pesticides in agriculture world-wide is still the most widespread method for pest control.

Farmers could not achieve the same yields without the use of synthetic pesticides. It is estimated that without them 30 per cent of world crops would be destroyed by harmful organisms such as insects, weeds, or plant diseases even before they were harvested. Pesticides prevent these loses by controlling such pests. Pests can also render food unfit for consumption, harm human health and also harm the environment. In crops like cotton, tomatoes or melons, difficulties in pest control have forced farmers in some countries to abandon these crops completely.

CHEMICAL CONTROL

Chemical control is the use of pesticides to reduce pest populations. The three major pesticide categories are herbicides, insecticides, and fungicides. Herbicides are chemicals that are used to control weeds, insecticides are chemicals used to control insects and fungicides are chemicals used to control plant diseases caused by fungi. There are many different types of these pesticides which affect pests in many different ways.

Within each pesticide group are numerous chemicals for use in specific applications. Depending on the chemical structure, the pesticide may interfere with the growth, reproduction or development of the pest or may induce damage to the internal systems, resulting in death. Some pesticides are more popular than others for reasons such as potency, availability, cost, and safety. There are usually a few brand names of pesticides which dominate the market. It is important to remember that no pesticides kill 100 per cent of the pests all the time. They simply kill enough pests so that the few remaining are no longer a problem. Most pesticides commonly used for agricultural, home and garden use belong to chemical families that are not toxic to humans or animals, do not pose cancer risk to humans and do not accumulate in the environment.

Pollutant

After a pesticide is applied not all of the pesticide remains where it is needed. When a pesticide leaves the target area, it is often considered to be an environmental pollutant. Movement of pesticides can occur in several ways.

Drift, soil leaching, run-off, improper disposal and storage, and improper application are some of the major causes. Pesticide drift is a major cause of soil and air contamination. Drift occurs at the time of pesticide application when small spray particles are moved by air currents to non-targeted areas. Most of these sources of contamination can be avoided with proper care and handling on the part of the farmer.

Pesticide Problems

There are various problems which can occur when using synthetic chemicals to control pests. Some pests can build up resistance to pesticides. Pesticide resistance is the ability of an organism to tolerate a lethal level of a pesticide. This results in larger applications and a reduction in the effectiveness of the pesticide. Many of the chemicals used are not selective. They may also kill beneficial species, such as bees which pollinate crops. Synthetic chemicals may end up hurting the plant the farmer is trying to preserve. Environmental pollution is another problem as pesticides may build up in the soil or run off in the water supply. Pest resurgence may also occur. Pest resurgence occurs when a pest population level can establish itself after control measures against the population have been eliminated or reduced. In some countries other than Canada, the irrational use of pesticides and equipment has led to serious problems.

In some crops, the costs of pesticide use are already higher than the benefits. While modern pesticides have reached the most remote parts of the world, the application equipment matches that which existed 40 years ago. This has resulted in large wastes of pesticides and unnecessary environmental contamination. Farmers and application equipment operator's knowledge of pesticides and the correct method of application is usually deficient or non existent. In many cases they do not receive any training on the issue.

Sometimes, the only specialists are representatives of pesticide companies. Of course it would be a poor business move for them to show farmers how to save money by using less of their product. Some farmers and spray equipment operators still believe that the more pesticides you use, the better. The lack of technical knowledge and awareness of farmers and manufacturers are the main reasons for the low quality of equipment found in some countries. There are also many cases of poor equipment maintenance. Many products which are considered to be highly or even extremely toxic and can lead to a considerable amount of poisoning are still used in some countries.

New Technology

Pressure for the use of pesticides in a manner that is safer for environment and humans has led to new approaches to pest control. Modern application equipment allows a fairly safe and efficient application of pesticides of all kinds.

Legal requirements and farmer's demands have forced equipment manufacturers in Europe to offer only state of the art machinery. The equipment is designed to help prevent unnecessary contamination, accidents, losses and spills. The use of modern electronics has improved the accuracy of pesticide spraying. The Global Positioning Systems (GPS) for precise tracking has improved the safety of aerial pesticide spraying. GPS technology also allows patch spraying against certain weeds. If a weed is detected by the GPS system, the nozzle on the sprayer is opened only at that point. Therefore, lower amounts of pesticides are used. The entire field does not need to be dosed, which saves money for the farmer.

Application

When farmers spray pesticides, they try to distribute a small amount of active ingredient to the appropriate biological target without contaminating other areas. There are many different types of targets for pesticides. These include insects, plants, or soil. A variety of application techniques are required to affect these targets while using as little pesticide as possible. The equipment used for pesticide application ranges in price and complexity.

Some application techniques include:

- Release from the air in the solid state as dusts or granules or in the liquid state as sprays
- Injection into the plant
- Injection into the soil
- Release into water which is then used for irrigation
- Release from aircrafts when the areas to be treated are large and when operations on the ground are either not possible or will cause damage

Not Just on the Farm

Pesticides are more widely known for their use on farms to protect crops. Chemicals are used by people everyday to get rid of pests. Remember the last time you used Raid or Off?

Pesticides are used mostly to control insects, in homes, office buildings, schools, hospitals, supermarkets, and many other common areas. Most of these products are in the form of aerosols, pest strips, baits, pet products, and insect repellents. Pesticides are also used on lawns, gardens, golf courses, parks, and recreational areas.

Alternatives

Non- chemical alternatives for pest control have existed for quite some time. Some non-chemical alternatives include cultural practices such as crop rotation, choice of resistant plant varieties and use of biological products and agents. The concept of pest management, where synthetic pesticides are only applied as a last resort, is now considered common practice in professional agriculture. The development of plants with built-in pest resistance is a valuable alternative. This can be done by selecting plants with pest resistance, by crossbreeding a crop plant and a plant with a desirable characteristic, or by using plant biotechnology. However, some of these new developments have also encountered significant opposition.

Biological Control

Biological control uses natural agents such as predators, parasites, and pathogens to control pests. The natural enemy of an insect can be introduced to the crop. For example, the aphid is a pest which eats many crops. Lady bugs eat aphids. Animals such as this which control pests and help a crop are called beneficial species. There can be some problems with this method. It might take more time than is needed for the farmer to save his crop, it may not be affordable, and it is sometimes hard to predict if the beneficial species will have other adverse effects on the environment. Biological control requires a thorough understanding of the biology and ecology of the beneficial organism, as well as the pest.

Organic Farming

Organic farming involves growing crops without applying synthetic fertilizers, herbicides, or pesticides. Only natural products, such as composted manure, are used to improve soil quality. Many natural products are used to control pest populations. For example, insecticidal soaps and oils are effective at controlling many different insects. These soaps and oils are non- toxic to humans and pets. Botanical pesticides are derived from raw plant materials like flowers, roots, and stems. They are also safe to humans. Both types of chemicals can damage plants and kill beneficial organisms if care is not taken in their use. Here are two recipes for making these chemicals:

Soap and oil	Garlic Oil
1 Tbsp Liquid Dish Soap	1/2 Cup Minced Garlic
1 Cup Oil (Peanut, Corn, Soybean, or Sunflower)	2 Tsp Mineral Oil
1 Cup Water	1200 ml Water
	1 Tsp Liquid Dish Soap

Organic agriculture appears to be gaining acceptance on Prince Edward Island. There are more than 30 certified organic growers on the Island. More than 1,000 acres of land are currently being used to grow crops such as apples, blueberries, vegetables, herbs, soybeans, potatoes, hay, buckwheat and grain.

Pesticide Safety

Because pesticides can kill pests, they also pose a potential risk to other organisms such as humans. For this reason, they are not allowed to be used without the most thorough checks and safeguards to ensure that they do not harm humans, other species, or the environment. Because of the concerns, synthetic pesticides have undergone a development process to match with today's requirements.

They have become less toxic for humans, they have become more specific, and they have become more powerful. For example, 40 years ago pesticides were applied in kilograms of litres of active ingredients per hectare. Modern pesticides only require grams or millilitres to achieve the same of better results. Remember that pesticide residues remain only on the surface of produce. Rinsing with clear water will remove most of any trace residues.

MICROBIAL PESTICIDAL AGENTS

Baculoviruses are a group of viral pathogens that cause fatal disease in arthropods. More specifically, these baculoviruses infect primarily only members of the class Insecta and even within this class limit themselves to only a few orders, *i.e.* Lepidoptera, Diptera, Hymenoptera and Coleoptera. Baculoviruses have been studied extensively dating back to the 1960s. The literature is filled with a plethora of information on their biology and safety. The wild-type baculoviruses have been evaluated for use as insecticides numerous times, especially during the late 1960s and up into the early 1970s. The first baculoviral product introduced in the commercial arena was Elcar. Three other non-commercial preparations produced and used by the US Forest Servicc for control of forestry pests were developed and registered under the names of Gyp-Chek (Lymantria dispar NPV, LdMNPV), EM Biocontrol-1 (Orgyia pseudotsugata NPV, OpMNPV) and Neo-Check-S (Neo-diprion sertifer NPV, NsMNPV).

These products have received only limited use. However, recently the Elcar product was reborn under the trade name of GemStar™ and is being sold in various countries around the world. Intricate evolutionary relationships have developed between the baculoviruses and the arthropods that they exploit. This close evolutionary relationship has resulted in the restriction of host range. Over the past 40 years, extensive studies evaluating the safety of baculoviruses against vertebrate species have been carried out.

A summary of the data indicates that over 26 different insect baculoviruses have been tested for pathogenicity against 10 different mammalian species, including rats, dogs, mice, guinea-pigs, monkeys and humans. The methods of administration in these tests varied, including oral, intravenous injection, intercerebral injection, intramuscular injection and topical application. In no case was there any indication of toxicity or allergenicity. These results have been summarized in a number of reviews. Specificity is also a major factor in making baculoviruses safe to non-target insects. Baculoviruses generally only infect insects in a few families. In some cases only one species in a genus may be susceptible (*e.g.* LdMNPV). Those insect species that are considered beneficial (control pest insect species) are generally in families other than those targeted by baculoviruses. There are a few exceptions to this rule, however, as they can also infect some desirable lepidopterans.

This insect specificity makes baculoviruses excellent candidates for use in integrated pest management (IPM) programmes as they are compatible with other insect control technologies (bioinsecticides, chemical insecticides and transgenic plants), and have no effect on any non-target insects such as honeybees. If baculoviruses have all of these attractive parameters, why are they not widely used as insecticides? Baculoviruses are unable to rapidly kill a susceptible insect.

It may take 5-15 days to kill a host larva following infection. This period is affected by such factors as temperature, dosage and larval age. Younger larvae are generally more susceptible to infection and die sooner, but even though they are infected they continue to eat, usually up until the day they die. Obviously a crop with a low economic threshold cannot tolerate this continued feeding damage until the insect dies. Growers want to see dead insects soon after the application of any insecticide.

Fortunately, recombinant technology offers a new life to the old concept of utilizing baculoviruses as insecticides. It is now possible to genetically engineer a baculovirus with a toxin gene, giving it improved insecticidal properties and increasing the speed of action. These toxin genes usually code for an insect-specific, proteinaceous toxin. The genes can be isolated from the venom of certain arthropods (*e.g.* scorpions or spiders).

Typically, when a baculovirus containing a toxin gene infects an insect, the feeding is rapidly halted and causes a quick death. Insertion of the insecticidal gene does not affect the LD50 of the baculovirus but does significantly hasten the killing speed through the pharmokinetic interaction of the toxin within the insect. Even before actual death, the toxin causes cessation of feeding resulting in less plant damage, then paralysis, and eventually death.

Diet-overlay assays using an LD99 quantity of recombinant NPV have shown that, when second-instars of susceptible insect species are exposed to the wild-type virus versus the genetically engineered virus, at 10 days the LD50 is the same for both groups; however, the wild-type infected larvae characteristically take as long as 5.1 days to die versus 2.6 days for larvae infected with genetically engineered virus. Greenhouse studies have shown that this speed of kill translates to significantly better plant protection. In one greenhouse study, cotton plants sprayed with a genetically engineered virus containing a scorpion toxin gene were compared with plants treated with a chemical standard, esfenvalerate. The plants were then artificially infested with target insect species and then evaluated 4-5 days after treatment. The cotton incurred 18.9 per cent versus 13.0 per cent damaged squares for the two treatments, respectively. Statistically, both treatments were equal. Numerous US field trials have been conducted, again showing that the speed of kill is influential in reducing the damage.

One study conducted in North Carolina (USA) on tobacco against the tobacco budworm, showed that the genetically engineered virus performed as well against this insect as did the standard, acephate. Additionally, toxin production is limited only to the body of the insect (*i.e.* there is no release of toxicant into the environment as with conventional insecticides).

This makes the genetically engineered baculovirus a prime candidate for insecticidal use, and in some situations makes it competitive with chemical insecticides. Block and Zhihong have given a good summary of the progression of the technology in engineering of baculoviruses, culminating with the actual field release of a toxin gene carrying baculovirus by a commercial company in 1995.

RISK ASSESSMENT

In addition to evaluating the insecticidal properties of baculoviruses against targeted pest species, it is essential also to assess the potential environmental impact of such gene-inserted pathogens prior to, as well as during, initial small-scale field trials. In their review of ecological considerations pertaining to recombinant baculoviruses, Richards *et al.* noted that an environmental impact evaluation must address the effect of the recombinant entity on non-target species and populations, as well as the impact of the genetic modification on the ability of the virus to persist in the environment, relative to that of the wild-type or feral counterpart.

It is important to determine whether the addition of a foreign gene alters the host-range of the virus, and whether the insecticidal protein produced in the virus-infected insect will present a hazard to other species. Additionally, comparative testing of recombinant and feral baculoviruses for physical characteristics, genetic fitness and dispersal properties can provide insight as to relative stability in the environment.

Recombinant NucleopolyHedroviruses

Nucleopolyhedroviruses isolated from alfalfa looper, Autographa californica (AcMNPV), and cotton bollworm, Helicoverpa zea (HzNPV), have been engineered as vectors for carrying and expressing insecticidal genes into the body cavities of pestiferous insect species. AcMNPV is known to infect about 39 species of Lepidoptera. Larvae of tobacco budworm, Heliothis virescens (F.), and cabbage looper, Trichoplusia ni (Hubner), are among the most permissive to infection by AcMNPV, whereas others such as H. zea are only moderately sensitive, or semipermissive, to this virus.

Conversely, HzNPV is highly virulent against larvae of the entire Heliothine spp. complex, *i.e.* H. virescens and Helicoverpa spp. A number of insect-selective toxin genes have been inserted into the genomes of AcMNPV and HzNPV in an attempt to make them more effective as biocontrol agents in cropping systems. Two such genes, AaIT and LqhIT, which were isolated from venoms of the Algerian scorpion, Androctonus australis (Hector), and Israeli yellow scorpion, Leiurus quinquestriatus hebraeus Birula, respectively, are known to inappropriately modulate neuronal sodium channels.

The protein encoded by AaIT acts as an excitatory sodium channel agonist, causing repetitive firing of the insect's motor nerves and overstimulation of skeletal muscle. Symptoms exhibited by insects intoxicated by AaIT include cessation of feeding, paralysis and death. Conversely, LqhIT is a depressant neurotoxin, causing a blockade to sodium conductance and suppression of axonal depolarization and neuronal action potentials. LqhIT blocks neuromuscular transmission and induces progressive paralysis in affected insects.

A third invertebrate toxin gene which has been incorporated into selected baculoviral genomes is tox34, which was cloned from the genomic array encoding venom of the straw-itch mite, Pyemotes tritici. In assays conducted on primary cultures of motor neurons isolated from selected lepidopteran species, it has been shown that the product of the tox34 gene, a 33-kDa protein designated as TxP-I, causes a blockade of presynaptic inward calcium currents, which likely results in reduced secretion of the neurotransmitter acetylcholine.

The recombinant baculoviruses AcMNPV-AaIT, AcMNPV-LqhIT, AcMNPV-tox34, HzNPV-AaIT and HzNPV-LqhIT have been evaluated for insecticidal efficacy in open field trial settings. Considerable research has been devoted to assessing mammalian and environmental safety of these aforementioned baculoviruses.

Selectivity of Isolated Toxins

Studies conducted to define the pharmacology of purified AaIT and LqhIT have shown these toxins to possess remarkable specificity. Although both of these toxins are very potent against insects, they appear to have no effect on mammals even at very high dosages. Using hemocoelic injection methodology, DeDianous *et al.* determined that LD50 values for AaIT against selected species of Diptera and Orthoptera ranged from 20 to 3770 ng/kg.

Conversely, these same researchers found that subcutaneous injection of AaIT, at dosages as high as 50 g/kg, caused no adverse effects in mice. In separate oral and inhalation dosing studies conducted on labouratory mice, Possee *et al.* showed that AaIT at 1 μg/animal had no effect on the animals' body weight, dietary intake, survival or appearance of internal organs at necropsy. Experiments in vitro and in situ have demonstrated that AaIT causes convulsive firing and overstimulation of insect motor neurons and skeletal muscle, whereas it caused no effects in neuromuscular preparations isolated from mammals and selected species of Crustacea and Arachnida. Further, in experiments conducted on isolated nervous tissue, radiolabeled AaIT readily bound to such tissue from insects but not to nervous tissue from crustaceans or mammals.

LqhIT has also demonstrated a high affinity for insect neuronal receptors and a high degree of potency against species within this class, but no detectable toxicity to mammalian or crustacean species. Researchers at DuPont Agricultural Products intravenously injected individual mice with LqhIT at a dosage of 0.5 mg/kg and then observed them for 14 days. None of the mice exhibited any clinical signs of toxicity over the duration of the study, and necropsies performed on the animals at the conclusion of the test revealed no gross lesions on tissues and organs as a result of exposure to LqhIT.

Compared with AaIT and LqhIT, studies pertaining to pharmacology of TxP-I are limited. However, Tomalski *et al.* reported that the PD50 (paralysis dosage) of this straw-itch mite toxin in larvae of wax moth, Galleria mellonella, was about 0.5 mg/kg, whereas TxP-I had no effect on labouratory mice at dosages as high as 50 mg/kg.

Environmental Stability of Feral and Recombinant NPV

The potential for recombinant NPVs to displace their feral counterparts in the environment would be dependent upon expression of an advantage in fitness by the genetically altered virus. Fitness of a virus is determined by its physical characteristics and capacity for production of active progeny. Evidence gathered to date indicates that NPVs which have been engineered for enhanced insecticidal speed without significant alteration in host-range are likely to be less competitive in the environment than corresponding feral viruses.

Insertion of a gene encoding an insecticidal protein does not impact the physical or biochemical properties of viral occlusion bodies. Thus, such recombinant and feral NPVs are equally sensitive to abiotic and biotic degradation factors. Photolysis and photo-mediated oxidation are the primary mechanisms for inactivation of baculoviruses. It is thought that ultraviolet (UV) sunlight in the UV-B spectrum can directly cause strand breakage in viral DNA, or lead to production of highly reactive radicals (*e.g.* peroxides) which, in turn, degrade viral particles. Field studies have shown that selected feral NPVs which were applied to cotton foliage lost 70-80 per cent of their initial biological activity in less than four days.

Similarly, following spray-application of a wettable powder formulation of AcMNPV onto field-grown cotton, American Cyanamid Company researchers found that 3-hour-old foliar residues of the virus caused 98 per cent mortality in populations of H. virescens larvae, whereas larval mortality was only 5 per cent on cotton leaves that had been aged in the field for two days. Although

NPVs are rapidly inactivated when exposed to sunlight, their persistence when buried in soil can be for months or years.

In their review of viral persistence in soil, England *et al.* highlighted the following as factors that can impact stability of viral OBs and DNA:

- Proteolytic microorganisms (*e.g.* bacterium);
- Soil pH;
- Soil temperature (*e.g.* freezing and thawing);
- Soil-binding capacity;
- Soil moisture;
- Predation by soil-dwelling protozoa.

A characteristic that limits the survivability of recombinant NPVs is the reduced ability to produce viable progeny. The yield of progeny occlusion bodies from a baculovirus infection is a key factor controlling viral population dynamics in the field. The slow killing speed imparted by a feral NPV on its host is strategic, because the virus can continue to replicate and maximize the progeny numbers while the infected larva stays alive.

Conversely, recombinant NPVs containing genes that encode insecticidal protein rapidly kill the infected insect, thereby reducing the amount of time available to produce occlusion bodies. In a labouratory study conducted at American Cyanamid Company's Agricultural Research Centre, third-instar H. virescens were dosed with OBs of feral NPV or any of seven different AcMNPV-AaIT clones (differing in promoter-signal sequences within the chimeric toxin-gene cassette).

On average, each of the AcMNPV-AaIT clones killed H. virescens larvae in half the time required by wild AcMNPV, and when cadavers were weighed and assessed for quantity of viral progeny, larvae infected with any of the AcMNPV-AaIT isolates had about five-fold less body mass and contained approximately eight-fold fewer OBs than larvae that were infected with feral NPV. Numerous published studies have indicated that production of progeny by AcMNPV-AaIT and AcMNPV-LqhIT clones in larvae of H. virescens and T. ni is markedly reduced, in some cases by more than 10-fold, compared with the number of OBs produced by feral AcMNPV in these same host species.

Fuxa *et al.* conducted a series of labouratory and greenhouse tests to compare wild and recombinant forms of AcMNPV for their abilities to propagate and disperse following infection and death of insect hosts. Similar to observations made by other researchers, Fuxa *et al.* found that T. ni survival time and viral reproduction was significantly reduced for insects infected with AcMNPV-AaIT than for those infected with feral AcMNPV. Further, one week after placement of virus-infected T. ni onto non-treated collard plants in the greenhouse, it was found that larvae infected with wild AcMNPV deposited about 25 times more OBs onto collard foliage than T. ni infected with AcMNPV-AaIT.

This difference in release of OBs onto crop foliage was attributed to:

- Fewer OBs being produced in AcMNPV-AaIT infected larvae
- Reduced incidence of cadaver disintegration by AcMNPV-AaIT diseased hosts and a subsequent decrease in liberation of OBs onto collard leaves.

Based upon their findings, Fuxa *et al.* suggested that AcMNPV-AaIT would not be as capable as feral AcMNPV in establishing infections in subsequent hosts, and would therefore be selected against in competition with its feral counterpart in the environment. Indeed, in a caged-plot field trial, Cory *et al.* determined that transmission of virus between successive cohorts of T. ni was significantly lower on cabbage sprayed with AcMNPV-AaIT than in plots of cabbage treated with feral AcMNPV.

A recent labouratory experiment conducted by entomologists and molecular biologists at American Cyanamid Company showed that AcMNPV-AaIT was not competitive with feral AcMNPV in co-infected H. virescens larvae. In this experiment, individual larvae were initially exposed to a mixture (LD95) of AcMNPV-AaIT and AcMNPV at a 10:1 numeric ratio of OBs, respectively.

Viral OBs produced from this first larval population were extracted from cadavers, pooled and fed to a second cohort of H. virescens. This process of passing viral progeny onto subsequent host-insects continued through six cohort sequences. Restriction enzyme digestion and quantitative DNA blot hybridization were used to determine relative abundance of AaIT-containing viral genomes extracted from each larval passage. Analyses of virus extracted from the initial larval passage showed that 78 per cent of the viral DNA molecules contained the AaIT gene.

However, the viral DNA population collected from the sixth serial passage consisted entirely of feral AcMNPV genotype, *i.e.* AaIT-containing DNA molecules were not detected in the samples. Such results again suggest that AcMNPV-AaIT is at a competitive disadvantage with feral AcMNPV. Thus, such changes made to the viral genome to make it a more effective insecticide may also be responsible for increasing the likelihood of extinction of the recombinant NPV in the environment.

Vertebrate Safety of Recombinant NPVs

It is intuitive to expect that recombinant NPVs would have no impact on mammals, birds or fish, because the NPVs would have to infect cells of these vertebrate species in order to produce toxin. It is well documented in the scientific literature that NPVs are highly specific for insect hosts, with AcMNPV and HzNPV able to infect only certain species of Lepidoptera. However, it has been considered prudent to repeat such vertebrate pathogenicity/toxicity studies with the recombinant forms of NPVs.

Possee *et al.* conducted a series of studies with AcMNPV-AaIT to determine its acute oral, dermal and subcutaneous toxicity or pathogenicity to rats and guinea-pigs. Over a 14-day period following a single dosing of each animal with 1×106 occlusion bodies (OBs) of recombinant NPV, all rats and guinea-pigs remained healthy and showed no abnormalities in weight gain, dietary intake, physical appearance or motor skills.

Additionally, necropsies performed at the end of the study showed that no gross changes occurred in organs of treated animals. When used as a bioinsecticide, a recombinant NPV would be formulated and sprayed onto crops in the form of occlusion bodies. Since toxin would not be present in OBs of a recombinant NPV, the only route of exposure to the toxin by vertebrates would be through ingestion of infected insect larvae or herbivory of foliage contaminated with insect cadavers. In an attempt to mimic this route of exposure, researchers at American Cyanamid Company fed labouratory mice a diet which consisted partly of H. virescens larvae that had been previously infected with AcMNPV-tox34. Prior to use in the study, H. virescens were exposed to 10 times the larval LD95 of AcMNPV-tox34. Three days after exposure, larvae that exhibited paralysis (*i.e.* indicative of TxP-I production and intoxication) were homogenized and fed to mice by gavage.

Each mouse was fed the equivalent of four infected H. virescens larvae on a daily basis for five consecutive days. No clinical signs of toxicity or effects on survival were seen in treated mice during the study period. Dietary intake and body weight gains for treated and non-treated mice were equal throughout the five-day study. Macroscopic evaluation of tissues and organs in mice which were sacrificed at the end of the five days of dosing revealed no treatment-related pathological changes or abnormalities.

American Cyanamid Company evaluated AcMNPV-AaIT in a series of studies which were established by the US EPA to ascertain the safety of microbial pesticides. In an acute oral toxicity-limit test, AcMNPV-AaIT had no impact on survival of labouratory rats, nor did it impact internal organs of the animals. The budded form of the recombinant virus was also evaluated for its ability to infect mammalian cells (*i.e.* diploid human lung cell line MRC-5). AcMNPV-AaIT did not alter cell doubling time or cell morphology, and as expected, there was no evidence of viral replication or expression of AaIT. Mallard ducks, bobwhite quail and rainbow trout were fed diets consisting of AcMNPV-AaIT OBs as well as H. virescens larvae infected with AcMNPV-AaIT, and no adverse effects were observed in any of these three vertebrate species. Finally, no abnormalities were observed in the aquatic invertebrate, Daphnia magna, when it was maintained in aqueous suspensions of AcMNPV-AaIT. Researchers at DuPont Agricultural Products compared feral AcMNPV and AcMNPV-LqhIT for infectious and pathogenic behaviour in labouratory rats using oral, intravenous and inhalation routes of exposure.

Neither virus was infectious or toxic to rats by any of the aforementioned means of exposure at dosages ranging from 1 × 107 to 1 × 108 OB/animal. DuPont researchers also confirmed that AcMNPV-LqhIT was not infective or cytotoxic to human liver (Chang liver), intestine (strain 407) and lung cell lines. Additional studies showed that bobwhite quail receiving oral and intraperitoneal exposure to OBs, and trout and grass shrimp exposed to the NPV via food and/or aquatic environment, were unaffected by AcMNPV-LqhIT.

Invertebrate Selectivity of Recombinant NPVs

In order for toxin to be expressed by a recombinant baculovirus, the virus must first establish a productive systemic infection within the host (particularly if the toxin gene is under transcriptional control of a late promoter). If an organism is non-permissive to infection by the vectoring agent, the foreign gene will not be transcribed, *i.e.* host-range is determined by viral gene products rather than by those of the introduced gene.

Using either hemocoelic injection or a per os inoculation methodology, Huang *et al.* tested budded and pre-occluded forms of the following viruses for their abilities to establish an infection and replicate within selected non-lepidopteran insects: AcMNPV, LdMNPV, OpMNPV and Bombyx mori NPV (BmNPV). Each of the aforementioned viruses was engineered to express an inserted reporter gene (*e.g.* luciferase, β-galactosidase), so that even symptomless infections could be identified within the insects (*i.e.* viral replication could be based on detection of reporter gene product).

Huang *et al.* demonstrated that individuals from 13 species of insects, representing Blattodea, Coleoptera, Diptera, Hemiptera, Homoptera, Neuroptera and Orthoptera, did not support detectable replication of any of the four viruses included in the study. Maximum challenge labouratory studies have been conducted to determine the impact of certain recombinant NPVs on growth, behaviour and survival of numerous species of non-lepidopteran invertebrates. Depending on species or growth stage tested in these labouratory assays, animals were exposed to AcMNPV-AaIT, AcMNPV-LqhIT, AcMNPV-tox34 and/or HzNPV-AaIT via ingestion of contaminated artificial diet or foliage (OBs), consumption of virus-infected prey.

Results from these labouratory studies showed that none of the aforementioned recombinant NPVs directly caused adverse effects in individuals of the following predatory, parasitic, herbivorous or saprophytic invertebrate species: green lacewing (Chrysopa carnea Stephen), insidious flower bug (Orius insidiosus (Say)), ground beetle (Pterostichus modidus), honeybee (Apis mellifera), hymenopterous parasitoid (Microplitis croceipes), twospotted spider mite (Tetranychus urticae Koch), boll weevil (Anthonomus grandis grandis Boheman), western corn rootworm (Diabrotica virgifera virgifera LeConte), Japanese beetle (Popillia japonica Newman), earthworm (Lumbricus spp.),

Chinese mantid (Tenodera aridfolia), funnel web spider (Ixeuticus spp.), subterranean termite (Reticulotermes flavipes Kollar), convergent lady beetle (Hippodamia convergens), bigeyed bug (Geocoris spp.), red imported fire ant (Solenopsis invicta Buren), the social wasp (Polistes metricus (Say)), German cockroach (Blatella germanica) and common fruitfly (Drosophila melanogaster).

In a series of small-scale field trials conducted on cotton, tobacco and cabbage, it was determined that weekly foliar sprays of AcMNPV-AaIT, at rates as high as 2.0 × 1012 OB/ha, had no effects on diversity or density of non-target arthropods, relative to arthropod populations in non-treated plots of each crop. At least 45 species, representing 26 families of insects, spiders and mites, were found in collections taken from these small-plot field tests. Heinz *et al.* found that plots of field-grown cotton which were sprayed with either HzNPV-LqhIT or AcMNPV-LqhIT had similar populations of non-target predatory arthropods as those plots treated with feral forms of the viruses.

Further, cotton treated with the aforementioned LqhIT-inserted viruses had significantly greater numbers of beneficial arthropods than plots treated with conventional insecticides, such as the pyrethroid esfenvalerate and the semi-synthetic macrolide emamectin benzoate. Published and non-published studies have also shown that even within Lepidoptera, recombinant NPVs exhibit a host-range similar to that of their feral counterparts. For example, in a series of dose-response assays conducted on larvae of 52 different species of Lepidoptera, Bishop *et al.* found that AcMNPV-AaIT exhibited potency or virulence against each species at a level similar to that imparted by feral AcMNPV. There have been results from time-limited assays which would appear to suggest that AcMNPV-semipermissive H. zea might be more sensitive to gene-inserted forms of AcMNPV than the feral form of this virus. However, studies such as those reported by Treacy and All were designed to compare recombinant and feral viruses for speed of insecticidal activity against lepidopteran pest species, and thus focused on levels of larval mortality during the initial 8-10 days of viral exposure.

Although such studies can demonstrate potential of a recombinant virus to be an effective insecticide, they do not provide insight as to the overall impact of such a gene-inserted NPV, relative to is feral counterpart, on the life cycle of a semipermissive lepidopteran species. Therefore, entomologists at American Cyanamid's Agricultural Research Centre in Princeton, New Jersey recently conducted a labouratory study to compare wild-type AcMNPV, AcMNPV-AaIT and AcMNPV-tox34 for effects on survival of AcMNPV-semipermissive H. zea at selected points in time throughout its life cycle, *i.e.* middle and late larval development and adult eclosion (or 5, 10 and 35 days after placing second-instars on viral contaminated diet).

Based on temporal LD50 values and presence of non-overlapping fiducial limits, the two aforementioned recombinant AcMNPVs were about 800 and 25 times more lethal to H. zea than feral AcMNPV at 5 and 10 days after initial exposure to the viruses, respectively. However, when LD50 values were calculated from final test-population responses to treatments (*i.e.* 35 days after test initiation), it was found that AcMNPV-AaIT, AcMNPV-tox34 and AcMNPV were statistically equal in virulence against H. zea.

Results from this later study indicated that when the feral and recombinant forms of AcMNPV were allowed to complete their infection cycles in AcMNPV-semipermissive H. zea, there were little to no difference among the NPVs in virulence. The differential H. zea LD50 values between the two recombinant viruses and wild-type AcMNPV seen during the first 10 days for the study were likely due to pharmacology of the expressed toxins (per the recombinant NPVs), and not a result of any genome-related changes in viral infectivity.

REGULATORY AFFAIRS

The US EPA is charged with the regulation of pesticides under both the Federal Insecticide Fungicide and Rodenticide Act (FIFRA) and the Federal Food, Drug and Cosmetic Act as they were amended by the Food Quality Protection Act (FQPA) of 1996. This function is managed by the Agency's Office of Pesticide Programmes (OPP) which is tasked with assuring that the use of a pesticide does not cause unreasonable adverse effects to humans or the environment and that there is a reasonable certainty no harm will result from the aggregate exposure to the pesticide residue (under FQPA).

A pesticide, as defined by FIFRA, is 'any substance ... intended for preventing, destroying, repelling, or mitigating any pest'. This broad definition includes baculoviruses as well as other microbial products being used for control of insect pests. A separate group within OPP was established in 1994 to specifically facilitate the registration of biopesticides such as the baculoviruses and encourage their use.

This group, called the Biopesticides and Pollution Prevention Division (BPPD) performs risk/benefit and risk management functions for microbial pesticides among other responsibilities centreed around the use of safer pesticides and IPM programmes.

Within BPPD, responsibility is divided among a Microbial Pesticides Branch (which includes plant-pesticides), a Biochemicals Pesticides Branch, and a Pollution Prevention Staff. In order to register a pesticide for use under FIFRA, certain data requirements must be met which allow the Agency to assess the risk to humans and the environment from the use of the product. In general, biopesticides pose fewer risks than conventional pesticides; thus, EPA's data requirements for registration are typically less than those required to register a conventional product.

Since less data need be reviewed by EPA, the time from submission to registration of a biopesticide also is typically less than that needed to register a conventional product. Nonetheless, the Agency is tasked with the conduct of a thorough evaluation of data to ascertain the potential risks. In addition, any pesticide used on food or feed crops in the US must have an established tolerance or, due to the nature of the pesticide, be exempted from the requirement of a tolerance.

A tolerance is the legal maximum residue concentration of a pesticide allowed on food or feed. If a crop is found to have pesticide residues above the legal limit or if pesticidal residues are found on a crop which is not supported by a tolerance, the crop may be considered adulterated and be seized by Federal or State enforcement agencies. Again, since biopesticides are generally considered inherently less harmful than conventional pesticides, tolerance exemptions are typically granted by EPA for microbial products. The data requirements for the registration of microbial pesticide products can be found in the Code of Federal Regulations.

Data required for registration depends upon the proposed use of the microbial pesticide products (*i.e.* terrestrial food use, forestry use, indoor use, etc.) and includes, but is not limited to, the areas of product analysis, residues, mammalian toxicology, non-target organism toxicology, and environmental expression. Data requirements for an Experimental Use Permit (EUP) under FIFRA 5 are typically a subset of what is required for a full registration under FIFRA 3.

Examples of tolerance exemption regulations for microbial products can be found under 40 CFR 180. Prior to making a development decision and generating data for registration or EUP purposes, the efficacy of a microbial product is typically ascertained under field use conditions. According to 40 CFR 172 Subpart C, field efficacy testing can be accomplished on a small scale (*i.e.* less than 10 acres for terrestrial crops) without EPA approval for naturally occurring, indigenous microbial products as long as the treated crop is destroyed at the completion of the trial. However, for certain genetically modified microbial products, the Agency requires the submission of a notification of the intent to conduct field trials. Notification is required for small-scale field testing of microbial pesticides 'whose pesticidal properties have been imparted or enhanced by the introduction of genetic material that has been deliberately modified' (*i.e.* gene-inserted constructs) and for 'non-indigenous microbial pesticides that have not been acted upon by the US Department of Agriculture (*i.e.* either by issuing or denying a permit or determining that a permit is unnecessary; or a permit is not pending with USDA)'.

Although previously not exempt from the requirement of a notification under 40 CFR 172 Subpart C, a notification is no longer required for field testing genetically modified 'microbial pesticides resulting from deletions or rearrangements within a single genome' (*i.e.* gene-deleted constructs). The notification itself must include information and data on the microbial pesticide including but not limited to its taxonomy and natural habitat, means and limits of detection of specific analytical methods, techniques used to genetically modify the microbe, the identity and location of the gene segments inserted or deleted, its physical, chemical and biological features, its host-range and survivability in the environment, along with detailed information on the field trial programme such as rates, specific trial locations and application methods, target pests, site security measures, methods of sanitation of equipment, methods used to monitor movement of the microorganism, and a statement of composition of the experimental formulation.

In addition, the means of detecting potential adverse effects must be submitted along with methods of controlling the microbial pesticide if detected outside the test area. Basically, this information and data are a subset of what would be required for a full registration or EUP for the genetically modified microbial pesticide.

The Agency has established a mechanism in the notification process for protecting data claimed to be confidential business information (CBI). EPA also allows for the petitioning for exemption from the requirement of a notification under 40 CFR 172.52. Once the notification is submitted, the Agency has established a timeframe of 90 days to review the data and respond back to the submitter; however, approval of the notification by EPA must be granted before the field trials can take place.

In addition, the EPA announces receipt of the notification in the Federal Register and thereby opens a public comment period. Upon review of the notification and public comments the Agency can:

- Approve the proposed test;
- Approve the proposed test with certain modifications to the programme;
- Request additional information from the submitter;
- Require an EUP be submitted under FIFRA 5;
- Disapprove the proposed test and any EUP application. Following Federal approval of the field testing of the microbial pesticide, individual states (*i.e.* California) may have their own regulatory approval processes in place and the appropriate State regulatory authorities should be consulted prior to conducting field trials.

As part of its programme to develop recombinant baculovirus biopesticide, American Cyanamid Company first submitted a notification to conduct small-scale field testing of its egt gene-deleted Autographa californica MNPV construct in 1993. Although this gene-deleted construct does not require a notification under the current regulations, at the time of submission, notifications were indeed required for gene modifications of this type. Based on information submitted with the notification in conjunction with literature from the public domain, the Agency concluded that 'no unreasonable adverse effects on human health and environmental effects from the conduct of this small-scale field test are foreseen'. EPA approved the conduct of the field trial programme with a soil-monitoring requirement and feeding of the sampled soil to the highly permissive H. virescens larvae. The Agency required that any larvae showing signs of viral infection after being exposed to the treated soil must be examined for occlusion bodies (OBs) and any OBs found must be analysed by polymerase chain reaction assay (PCR) to determine if recombinant DNA is involved.

The following year, American Cyanamid Company submitted a notification of intent to conduct small-scale field testing of its AaIT gene-inserted Autographa californica MNPV (AcMNPV-AaIT) construct. Upon review of the submitted data and in response to public comment, the Agency requested more information concerning the environmental expression (*i.e.* persistence in the environment) and host-range of the gene-inserted baculovirus.

After several rounds of meetings and correspondence with EPA personnel aimed at addressing their concerns, the initial notification for field testing was ultimately approved by the Agency, albeit with substantially increased mitigation (*i.e.* limited acreage, site disinfection procedures, soil barriers, 'wild-type' baculovirus oversprays), increased monitoring (*i.e.* soil analyses for OBs, non-target insect surveys), and decreased acreage requirements in comparison to the gene-deleted construct. The mitigation and monitoring requirements were aimed at addressing the questions on persistence and host-range and more fully characterizing any potential risk. As the initial field trials were underway, scientists at American Cyanamid Company continued to conduct labouratory studies focused on the host-range determination by looking at various species of lepidopteran and non-lepidopteran insects.

Additionally, a 'competition' study was conducted which proved that the ecologically disadvantaged modified baculovirus (*i.e.* recombinant produces less OB progeny than the wild-type virus) would not be able to overcome the naturally occurring virus in the environment. These studies, in conjunction with the results from two years of field monitoring work, provided the Agency with sufficient data to ultimately reduce the mitigation and monitoring requirements in follow-on notifications.

By 1997, the Agency had monitoring data collected from the 1995 and 1996 trials which, along with numerous labouratory bioassays, showed that AcMNPV-AaIT had no effect on non-target arthropods (at least 26 families of Insecta and Arachnida representing 45 species were encountered in monitoring surveys with no ill-effect observed). In addition, soil monitoring data from the earlier trials showed that the recombinant 'does not vary in its environmental fate (*i.e.* persistence) profile in comparison to the naturally occurring Autographa californica MNPV'.

The Agency subsequently approved a small-scale field programme for the 1997 season and removed the requirements for soil/non-target insect monitoring and site disinfection. In fact, EPA no longer requires notification for field testing of AcMNPV-AaIT as long as certain minimal mitigation criteria are followed. Also in 1997, American Cyanamid Company submitted a notification for small-scale, crop-destruct field testing of an AaIT gene-inserted construct based on a H. zea NPV baculovirus.

This notification built on the experience gained from the modified AcMNPV field work and included labouratory-derived results which once again showed that neither the host-range nor the infectivity of the recombinant H. zea NPV changed in comparison with the wild-type baculovirus. The Agency approved the small-scale field trial programme and stated that no further notifications were necessary for similar constructs of AaIT gene-inserted H. zea NPV 'if there is no indication of increased host-range' as long as the tests are not conducted in areas adjacent to potentially susceptible endangered insects and that wild-type H. zea NPV baculovirus is introduced to the site as part of a standard treatment or overspray. According to EPA, this overspray of the trial site is a valid way to provide a source of wild-type baculovirus to compete with the genetically engineered baculovirus.

In a notification submitted in 1998, American Cyanamid Company requested approval of small-scale field testing of a tox34 gene-inserted AcMNPV (AcMNPV-tox34). Expression of the tox34 gene in recombinant AcMNPV significantly improved the efficacy of the baculovirus in the same fashion as expression of the AaIT gene in the earlier AcMNPV constructs.

This submission once again provided EPA with information on the host-range and infectivity of the construct. As this tox34 gene-insert construct introduced an insect-specific toxin not previously reviewed by EPA, labouratory data were submitted to ascertain the mammalian toxicity characteristics of the toxin expressed in infected target insect larvae. No adverse toxicological effects were seen. The notification was approved (and further notifications of the tox34 construct waived) with the same wild-type baculovirus overspray restriction and infectivity testing as the AaIT gene-insert H. zea NPV construct.

Overall, EPA concerns with small-scale field testing programmes of genetically modified baculoviruses focused on potential risks of an increase in host-range. Because of the novel technology and after a thorough review of the submitted host-range and environmental safety data, EPA approved the initial small-scale field programms of the original recombinant baculoviruses with extensive risk mitigation criteria. As the database on host-range, infectivity and environmental expression increased, the Agency was able to rely on past studies and extensive monitoring as a foundation to support newer construct introductions. This increase in familiarity with the science ultimately led to the approval of small-scale field testing of several different recombinant constructs with lessened mitigation requirements and laid the groundwork for further development work. Baculovirus safety in conjunction with recombinant technology should be perceived as having two levels: level 1, species specificity and level 2, target specificity. The safety of feral baculoviruses has been clearly established through several decades of high-quality research. The first level of safety is dictated by the fact that infection can occur only in the limited number of species susceptible to any baculovirus. In the recombinant form of the virus, a second level of safety is added, in that the inserted gene encodes a toxin which binds only to insect targets.c Research on the recombinant viruses is somewhat limited since their availability is a recent development. The constructs currently available meet the two-level safety thesis. The limited database does show that these constructs do not infect species outside their established host-range and, in fact the addition of the toxin gene does not even alter the LD50.

The major benefit is a faster kill and better crop protection (desirable changes). Very sophisticated cellular level experiments have shown that there is no replication in non-target hosts (vertebrate and invertebrate). Finally, the recombinant NPV is clearly at a competitive disadvantage to the feral NPV due to its limited ability to propagate. The growing database has been well received by rational environmental groups, academics and growers. USA regulatory agencies are becoming convinced about the safety of these recombinants and see them taking their place in the grower's arsenal of insect control tools.

PUBLIC RELATIONS ACTIVITY

These genetically manipulated baculoviruses obviously need approval by regulatory agencies prior to any field release and eventual registration. Using the scientific process to demonstrate safety of a genetically engineered organism to the public is only part of the 'proof of safety' process. This information has to be transmitted to the public to establish the true level of safety in their perception. Biotechnology has come under public scrutiny to varying degrees depending on where one is in the world. Environmental activists have attacked biotechnology from scientific, ethical and personal viewpoints. These activities have created doubts, which must be addressed, no matter how trivial or unscientific they may appear. Obviously the development of a plan is necessary.

Assuming that approval of regulatory agencies is adequate (to make a release) could be a major mistake, as demonstrated by the experience of the National Environment Research Council (NERC) Institute of Virology, Oxford, England. All the necessary regulatory clearances were obtained for a release, but no effort had been made to inform the public around the application site. The end result was that the public felt that 'a fast one was being pulled'. The researchers then had to backtrack and attempt to educate an irate public. American Cyanamid Company conducted field releases in 1995, 1996 and 1998 of baculoviruses containing an insect-specific toxin gene from either the scorpion Androctonus australis or the straw-itch mite Pyemotes tritici. Before any release was attempted, American Cyanamid Company scientists developed a safety database (extracted from the literature) and through carefully conducted experimentation.

This database was summarized and written into a brochure. Also, a video was produced, which explained the engineering process and the resultant recombinant baculovirus' impact on the susceptible host. Both the brochure and video were made available for public distribution. This information was utilized at numerous open meetings with representatives from many areas, academia, regulatory, mass media, environmental groups, politicians and the general public. Thus communication lines were opened and the dialog that developed served to mitigate any concerns. Early on, American Cyanamid Company scientists opened communication lines with responsible national environmental groups to explain the technology.

Gaining the support of these groups inevitably helped ease any concerns that may have been present. Using the basic elements of risk communication, a policy was adopted that all the safety information available would be made public. This policy was followed throughout the public education process and in all US EPA submissions (*i.e.* all safety data supplied to the US EPA were available to anyone for the asking). This strategy resulted in the generation of very little negative response from the aforementioned groups.

10

A Participatory Approach to Encourage Non-pesticide Methods

In the last few decades, plant protection has been dominated by reliance on pesticide methods. However, evidence from oil palm, tea, rice and crucifer crops provided clear evidence that reliance on chemical insecticides led to pest outbreaks. Indeed, contrary to popular beliefs, chemical insecticides often cause pest outbreaks and have to be avoided or at least minimized. The reliance on chemical insecticides often meant that efforts on investigation of non-pesticide methods were usually neglected. While not as common as chemical insecticides, non-pesticide methods are present and ecological studies have led to better understanding and even application.

The impact of crop resistance is often complicated by the impact of existing natural enemies, a phenomenon often referred to as natural biological control. This chapter will focus on two types of non-pesticide methods that farmers can engage in, namely; biological control and physical and cultural practices. It is difficult to include the use of attractants as a non-pesticide method as often the use of pheromones/attractants is associated with use of pesticides to kill off the trapped insects. In the case of using rotting meat to attract the rice ear bug, *Leptocorisa oratorius* (Fabricius) (Hemiptera: Alydidae), it was found that these attract mainly males. The chapter is largely based on the author's personal experience with non-pesticide methods and may not cover fully this large and interesting field.

The problems arising from use of chemical insecticides were realised as early as 1950s. By the early 1950s, resurgence of pest outbreaks and resistance to chemical insecticides were reported from all over the world. Initial attempts led to the first development of IPM as a combination of chemical and biological control. This did not go far enough to address the conflict of using two opposing approaches. Chemical insecticides by their very mode of action remove or inhibit the function of biological control and this poses a problem of integrating chemical and biological control. Recognizing this, steps were then taken in the 1960s to formulate a more ecological approach. In the 1970s, the approach recognizes the need to identify social and economic requirements in addition to ecological consideration. Despite all attempts at defining IPM, it remained very much at academic levels, particularly in the wet tropics. Part of this problem lies in the heavy reliance on the use of economic thresholds as a key element to decision-making. Economic threshold levels were insensitive to both social and economic situations of farmers and it was realised by 1980s that the use of economic threshold levels continued to maintain unnecessary use of insecticides due in large part to lack of understanding of underlying ecological principles. Realizing this dilemma, the FAO implemented IPM programme for Asia in the late 1980s initiated a non-formal education approach to help farmers understand these underlying ecological principles.

The process came to be known as the IPM Farmer Field School. The IPM FFS is the primary learning approach piloted in Indonesia. The IPM FFS is a season-long learning experience. In the FFS, farmers learn about agro-ecosystem management. The FFS makes use of the crop field as a field laboratory. In the laboratory, FFS participants learn about the ecology of the crop by means of regular observation and hypothesis testing. Discussions are based upon participant analysis of field conditions. The FFS uses a participatory learning process. The process emphasizes taking decisions and actions based on an open discussion of ideas which is free from the domination of any individual. These decisions are tested in the field laboratory. The FFS process, besides its emphasis on field ecology, provides participants with an opportunity to examine human social dynamics.

As a result, FFS participants not only learn about the cause and effect relationships that exist in the field; they also acquire a greater understanding of human relationships. The analytical processes employed in the FFS enhance farmers' capacities to examine the conditions, in which they live and work. Participants, having completed their FFS, are able to take decisions and take actions that would improve those conditions. The increased understanding of participants regarding human social dynamics enables them to develop collaborative efforts to ensure that planned actions are implemented. Even after the FFS, farmers continued to experiment and in some cases worked in groups. For example, farmers in the village of Kalensari continued to study a non-pesticide method to control the white stem borer. Examples of follow-up studies by farmers are reflected.

This confirms that farmers are able to carry out experiments and develop innovative non-pesticide methods of controlling pests. As noted in the first case study, encouraging IPM farmers to continue research is critical to the realization of the potential of a biological control agent. Results from a better understanding of non-pesticide methods have helped farmers to convince policymakers to take the right steps in preventing outbreaks. It is time that policymakers should make this approach *a priori*. An enabling policy to encourage farmer education in FFS and carry out field ecological studies will invariably promote the development of more non-pesticide methods to manage insect pests in Asia.

THE BENEFIT OF ERADICATION

Despite all the noble efforts of our quarantine services there are going to times when new pathogens are introduced. Even at this stage an effective emergency response by quarantine and plant protection personnel can reduce the chances of a pathogen establishing in a country. The cost-benefit analysis of this approach highlights the advantages. In 1984 a survey detected citrus canker on Thursday Island. This disease has the potential to destroy the citrus industry in the moist coastal production areas of Australia. Eradication of the disease from the island involved the destruction of several trees and detailed monitoring of the remaining trees for four years. Thursday Island was declared free of citrus canker in 1988 at a cost of US$20,000. Similar eradication programmes have been successful for black sigatoka.

A host plant will grow better and produce a higher yield if the planting material used is pathogen-free. For this reason every effort should be made to ensure that farmers get access to such planting material. In developed countries certification schemes are well developed for potatoes and certain fruit species and the provision of certified stock has led to significant improvements in general health and production. Similar certification schemes or programmes to produce pathogen-free planting material exist in developing countries. In the Asia-Pacific region, Taiwan has an extensive range of schemes covering staple crops such as rice and coarse grains such as corn, soybean and peanut.

Other propagation and inspection systems of virus-free seedlings are now in place for potato, strawberry, sweet potato, asparagus, passion fruit, banana, citrus, sugarcane, anthurium, oncidium and bamboo. Most other countries in Asia have seed certification schemes, backed up by appropriate seed laws, for one or more crops, which have a degree of seed health testing to control the possibility of seed transmitted pathogens. In the Pacific region the formal seed supply network is poorly developed and there are no seed laws or quality standards. Some farmers have access to certified seed through agricultural merchants but generally most use farmer-saved seed.

IMPACT OF VIRUS-FREE SWEET POTATOES

Sweet potatoes are currently grown by small farmers on 600,000 ha in Shandong province, China with an annual production of about 17 million mt. In the late 1980s, observations indicated high levels of virus infection and propagation of virus-free seed was identified as a possible means to improve yield. International Potato Centre (CIP) and Chinese scientists began a project to develop and transfer methods for propagating virus-free planting material using tissue culture and Enzyme-linked Immunosorbent Assay (ELISA).

Between 1994 and 1998, virus-free plants were extended to about 80 per cent of the hectares and contributed to a 30-per cent yield increase with no change in the use of inputs. The internal rate of return estimated at 202 per cent with a net present value of US$550 million. By 1998, yearly productivity increases were put at US$145 million, improving the income of the provinces seven million sweet potato growers by 3-4 per cent.

Compared to seed, vegetatively propagated planting material can potentially carry all pathogens that are present systemically in the mother plant in addition to pathogens that are carried externally. Pathogen-free plants may be obtained by meristem tip culture of the upper millimeter or so of the growing tip of the plant. Most viruses do not invade this portion of the plant for a number of possible reasons. Meristem tip culture is sometimes accompanied by heat treatment. This may involve putting plants in controlled temperature cabinets up to 40 °C for a number of weeks. Under these conditions the rate at which new meristematic tissue is synthesized may outstrip the rate of viral replication. Shoot-tips removed from these plants should be virusfree. Virus and virus-like diseases are serious constraints to tropical fruit production in the region. Several large citrus development programmes in the region have failed because of the quality of planting materials used. In Asia, a number of initiatives have been implemented for the production and cultivation of virus-free banana and citrus seedlings which have been integrated with other measures necessary for long-term control.

The development of sensitive serological and molecular diagnostic tools for these damaging viruses has improved greatly the impact of disease-free nursery schemes. In Taiwan, a national scheme for certification of citrus bud-wood has been underway since 1983 which employs an improved method of shoot-tip micro grafting, heat therapy combined with virus indexing that is conducted by the Plant Virology Lab (PVL) at the National Taiwan University (NTU). The stocks indexed as pathogen-free by the PVL/NTU are used as the pathogen-free citrus foundation stock collection for the programme and are maintained in insectproof screen-houses. In Malaysia, several states have been involved in citrus rehabilitation programmes and the use of pathogenfree planting materials has been recently implemented. The Malaysian Agricultural Research and Development Institute (MARDI) and Department of Agriculture, Sarawak maintain pathogenfree foundation stocks of popular citrus cultivars. Bud-woods from these plants are subsequently used for the propagation of pathogen-free planting materials. In Thailand, the disease-free Citrus Nursery Programme was implemented by the Thai-German Integrated Pest Management (IPM) project in collaboration with the Department of Agriculture (DOA). The National Citrus Foundation Block was established in 1994.

The DOA is responsible for the National Citrus Foundation Block and Certified Mother Tree Block. Disease-free bud-woods are produced from certified mother trees and released to approved citrus nurseries or farmer groups. A nursery accreditation scheme was introduced in 1993 in order to improve the quality control process. Those nursery owners who wish to be involved in the scheme must follow the conditions, regulations and recommendations indicated in the scheme and must reapply for accreditation every 12 months. Nurseries are inspected twice a year. Similar programmes also exist in Japan, Philippines (Phil-German Fruit Tree Project), Vietnam (Southern Fruit Research Institute [SOFRI] Citrus Disease-free Seedlings Programme) and Indonesia (Indonesian Citrus Variety Improvement Programme).

In Taiwan, since 1983 a total of 26 million banana plantlets have been produced *in vitro* for commercial planting. This has benefitted growers by making available pathogen-free plants and also in the rapid multiplication of new varieties. To meet the demand for large numbers of samples, virus-free stock foundations are kept in special facilities that are insect-proof and vector-free and the viral status is closely monitored. These virus-free plants are used to produce mother plants that are used for multiplication by commercial nurseries. Meristem culture of banana for the production of pathogen-free plantlets has been developed elsewhere in the Asia-Pacific region. This approach is used in the Philippines as an ongoing eradication and banana rehabilitation programme.

A project for the control and rehabilitation of BBTV affected areas was implemented in 1994 and focused on a communitybased approach for smallholder banana growers. This involved consultations with officials and banana growers and a rapid rural appraisal (RRA) with farmer cooperators to determine their understanding of disease management practices. This was followed by training on the early diagnosis of BBTV and proper eradication methods. Pathogen-free plantlets were distributed to farmers at four project sites. The results clearly indicate that areas previously devastated by BBTV can be made productive again. However, this can only be achieved with proper understanding of the disease by the farmers and appreciation of the need to integrate with other methods of control such as proper sanitation to remove alternative sources of inoculum and appropriate vector control.

Biological Control and Crop Disease Management

Biological control is increasingly recognized as the foundation of a many successful IPM programmes. Biological control has a number of benefits. It is environmentally benign and once established can give long-term control. After initial costs it is relatively inexpensive. It also ensures control over a large area compared to other control measures such as chemical spraying which only gives control in the area treated. It can contribute to healthy food, cleaner environment and increased profits for farmers. In Asia it has been demonstrated that outbreaks of the brown planthopper (BPH) in rice were caused by the destruction of effective natural enemies through the indiscriminate use of broad-spectrum insecticides.

Failure to recognize the importance of natural biological control led to continuous outbreaks of BPH in several countries in Asia until the FAO Inter-Country Programme encouraged countries to appreciate this ecological phenomenon. This Programme has encouraged extension staff to understand the role of natural enemies in their farming systems. This has been done by implementing Farmer Field Schools (FFSs) where biological control is the main focus of an integrated approach to pest control. Unfortunately, efforts at biological control of plant pathogens have lagged behind developments in the field of entomology and there are few examples of successful field implementation of biological disease control. However, significant development work would seem to imply that biological control has great potential for the sustainable management of plant pathogens. Biological control of plant pathogens usually refers to the use of one organism as an antagonist to another. This can take the form of suppressive soils that contain antagonistic microorganisms, using antagonistic and trap plants.

These biological control agents may be employed to keep the pathogen population below an acceptable threshold level, or may be used as a living barrier that prevents pathogen infection. Traditional farmers have been using biological control for millennia through the development of suppressive soils and this was highly developed in China and a number of Pacific Island countries. Examples of biological control of plant pathogens include the following: Bacteria such as *Bacillus cereus, B. subtilis* and fluorescent pseudomonads have also been successfully used to control fungal root pathogens. In Indonesia, fluorescent pseudomonads have been shown in the laboratory to reduce the severity of *Ralstonia solanacearum*, the cause of blood disease in banana. Endophytic fungi of durian may have some potential as an antagonist of *Phytophthora palmivora*.

In Taiwan, a *Bacillus subtilis* isolate was recently commercialized by Kuan-Hwa Chemical Co. for the control of powdery mildew of pea. *B. subtilis* was also found outstanding because of the superior antagonistic activity against *Rhizoctonia solani* AG4, *Sclerotium rolfsii*, and *Xanthomonas* sp. pv. *mangiferae-indicae* (Xcm). Its mass production technique has been developed in Taiwan. Commercial preparations of *Trichoderma* and *Streptomyces* spp. are also commercially available. Inoculation of a healthy plant with a mild, or less severe strain, of a particular virus, gives protection to the plant against other more severe strains of the same virus.

Cross protection technology has been used in some countries of the Asia-Pacific region. This includes work on cucurbits infected with zucchini yellow mosaic virus (ZYMV). The process involves mechanically inoculating the leaves of cucurbits with the mild strain of the virus before they are transplanted into the field. In one trial in the Pacific, cross protection resulted in a reduction of almost 50 per cent in the levels of disease compared to unprotected plants. Cross protection has been used in Taiwan and some other countries of Asia with mild strains of citrus triesreza closterovirus (CTV). One obvious problem with this technique is the need for labour to carry out the mild strain inoculations before plants are transplanted into the field.

Host Plant Resistance and Crop Disease Management

Traditional farmers in the Asia-Pacific region have been selecting, consciously and unconsciously, disease-resistant plants for centuries. These farmers often have knowledge of a large number of traditional varieties that they know will give good agronomic performance but also have a degree of resistance to the many injurious diseases that attack their fields. More recently, scientists have been breeding plants for improved resistance to plant disease. Since the 1960s, host plant resistance has been increasingly successful in the management of important diseases of staple food crops in developing countries with significant and sustained benefits to both smallholder farmers and their environment.

The gains from resistant varieties have been most significant in Asia. HYVs of wheat and rice of the Green Revolution with resistance to fungal and viral pathogens have contributed significantly to increased productivity, cheaper food and reduced poverty. Similar gains have been experienced in India with disease-resistant legumes. Banana BLS-resistant varieties released by the Honduran Agricultural Research Foundation (FHIA) breeding programme have had some positive impact in Pacific countries. These FHIA hybrids are currently undergoing screening for resistance to other pathogens including nematodes and are being evaluated in over 50 locations. Agricultural Research for Developing Countries (CIRAD) have plans to screen bananas from their breeding programme for BLS resistance in New Caledonia. In the Pacific, the TaroGen project has supported taro breeding programmes in PNG and Samoa. These programmes, based on durable resistance to TLB, have resulted in the release of nine new resistant varieties. There are plans to distribute these new varieties to other countries in the Pacific with TLB and also to countries still free from the disease. Resistance conferred on a plant by one or a few genes is often referred to as monogenic/oligogenic or race-specific resistance.

This type of resistance usually results in the hypersensitive response and gives complete control against the pathogen. This is easy to work with in breeding programmes because it is easily incorporated into plants and gives a high level of resistance. Unfortunately it is this type of resistance that is easily overcome by new races of the pathogen. For this reason, resistance based on one or a few genes has generally not proved to be long-lasting. The breakdown of resistance in potato to *Phytophthora infestans* clearly illustrates this point.

When many genes control resistance it is referred to as polygenic or non-specific resistance. Unlike specific resistance this type of resistance does not give complete control but the host plant is able to reduce or restrict the colonization or spread by the pathogen or limit the rate at which it can produce spores. Because none of the genes involved confer resistance if acting in isolation they are referred to as minor genes.

Non-specific resistance has the advantage of being long-lasting or durable, and it is being increasingly utilized in breeding programmes such as the TaroGen taro breeding projects. Robinson describes a number of innovative participatory approaches to plant breeding based on horizontal or durable resistance aimed at 'demystifying' plant breeding and taking control away from professional breeders and putting it in the hands of university students and farmers. He describes the formation of university breeding clubs and decentralized farmer participation schemes. The first university breeding club in the world was started in 1995 in Mexico. A university taro-breeding club was initiated at the University of the South Pacific (USP), Samoa in 1999. This was the first club to be inaugurated outside of Latin America. The club represents an innovative approach to teaching and learning at USP. It is a cheap and easy approach to breeding. It ensures that there are many hands to do breeding work and has resulted in increased taro breeding activity.

Robinson has proposed university breeding clubs as a "hands-on" approach for students to learn about breeding for horizontal resistance and a way of "scaling-up" farmer participation in plant breeding. Robinson envisaged student members of breeding clubs returning to their family farms with potential new cultivars for evaluation. After a few decades, there could be hundreds, or even thousands, of former club members testing new lines as they emerge from clubs. Additional breeding clubs would increase the output even more, providing the widest extent and the highest possible quality of farmer participation in plant breeding.

Cultural Practices and Crop Disease Management

Many of the practices classified as cultural control have been practiced by farmers for centuries in the Asia-Pacific region and form a major component of their indigenous knowledge. Thurston has extensively documented many of these traditional practices and there are increasing demands that such practices and knowledge are integrated to ensure sustainable crop disease management. Traditional smallholder farmers in the tropics can rarely afford the cost of pesticides and labour is relatively inexpensive. This has meant that cultural control has and continues to play a major role in the management of crop diseases.

In the Asia-Pacific region cultural control methods represent locality-specific options for disease management. They have proven their adaptability and worth in a particular environment. They have often arisen as a result of farmers' insightful knowledge of their farming systems and the pests within those systems.

Dent has recognized the value of cultural control and believes that it should be the first line of defence in pest management programmes. This can act as a foundation on which other control strategies can be implemented. Above all, cultural control is environmentally safe and is compatible with other control measures.

The following list of cultural methods has been shown to have some benefit for disease management in the Asia-Pacific region:

- *Destruction of Plant Disease Inoculum:* Many pathogens can survive in soil or on plant residues, and readily infect newly-planted crops. Such practices as clearing land and burning residues is believed to have a number of beneficial disease control effects. Soil-borne fungal pathogens are destroyed in the upper soil levels during burning. Flooding with water is a common feature of taro cultivation and may reduce pathogenic fungi in soil. The early removal (rouging) of virus-infected plants may help reduce the incidence of disease. In some countries, such as Hawaii and Guam, removal of watermelon mosaic virus infected plants has led to a reduction in disease incidence by providing less inoculum for insect vectors. The early removal of dasheen mosaic virus and *alomae*- and *bobone*-infected taros may provide good control. Banana bunchy top may be controlled in a similar way. It is important to be careful when removing infected plants as there is the risk that the vectors will disperse to uninfected plants. Plants may be doused in kerosene and burnt. Roguing of *Pythium*-infected taros may be useful and care should be taken to remove all decayed roots and corms. In Palau, the traditional approach is to remove the plant and all the soil surrounding the roots.
- *Crop Rotations:* Asian and Pacific farmers have used rotations for millennia. Particular plant pathogens are associated with specific crops. If these crops are continuously grown then these pathogens can reach very high levels. By planting unrelated crops the buildup of disease is halted.
- *Plant Density and Spacing:* Low plant densities contribute to a reduction in humidity in the plant microclimate that lowers the incidence of a number of plant diseases such as Sigatoka of banana. In some Pacific countries increased plant spacing has been observed to reduce the incidence of TLB. In Asia, traditional spacing practices have been useful in managing downy mildew in maize. Sheath blight is now one of the major fungal diseases on highyielding varieties (HYVs) of rice in Asia. The HYVs tend to be planted in high densities

which make them more susceptible to the disease. Prior to the release of IR8 in 1966 sheath blight was considered a minor rice disease. There is evidence that planting at low densities reduces the incidence of disease.

- *Depth of Sowing:* The perfect stands found in many crops grown by traditional farmers in Asia reflects their knowledge in using this practice.
- *Time of Planting:* Plant diseases can often be avoided by planting at times of the year that are unfavourable for disease development. In Asia, early planting of rice and maize has been reported to reduce the incidence of rice blast and downy mildew, respectively.
- *Flooding:* The paddy system of rice production represents one of the oldest practices that uses flooding for plant disease management. Severity of rice blast is reduced on flooded paddy rice compared to upland rice due to reduced infection periods. A similar system for taro production in the Pacific may contribute to disease management.
- *Mulching:* The placement of a mulch of old leaves or weeds around the base of beans has been shown to reduce infection by web blight. Mulching is believed to prevent the dissemination of soil-borne fungi by not allowing soil splashing by rain to occur. This has been demonstrated for yam anthracnose in the Pacific.
- *Raised Beds:* The use of raised beds for modification and cultivation of wetland areas has been described for a number of countries in Asia. In the Pacific, raised bed agriculture has been, and continues to be, effectively employed for the cultivation of mainly root crops. Raised bed systems have been described for a number of different environments and represent a highly productive form of cultivation. Thurston and Parker report that, in addition to their obvious irrigation and drainage value, there is little doubt that disease management is an added benefit of raised bed systems. They cite examples of the suppressive nature of raised bed soils against damping-off fungi (*Pythium* spp.) This is attributed to the increased biological activity of known *Pythium* spp. antagonists such as *Trichoderma* spp., *Pseudomonas* spp., and *Fusarium* spp. in raised bed soils due to the addition of large amounts of organic matter. These antagonists are also likely to be active against other soil-borne plant pathogens. Raised bed soils were also found to suppress plant parasitic nematodes, containing nine organisms that had anti-nematodal activity. In addition, the high organic content of these soils was thought to contribute to the low numbers of nematodes in these soils.
- *Crop Diversity:* Polyculture or mixed cropping (the growing of two or more crops at the same time) has been observed to reduce the

incidence of disease. It is believed that such environments represent a more stable ecosystem and offer a greater degree of protection compared with monocultures. A diversity of crops helps to delay the buildup of plant pathogens. It is interesting to note that the incidence of the disease kava dieback, caused by cucumber mosaic virus (CMV), is more severe in places such as Fiji where it is planted mainly as a monoculture. The incidence in Vanuatu where kava forms a component of the traditional mixed cropping system is lower. The virus is transmitted by an aphid vector and in the traditional mixed cropping system the aphid may have difficulty locating kava plants and along the way it will feed on other non-kava plants and quickly loses the ability to transmit the virus. There is abundant evidence to show that crop diversity contributes to the management of other pathogens also.

- *Destruction of Alternate Hosts:* Weed hosts act as carriers for a number of plant pathogens, especially viruses, over successive seasons. The control of these weeds is complicated, given that they are often symptomless carriers of viruses and root pathogens but their destruction has been demonstrated as contributing to a reduced incidence of certain diseases. Weed control is important for the control of a number of cucurbit viruses that cause damage in Pacific Islands and for control of CMV in bananas.

NON-PESTICIDE METHODS FOR MANAGING CROP DISEASES

The problems highlighted above can be greatly reduced by using an integrated approach based on some of the non-pesticide methods of disease control. This will start with a discussion of methods used to prevent the entry of a pathogen to a new area or attempts to quickly eradicate such an incursion. Both approaches highlight the cost benefits achieved in preventing disease establishment and they should be the cornerstone of non-pesticide approaches to disease management in all countries of the region. This is followed by a description of certification schemes to provide pathogenfree plants to farmers and an in-depth examination of various cultural and physical practices and biological control methods that may reduce disease levels before discussing the importance of host plant resistance for achieving durable and sustainable disease control.

An examination of the role of biotechnology and its application to disease control. It is stressed that no one method or approach to disease management holds the key to future success rather that solutions will be location-specific and will depend on a basket of options in order to achieve sustainable disease management. The problems and issues specific to all methods are discussed with possible strategies for future improvement. The chapter highlights a number of case studies that highlight the impact of some non-pesticide approaches to disease management.

Regulatory Methods for Managing Crop Diseases: The Role of Quarantine and Eradication

Regulatory methods of disease management have as their objective measures that achieve exclusion of the pathogen from its host or a specific geographic area. This could be an entire country or a region within a country. These methods are enforced by regulations or acts of quarantine. Preventing the pathogen from coming into contact with its host will ensure that disease does not develop. Such an approach should be the foundation for the effective control of plant disease. It is the most cost-effective method for disease control and benefits all farmers in the region. The exclusion of banana bunchy top virus (BBTV) from PNG or TLB from Fiji saves both countries millions of dollars yearly. Unfortunately, exclusion is becoming increasingly difficult with the escalation in tourism, travel and commerce.

There are growing demands on an effective quarantine service to support this objective. However, at the same time detection methods are becoming more reliable as sensitive diagnostics become available. The island countries of the Pacific have a comparative advantage when it comes to exclusion. The island nature and isolation of many of the countries provides natural barriers to the movement of plant pathogens and make it easier for these countries to police the movement of pathogens and plants compared to countries in Asia that share a common border. Consequently, many of the island nations of the Pacific such as Fiji, Samoa, Tonga grow crops that are still free from major diseases.

However, this comparative advantage is only as good as the enforcement and there are a number of constraints that face quarantine services in the Asian-Pacific region in the face of increasing trade and commerce. In recent years there have been a few notable cases of exotic diseases introduced to certain countries in the region with devastating impact. TLB is a good example. Most recently, TLB spread to Samoa and American Samoa in 1993, most likely from Hawaii, devastating production. Within a matter of months the disease had caused millions of dollars in lost production. Pathogens introduced into an area in which they did not previously exist often have a dramatic impact. This is because the plants that the pathogens attack have not had the opportunity to develop any significant level of resistance in the absence of the pathogen. The nearby islands of Fiji, Cook Islands, Tonga, Wallis and Futuna and Niue have important taro industries and for the time being remain free from the disease but constant vigilance is required by quarantine services.

At the western side of PNG, along the border of West Papua, villagers are dealing with many plant diseases that seriously constrain smallholder production across the country. In addition, they are facing the added difficulties of having to deal with new diseases and threats to food security that are coming over the border from West Papua and further afield. Of great importance are blood disease of banana and citrus huanglongbing (greening) disease. One other banana disease, fusarium wilt, has recently been introduced to PNG and could be catastrophic. Every effort needs to be made to prevent incursions of blood disease and citrus huanglongbing to PNG and the further spread of fusarium wilt to other parts of PNG. Likewise, devastating taro virus diseases such as *alomae* and *bobone*, currently confined to Melanesian countries pose a major quarantine threat to Indonesia and other Asian countries where this crop is important.

SUSTAINABLE CROP DISEASE MANAGEMENT

The Asia and Pacific regions represents almost 23 per cent of the world's total land area and about 30 per cent of the world's cultivated land but 56 per cent of total world population and 72.5 per cent of the world's agricultural population. Agriculture has been the important economic activity that has sustained populations in the Asia-Pacific region for thousands of years and it continues to be the most important economic activity for the majority of countries in the region. In some countries its share of GDP is as high as 60 per cent.

Although Asia and the Pacific are often grouped together as a political region there are stark contrasts between them and the statistics hide major discrepancies and differences of scale in terms of land area, natural resources, population, culture, access to information, production, commerce, human capacity and so forth. For example, in 1998 the combined population of Melanesia, Micronesia and Polynesia accounted for only 0.12 per cent of the world total of six billion compared to the massive 56 per cent contribution of the Asian region. Many of these physical and socioeconomic differences have important implications for plant protection and the search for appropriate agricultural technologies to combat pests.

The Pacific region is composed of a vast number of islands which vary in size from Papua New Guinea (PNG) with a population of around four million, to Niue with a population of less than 2,000 and decreasing. This gives the Pacific a certain comparative advantage in terms of isolation and minimizing the number of exotic pests and diseases. In terms of agriculture, Asia is responsible for more than 90 per cent of the world's production of rice, 30 per cent of wheat and about 20 per cent of coarse grain.

The area devoted to vegetables in Asia is about 21 million ha. The major perennial crops in Asia are coconuts (9 million ha), fruits (17 million ha), sugarcane (8.6 million ha), rubber (6.7 million ha), oil palm (4.7 million ha), tea (2 million ha) and coffee (1.6 million ha). In the Pacific, agriculture is mainly subsistence-based, though export markets have been developed for certain crops. Compared to Asia the scale is very small. Unlike Asia, the important crops grown are dominated by vegetatively propagated species such as taro, sweet potato, yams, cassava, banana, sugarcane, breadfruit, and kava. In terms of world production the contribution from the Pacific is minor except for crops such as kava and possibly breadfruit. Other important crops include coconut, vanilla, squash, Asian vegetables, pawpaw and cucurbits. Agricultural intensification is a problem in the Asia-Pacific region, and is largely due to pressures resulting from rapid population growth and increasing demand from limited land.

This is manifested by increased soil erosion, forest clearing, agriculture on steep slopes, declining fallow periods, pest and disease problems, erosion of crop genetic resources, loss in forest biodiversity, and loss of traditional knowledge once employed in agriculture and natural resource management. These negative ecological and sociocultural impacts have contributed to increasing farmer marginalization and increased rural poverty. With increasing population in the region we are faced with the challenge of producing more food from less land.

Such increased productivity will have to be based on agricultural research and technologies that have minimal adverse environmental and ecological impacts and that contribute to social equity and poverty alleviation. Improved technologies for pest management will be a significant component of this new approach. Non-pesticide control technologies such as host plant resistance integrated with other appropriate cultural practices and biological control will play a vital role in achieving this goal. Poor farmers will benefit from increased yields, the rural and urban poor will gain from better availability of cheaper, safer and higher quality food that has little adverse environmental impacts. Employment opportunities will increase as a consequence of improved productivity and the employment of labour-intensive technologies. Consumers will benefit from improved health as a result of safer technologies that minimize the need for pesticides.

THE IMPACT OF AGROCHEMICAL CONTROL MEASURES

The second half of the 20th century was a buoyant time for the development of pesticides and very quickly became, and still is, a huge industry dominated by multinationals. People now talked of eradicating pests and initially the pesticides were so successful that there seemed to be no reason for carrying on the old preventative methods of pest control such as crop rotation and sanitation. Many farmers started to spray routinely even if pests were not present. This dependence on chemical sprays very quickly led to a number of ecological problems. The first sign that something was going wrong was the development of resistance among the target pests. Pesticides were also destroying the pest's natural enemies leading to problems of pest resurgence and secondary pests. Very often the response to these ecological problems was to spray more frequently, often at higher doses, in order to achieve control, commonly referred to as the "pesticide treadmill". The final group of problems observed with this intensive use of pesticides were problems related to the environment and health, which received wide coverage with the publication of Rachael Carson's *Silent Spring*. Problems arising as a result of excessive pesticide use included environmental pollution, contamination of soils and waterways and the destruction of many non-target species of wildlife.

The health of users and other members of society were also placed at risk. The growing awareness of these problems in certain Asian countries has recently led to bans on a number of damaging pesticides accompanied with policy support for non-pesticide approaches. Overall, fungicides are considered less toxic than the majority of other pesticides with the possible exception of the tin-based and organophosphate fungicides.

They are also considered to pose less threat to the environment compared to herbicides and insecticides. Fewer non-target effects using fungicides have been reported. However, chemical fungicides, especially the older fungicides, have the potential to contribute to some of the negative effects outlined above and should be treated with caution. The major problem with fungicides in recent years has been the rise of fungicide resistance in the target pathogen. This has arisen largely to systemic fungicides with a very specific mode of action. By comparison, fungicides with multi-site activity are much less prone to resistance problems. Resistance to benzimidazole fungicides (benomyl) has been found in the banana black leaf streak (BLS) pathogen in the Pacific region. Similar resistance problems can be anticipated with other dicarboximide or sterol inhibitor fungicides.

Pesticide use in developing countries is small compared to developed countries. However, the bulk of pesticides used in developing countries belong to those classified as highly hazardous such as organochlorines and organophosphates. In many instances these pesticides are banned in developed countries. It is predicted that pesticide use in developing countries will increase in the next decade or so. In 1997, the Food and Fertilizer Technology Centre (FFTC) and the Asian Productivity Organization (APO) undertook a pesticide production and use survey in Asia.

The results indicated an increasing trend in pesticide usage in the region. No such survey has been undertaken in the Pacific region but anecdotal evidence supports continued increasing trends in pesticide usage in the region. Clearly the challenge ahead is to develop alternative viable pest management technologies that will reduce the need for pesticides. It is interesting to note that US$570 million was spent on fungicides for rice cultivation globally in 1998, with over 50 per cent being used in Japan. Crop protection inputs in Japan for rice totalled US$680 per ha compared to US$3 per ha in other Asian countries.

THE IMPACT OF CROP DISEASE

Plant disease epidemics have influenced the course of history in many Asian-Pacific countries and continue to be of great importance especially to those people whose day-to-day survival depends on crop production. By reducing productivity, plant diseases contribute to the decline in rural income and diminish food security, both of which are a major determinant of poverty. Many times in history, plant diseases have resulted in thousands of deaths due to the destruction of staple food crops and consequent starvation.

Worldwide, it is believed that plant diseases reduce agricultural productivity by more than 10 per cent which is the equivalent of 500 million mt of food per year. Not only does the farmer have to cope with reduced levels of production he or she is also faced with the additional costs involved in trying to control the problem.

Reduced production levels of diseased crops means increased product prices for consumers. Losses in terms of exports and foreign earnings can be devastating as Samoa realised in 1993 when taro leaf blight (TLB) arrived. If attempts at control involve excessive use of pesticides there are hidden environmental and health costs. Plant diseases can also reduce the level of biodiversity or limit the variety of plants grown in an area. The devastating impact of plant disease on human affairs is vividly illustrated by the effect of late blight (*Phytophthora infestans*) of potato in Ireland from 1845 to 1946. At that time the potato plant was the staple food of hundreds of thousands of people. As a result of the blight there was mass starvation and three million deaths although other socioeconomic factors were also involved. There was a major exodus from the country as desperate people sought refuge in more 'prosperous' lands. The lingering impact of the potato famine in Ireland continues to this day and is deeply etched on the psyche of the people of that country. Interestingly, late blight of potato is once again becoming prevalent in some countries.

The impact of a plant disease epidemic on a national economy was vividly illustrated by the arrival of the coffee rust fungus (*Hemileia vastatrix*) in Sri Lanka (then Ceylon) in 1875. At that time there were around 160,000 ha of coffee planted and 45 million kg exported yearly. Within 20 years, production of coffee had virtually ceased and had fallen to 2,000 mt. The impact of the South American leaf blight *(Microcyclus ulei)* on rubber in Latin America is a similar scenario. Epidemics of groundnut rosette virus disease in Sub-Saharan Africa in the 20th century have resulted in the loss of billions of dollars. African cassava mosaic virus disease in Uganda is estimated to cause losses of US$70 million annually. In the Asia-Pacific region, plant diseases have, and continue to, pose a serious threat to rural livelihoods.

The Bengal famine of 1942-43 that resulted from *Biploris* brown spot of rice and contributed to the death of two million people is an important reminder of the power of disease epidemics in the region. Wheat rust in India from 1850 to 1950 contributed to 27 major famines and widespread starvation. In the Pacific region the impact of TLB and the threat it poses to countries not yet affected by the disease illustrates this point clearly. The TLB fungus (*Phytophthora colocasiae*) arrived in Samoa in 1993 and immediately had a devastating effect on the country.

Taro (*Colocasia esculenta*) was the principal staple crop at the time and of considerable economic, cultural and ceremonial importance. It was the major export commodity and income earner for many farmers. It has been estimated that this small island country lost approximately US$10 million as a consequence of this disease. In comparison to the economies of Asia such losses may not be significant but in small island states the impact can be devastating. The threat that this disease poses to other islands of the Pacific region continues to be of major concern and requires constant vigilance. Although plant diseases have contributed to substantial global crop losses, these would have more than doubled, and many of the world's poor would have been poorer, without significant advances in crop protection research. Continued crop protection research is essential in order to respond to future changes and challenges that will arise.

BIOLOGICAL CONTROL OF NATURAL ENEMIES

For the context of this chapter, biological control is defined as the use, or encouragement, of natural enemies of a pest to reduce pest abundance. Four types of biological control are recognized, namely; classical biological control, inundative method of biological control, augmentation method and a management method. Classical biological control in Asia is exemplified by the first case study. This case study showed that it is not enough to introduce an efficient parasitoid. It is equally important that farmers are educated on the subject so that they can be equal partners in developing this non-pesticide method. As shown by Ooi, farmers had to wait up to 12 years to realise the impact of the parasitoid.

Lessons learned from this case study were transferred to the highlands of Dalat in Vietnam and cabbage farmers studied the biology of the parasitoid. Farmers also studied the use of a microbial insecticide (*Bacillus thuringiensis* [Berliner]) to replace chemical pesticides. When cabbage farmers in Dalat learned about the biology of the parasitoid and the action of *B. thuringiensis*, they stopped using chemical pesticides.

The parasitoid was so effective that no further intervention was necessary. Inundative biological control involved rearing large numbers of biological control agents and releasing them into the field. Lim and Chong reported the periodic release of *Trichogrammatoidea bactrae fumata* (Nagaraja) (Hymenoptera: Trichogrammatidae) to control the cocoa pod borer. While it appeared to be able to reduce cocoa pod damage, many farmers find the process both tedious and expensive. Moreover, rearing large numbers of parasitoids for timely releases can be very challenging.

In some cases, the inundative releases appeared to have a placebo effect, allowing other natural enemies to control the pests when farmers stopped spraying chemical insecticides following periodic releases of the agent. In the case of augmentative biological control, beneficial insects are introduced at the time when pest populations are about to build up. An example of augmentative biological control is the encouragement of the black ant, *Dolichoderus thoracicus* (Smith) (Hymenoptera: Formicidae) to manage cocoa mirid pest. This involves establishing nests of the ant in cocoa plants. Liew also found that the black ants could control the cocoa pod borer.

Case study demonstrates that a larger biological control exists as a naturally occurring phenomenon. However, we often do not appreciate this and rice farmers were not aware of natural enemies in the field until they attended Farmer Field Schools (FFSs). Farmers studied and learned that friendly arthropods can keep a serious pest like the brown planthopper (BPH) in check. In the process, they soon discover that other pests like the rice stemborers are also kept at low populations by natural enemies. The most frequently asked question at rice FFSs is why farmers are not aware of such a ubiquitous phenomenon as natural biological control. It suggests that farmers were denied this information. It is ironical that farmers who live next to the rice fields are not aware of friendly insects that protect them from pest species. Hence, there is a need to educate farmers so that they can make more informed decisions about growing a healthy crop.

PHYSICAL AND CULTURAL CONTROL OF INSECT PESTS

Prior to the availability of chemical insecticides, farmers developed various means to prevent insect pest infestation. Some of these have been refined through research both by farmers and researchers. One such example of how farmers rediscovered an old method of avoiding pest attack. By studying the biology of the insect pest, farmers were able to confirm the findings made some 60 years ago by entomologists. More importantly, the information is now used by farmers to manage an important pest of rice.

In Bangladesh, farmers who participated in vegetable FFS organized by an Integrated Pest Management (IPM) programme funded by DANIDA (Danish International Development Agency) learned that the brinjal fruit and shoot borer, *Leucinodes orbonalis* (Guenée) (Lepidoptera: Pyralidae) pupated in leaf debris on the ground.

Using this information, farmers would maintain clean fields and destroy all infested brinjal fruits. This dramatically reduced the incidence of the borer without using insecticides. In the past, farmers sprayed as many as 50 times per crop season! Cocoa is an introduced crop in Southeast Asia. The key pest is the cocoa pod borer, *Conopomorpha cramerella* (Snellen) (Lepidoptera: Gracillariidae).

It is one crop that physical and cultural control received much attention. Shah described several methods that were used to control the cocoa pod borer. These included bagging of all husks following extraction of cocoa beans and it was found that a minimum of nine days period was required to effectively kill all the larvae, pupae and adults. Besides bagging the husk, it was suggested that harvesting pods at the first hint of yellow may leave a large number of larvae inside the pod and bagging the husks will kill them off.

To reduce pest population, it was recommended that a high standard of harvesting be maintained and that harvested pods be split the same day to extract the beans. Ecological harvesting refers to adoption of four aspects essential to effectiveness, *i.e.*, regular harvesting, complete harvesting, a low minimum ripeness standard for plucking, and continuation throughout the year. A more draconian measure to manage the cocoa pod borer was reported by Roepke which involved complete removal of all cocoa pods and related fruits attacked by the cocoa pod borer, a method named "rampasan". If carried out efficiently, it would encourage a pest-free period the following season. For smallholders, the use of a transparent plastic bag to sleeve young pods has been practised in the Philippines and Indonesia.

IMPORTED PARASITOID MAKE DIAMONDBACK MOTH A NON-PEST

A review of diamondback moth (DBM) management in the early part of 1970s would reveal that the literature was dominated by chemical evaluations for poisons to control this cabbage pest.

The preoccupation with chemical control resulted in many reports of insecticide resistance. The response by the research community was to find more toxic chemicals with similar results – development of insecticide resistance. At that point, it was obvious that the only option for DBM management was chemical control and the frustrations of cabbage farmers drove them to use cocktails which increased the cost of production. This suggests that farmers want to be involved in DBM management, which hitherto was left to researchers and scientists with very little change from the 1950s.

An ecological approach was initiated in the 1970s in the main cabbage growing area of Malaysia. This served to widen the options available to farmers when some parasitoids were imported. However, these remain options until a chain of events resulted in a severe reduction in use of chemical insecticides in Malaysia. When farmers reduced the use of cocktails and focused on *B. thuringiensis*, the impact of *Diadegma semiclausum* (Hellen) (Hymenoptera: Ichneumonidae) was realised. However, farmers had to wait 12 years to appreciate this! The lesson learned is that farmers should be involved in biological control from the beginning. Educating farmers about ecology should take precedence in pest management programme to ensure a sustainable programme. Farmers would not stop using chemical insecticides unless they understand biological control. At the heart of this is the lack of appreciation of ecology. Farmers, hitherto, had been exposed only to a unilateral approach to pest control and it would take a great effort to unlearn this and understand the need for a more holistic approach.

ECOLOGICAL UNDERSTANDING OF THE WHITE STEM BORER

Despite annual outbreaks of the white stem borer, *Scirpophaga innotata* (Walker) (Lepidoptera: Pyralidae) in the district of Indramayu, province of West Java, rice farmers were not aware of non-pesticide methods for pest control. Equally intriguing was the lack of understanding by the authorities of studies to control this pest carried out some 60 years ago. It was not surprising that farmers did not understand the biology of the white stem borer, despite long association with the insect.

Farmers observed that the white stem borer would occur in the first season following a three months drought. The farmers reported often seeing swarms of them in this first season. The only approach adopted by farmers and recommended by agricultural authorities in 1995 was to spray with chemical insecticides. Despite numerous sprays, the rice crops were often devastated by the white stem borer. The IPM facilitator assigned to the study used this observation as an entry point to introduce studies that help farmers find out what happened.

Effects of Insecticides

The IPM facilitator suggested that farmers carry out studies to determine the effects of insecticides. So, in July 1995, some white stem borer moths were collected from the fields and farmers sprayed them with insecticides. To their surprise, the moths spawned eggs before they died. Farmers then concluded that spraying insecticides did not prevent egg laying. The farmers' group in the village of Kalensari were graduates of IPM FFSs. This group of 25 farmers decided to continue to work together to discover why they had the white stem borer problem.

Regular meetings were held. Farmers went into the field to make observations and carry out field experiments. The IPM facilitator helped the farmers in experiment design and implementation. In one study, farmers evaluated the official recommendation of using an insecticide (carbofuran) coated on sand particles and applied in the seedbed. Results were variable, depending on the time of oviposition of the eggs. Hence, farmers realised that application of carbofuran did not guarantee an insect-free nursery.

Armed with this information, farmers went further, exploring why there were still outbreaks despite spraying insecticides on egg masses. They collected freshly-laid egg masses and sprayed these using the recommended dosage. Farmers noticed that after about a week, healthy larvae emerged. The results convinced farmers that spraying insecticides really did not stop an outbreak. They had to look for another approach and subscribed to the principle that the more they understood the biology and ecology of the stemborer, the greater the potential for management. The next study concerned learning about the origin of early season flights. Farmers suspected that the moths came from the fields and decided to survey rice stubbles left in the field from the previous season.

To their amazement, they found live larvae within some of the stubbles. Discussions with resource persons helped them understand the concept of insect aestivation (a form of hibernation during adverse weather conditions), observed by Goot. This discovery led to different experiments to kill off the aestivating larvae. This showed that burning and flooding did not kill many larvae, and that the best approach was to remove egg masses. They organized a local campaign to do this. This campaign proved successful, as the incidence of stem borer in the locality was about 5 per cent compared to neighbouring areas which had about 25 per cent bored tillers with prophylactic spraying of chemical insecticides. Continued discussion among farmers led to the design of experiments for avoiding the peak flights of the stem borer moths. Farmers carried out studies to understand when the stem borer moths took flight. They were able to duplicate Dammerman's study with light traps, using them to monitor the moth flights. Farmers also learned the role of water in breaking aestivation. Combining this information, farmers set up a strategy to 'forecast' peak flights and organized other farmers in their villages to only plant nursery beds after peak flights.

STRATEGIES FOR FACILITATING THE SUCCESS OF NON-PESTICIDE METHODS

Countries of the region need to appreciate the real cost associated with the past use and reliance on pesticides for pest control. These costs have been highlighted in depth in this chapter. There is an urgent need to actively support and promote non-pesticide methods if we are to change from the prevailing approach to pest control to a more ecological approach.

It is recommended that:

- Governments put in place policies to promote all aspects of non-pesticide and a more ecological approach to crop disease control;
- Farmers be educated and trained to enhance their skills and knowledge of non-pesticide methods. Over the last decade, there has been a significant shift in the emphasis of IPM programmes

throughout Asia, towards a more participatory, decentralized, community-based approach, termed community IPM, in which the farmers become the initiators, implementers and promoters of IPM and not just the recipients. In community IPM, farmers organize, manage and implement their own IPM activities, analyse problems, design field studies and carry out experiments and undertake efficient farming practices. This involves a range of approaches such as FFSs and farmer-to-farmer training, considered to be a costeffective, decentralized, community-based, farmer-first approach that will promote sustainability and expansion of IPM and reduce reliance on pesticides. Although community IPM and FFS initially grew out of problems related to insect pest management there is a growing awareness among plant pathologists of their relevance to sustainable crop disease management in the region. It is recommended that the FFS approach continue to be explored to enhance the use of non-pesticide methods;

- Linkages are developed between all stakeholders (farmers, extension workers, researchers, NGOs and other community-based groups including consumers) to enhance awareness and knowledge of non-pesticide methods through more effective networking;
- Countries strengthen regulation of the pesticide industry to reflect its real cost;
- There is further cooperation and coordination among national/regional/ international organizations and donors to highlight the problems with pesticide use and to promote a more ecological approach to crop disease control.

FARMER FIELD SCHOOLS AND CROP DISEASE MANAGEMENT

Small farmers understanding of disease is limited and management often ineffective. Farmers in Central Vietnam expressed interest in learning more about rice blast and disease management. A seasonlong FFS on management of rice blast was developed based on the approach used by FAO IPM programme. Farmers implemented field experiments including testing promising varieties, breeding lines, varietal mixtures, along with cultural methods. Farmers learned about host resistance, disease epidemiology. Farmers indicated the FFS very worthwhile. Farmers had a greater understanding of plant disease and the capacity to manage rice blast more effectively. Farmers reduced inputs and had healthier crops. Other approaches to help farmers learn about plant disease are worth exploring.

This chapter has identified a broad range of methods and approaches that can be used to achieve non-pesticide crop disease management. It has identified many problems and constraints that have limited the impact of these methods in the Asia-Pacific region. A range of broad strategies or solutions are outlined that might contribute to improving the success or wider adoption of non-pesticide approaches. However, it is important to remember that no one group of methods will solve crop disease problems. There is no blanket solution. Farming systems in the Asia-Pacific region are highly diverse and will require a range and diversity of solutions to crop disease problems. What is required is to ensure that countries have access to the widest possible basket of options. The challenge ahead for researchers is to identify and integrate sustainable disease management technologies that can contribute to increased productivity, such as breeding for durable resistance, biological control and cultural methods and which will have minimal adverse environmental and social impacts.

CARRYING CAPACITY OF FARM SYSTEMS FOR PEST POPULATIONS

In a non-farmscaped system, where pests have fewer natural controls and thus reach higher average populations, they are more likely to approach or exceed the economic threshold level for the crop, making pesticide treatments likely. In a farmscaped system, greater and more consistent populations of beneficial organisms put more ecological pressure on the pests, with the result that pest populations are less likely to approach the economic threshold. In other words, the ecological carrying capacity for a pest will probably be lower in a farmscaped system.

Maintaining and increasing biological diversity of the farm system is a primary strategy of cultural control. Decreased biodiversity tends to result in agroecosystems that are unstable and prone to recurrent pest outbreaks and many other problems. Systems high in biodiversity tend to be more "dynamically stable"—that is, the variety of organisms provide more checks and balances on each other, which helps prevent one species (*i.e.*, pest species) from overwhelming the system.

There are many ways to manage and increase biodiversity on a farm, both above ground and in the soil. In fact, diversity above ground influences diversity below ground.

Research has shown that up to half of a plant's photosynthetic production (carbohydrates) is sent to the roots, and half of that (along with various amino acids and other plant products) leaks out the roots into the surrounding soil, providing a food source for microorganisms. These root exudates vary from plant species to plant species and this variation influences the type of organisms associated with the root exudates.

Factors influencing the health and biodiversity of soils include the amount of soil organic matter; soil pH; nutrient balance; moisture; and parent material of the soil. Healthy soils with a diverse community of organisms support plant health and nutrition better than soils deficient in organic matter and low in species diversity. Research has shown that excess nutrients (*e.g.*, too much nitrogen) as well as relative nutrient balance (*i.e.*, ratios of nutrients—for example, twice as much calcium as magnesium, compared to equal amounts of both) in soils affect insect pest response to plants. Imbalances in the soil can make a plant more attractive to insect pests, less able to recover from pest damage, or more susceptible to secondary infections by plant pathogens. Soils rich in organic matter tend to suppress plant pathogens. In addition, it is estimated that 75 per cent of all insect pests spend part of their life cycle in the soil, and many of their natural enemies occur there as well. For example, larvae of one species of blister beetle consume about 43 grasshopper eggs before maturing. Both are found in the soil. (Unfortunately, although blister beetle larvae can help reduce grasshopper populations, the adult beetles can be a serious pest for many vegetable growers.) Overall, a healthy soil with a diversity of beneficial organisms and high organic matter content helps maintain pest populations below their economic thresholds.

Genetic diversity of a particular crop may be increased by planting more than one cultivar. For example, a recent experiment in China demonstrated that disease-susceptible rice varieties planted in mixtures with resistant varieties had 89 per cent greater yield and a 94 per cent lower incidence of rice blast (a fungus) compared to when they were grown in monoculture. The experiment, which involved five townships in 1998 and ten townships in 1999, was so successful that fungicidal sprays were no longer applied by the end of the two-year programme.

Species diversity of the associated plant and animal community can be increased by allowing trees and other native plants to grow in fence rows or along water ways, and by integrating livestock into the farm system. Use of the following cropping schemes offers additional ways to increase species diversity.

Crop rotations radically alter the environment both above and below ground, usually to the disadvantage of pests of the previous crop. The same crop grown year after year on the same field will inevitably build up populations of organisms that feed on that plant, or, in the case of weeds, have a life cycle similar to that of the crop. Add to this the disruptive effect of pesticides on species diversity, both above and below ground, and the result is an unstable system in which slight stresses (*e.g.*, new pest variety or drought) can devastate the crop.

An enforced rotation programme in the Imperial Valley of California has effectively controlled the sugar beet cyst nematode. Under this programme, sugar beets may not be grown more than two years in a row or more than four years out of ten in clean fields (*i.e.*, non-infested fields). In infested fields, every year of a sugar beet crop must be followed by three years of a non-host crop. Other nematode pests commonly controlled with crop rotation methods include the golden nematode of potato, many root-knot nematodes, and the soybean cyst nematode.

When making a decision about crop rotation, consider the following questions: Is there an economically sustainable crop that can be rotated into the cropping system? Is it compatible? Important considerations when developing a crop rotation are:

- What two (or three or several) crops can provide an economic return when considered together as a biological and economic system that includes considerations of sustainable soil management?
- What are the impacts of this season's cropping practices on subsequent crops?
- What specialized equipment is necessary for the crops?
- What markets are available for the rotation crops?

A corn/soybean rotation is one example of rotating compatible economic crops. Corn is a grass; soybean is a leguminous broadleaf. The pest complex of each, including soil organisms, is quite different. Corn rootworm, one of the major pests of corn, is virtually eliminated by using this rotation. Both crops generally provide a reasonable return. Even rotations, however, create selection pressures that will ultimately alter pest genetics. A good example is again the corn rootworm: the corn/bean rotation has apparently selected for a small population that can survive a year of non-corn (*i.e.*, soybean) cropping.

Management factors should also be considered. For example, one crop may provide a lower direct return per acre than the alternate crop, but may also lower management costs for the alternate crop (by reducing weed pressure, for example, and thus avoiding one tillage or herbicide application), with a net increase in profit.

Biological Controls

Biological control is the use of living organisms—parasites, predators, or pathogens—to maintain pest populations below economically damaging levels, and may be either natural or applied. A first step in setting up a biointensive IPM programme is to assess the populations of beneficials and their interactions within the local ecosystem. This will help to determine the potential role of natural enemies in the managed agricultural ecosystem. It should be noted that some groups of beneficials (*e.g.*, spiders, ground beetles, bats) may be absent or scarce on some farms because of lack of habitat. These organisms might make significant contributions to pest management if provided with adequate habitat.

Natural biological control results when naturally occurring enemies maintain pests at a lower level than would occur without them, and is generally characteristic of biodiverse systems. Mammals, birds, bats, insects, fungi, bacteria, and viruses all have a role to play as predators and parasites in an agricultural system. By their very nature, pesticides decrease the biodiversity of a system, creating the potential for instability and future problems. Pesticides, whether synthetically or botanically derived, are powerful tools and should be used with caution. Creation of habitat to enhance the chances for survival and reproduction of beneficial organisms is a concept included in the definition of natural biocontrol. Farmscaping is a term coined to describe such efforts on farms. Habitat enhancement for beneficial insects, for example, focuses on the establishment of flowering annual or perennial plants that provide pollen and nectar needed during certain parts of the insect life cycle. Other habitat features provided by farmscaping include water, alternative prey, perching sites, overwintering sites, and wind protection. Beneficial insects and other beneficial organisms should be viewed as mini-livestock, with specific habitat and food needs to be included in farm planning.

The success of such efforts depends on knowledge of the pests and beneficial organisms within the cropping system. Where do the pests and beneficials overwinter? What plants are hosts and non-hosts? When this kind of knowledge informs planning, the ecological balance can be manipulated in favour of beneficials and against the pests.

It should be kept in mind that ecosystem manipulation is a two-edged sword. Some plant pests (such as the tarnished plant bug and lygus bug) are attracted to the same plants that attract beneficials. The development of beneficial habitats with a mix of plants that flower throughout the year can help prevent such pests from migrating *en masse* from farmscaped plants to crop plants.

Applied biological control, also known as augmentative biocontrol, involves supplementation of beneficial organism populations, for example through periodic releases of parasites, predators, or pathogens. This can be effective in many situations—well-timed inundative releases of *Trichogramma* egg wasps for codling moth control, for instance.

Most of the beneficial organisms used in applied biological control today are insect parasites and predators. They control a wide range of pests from caterpillars to mites. Some species of biocontrol organisms, such as *Eretmocerus californicus*, a parasitic wasp, are specific to one host—in this case the sweetpotato whitefly. Others, such as green lacewings, are generalists and will attack many species of aphids and whiteflies.

Information about rates and timing of release is available from suppliers of beneficial organisms. It is important to remember that released insects are mobile; they are likely to leave a site if the habitat is not conducive to their survival. Food, nectar, and pollen sources can be "farmscaped" to provide suitable habitat.

The quality of commercially available applied biocontrols is another important consideration. For example, if the organisms are not properly labeled on the outside packaging, they may be mishandled during transport, resulting in the death of the organisms. A recent study by Rutgers University noted that only two of six suppliers of beneficial nematodes sent the expected numbers of organisms, and only one supplier out of the six provided information on how to assess product viability.

While augmentative biocontrols can be applied with relative ease on small farms and in gardens, applying some types of biocontrols evenly over large farms has been problematic. New mechanized methods that may improve the economics and practicality of large-scale augmentative biocontrol include ground application with "biosprayers" and aerial delivery using small-scale (radio-controlled) or conventional aircraft.

Inundative releases of beneficials into greenhouses can be particularly effective. In the controlled environment of a greenhouse, pest infestations can be devastating; there are no natural controls in place to suppress pest populations once an infestation begins. For this reason, monitoring is very important. If an infestation occurs, it can spread quickly if not detected early and managed. Once introduced, biological control agents cannot escape from a greenhouse and are forced to concentrate predation/parasitism on the pest(s) at hand.

Mechanical and Physical Controls

Methods included in this category utilize some physical component of the environment, such as temperature, humidity, or light, to the detriment of the pest. Common examples are tillage, flaming, flooding, soil solarization, and plastic mulches to kill weeds or to prevent weed seed germination.

Heat or steam sterilization of soil is commonly used in greenhouse operations for control of soil-borne pests. Floating row covers over vegetable crops exclude flea beetles, cucumber beetles, and adults of the onion, carrot, cabbage, and seed corn root maggots. Insect screens are used in greenhouses to prevent aphids, thrips, mites, and other pests from entering ventilation ducts. Large, multi-row vacuum machines have been used for pest management in strawberries and vegetable crops. Cold storage reduces post-harvest disease problems on produce.

Although generally used in small or localized situations, some methods of mechanical/physical control are finding wider acceptance because they are generally more friendly to the environment.

Other Cropping Structure Options

Multiple cropping is the sequential production of more than one crop on the same land in one year. Depending on the type of cropping sequence used, multiple cropping can be useful as a weed control measure, particularly when the second crop is interplanted into the first. Interplanting is seeding or planting a crop into a growing stand, for example overseeding a cover crop into a grain stand. There may be microclimate advantages (*e.g.*, timing, wind protection, and less radical temperature and humidity changes) as well as disadvantages (competition for light, water, nutrients) to this strategy. By keeping the soil covered, interplanting may also help protect soil against erosion from wind and rain.

Intercropping is the practice of growing two or more crops in the same, alternate, or paired rows in the same area. This technique is particularly appropriate in vegetable production. The advantage of intercropping is that the increased diversity helps "disguise" crops from insect pests, and if done well, may allow for more efficient utilization of limited soil and water resources. Disadvantages may relate to ease of managing two different crop species—with potentially different nutrient, water, and light needs, and differences in harvesting time and method—in close proximity to each other.

Strip cropping is the practice of growing two or more crops in different strips across a field wide enough for independent cultivation (*e.g.*, alternating six-row blocks of soybeans and corn or alternating strips of alfalfa and cotton or alfalfa and corn). It is commonly practiced to help reduce soil erosion in hilly areas. Like intercropping, strip cropping increases the diversity of a cropping area, which in turn may help "disguise" the crops from pests. Another advantage to this system is that one of the crops may act as a reservoir and/or food source for beneficial organisms. However, much more research is needed on the complex interactions between various paired crops and their pest/predator complexes.

The options described above can be integrated with no-till cultivation schemes and all its variations (strip till, ridge till, etc.) as well as with hedgerows and intercrops designed for beneficial organism habitat. With all the cropping and tillage options available, it is possible, with creative and informed management, to evolve a biologically diverse, pest-suppressive farming system appropriate to the unique environment of each farm.

Other Cultural Management Options

Disease-free seed and plants are available from most commercial sources, and are certified as such. Use of disease-free seed and nursery stock is important in preventing the introduction of disease.

Resistant varieties are continually being bred by researchers. Growers can also do their own plant breeding simply by collecting non-hybrid seed from healthy plants in the field. The plants from these seeds will have a good chance of being better suited to the local environment and of being more resistant to insects and diseases. Since natural systems are dynamic rather than static, breeding for resistance must be an ongoing process, especially in the case of plant disease, as the pathogens themselves continue to evolve and become resistant to control measures. Sanitation involves removing and destroying the overwintering or breeding sites of the pest as well as preventing a new pest from establishing on the farm (*e.g.*, not allowing off-farm soil from farm equipment to spread nematodes or plant pathogens to your land). This strategy has been particularly useful in horticultural and tree-fruit crop situations involving twig and branch pests. If, however, sanitation involves removal of crop residues from the soil surface, the soil is left exposed to erosion by wind and water. As with so many decisions in farming, both the short-and long-term benefits of each action should be considered when tradeoffs like this are involved.

Spacing of plants heavily influences the development of plant diseases and weed problems. The distance between plants and rows, the shape of beds, and the height of plants influence air flow across the crop, which in turn determines how long the leaves remain damp from rain and morning dew. Generally speaking, better air flow will decrease the incidence of plant disease. However, increased air flow through wider spacing will also allow more sunlight to the ground, which may increase weed problems. This is another instance in which detailed knowledge of the crop ecology is necessary to determine the best pest management strategies. How will the crop react to increased spacing between rows and between plants? Will yields drop because of reduced crop density? Can this be offset by reduced pest management costs or fewer losses from disease?

Altered planting dates can at times be used to avoid specific insects, weeds, or diseases. For example, squash bug infestations on cucurbits can be decreased by the delayed planting strategy, *i.e.*, waiting to establish the cucurbit crop until overwintering adult squash bugs have died. To assist with disease management decisions, the Cooperative Extension Service (CES) will often issue warnings of "infection periods" for certain diseases, based upon the weather.

In some cases, the CES also keeps track of "degree days" needed for certain important insect pests to develop. Insects, being cold-blooded, will not develop below or above certain threshold temperatures. Calculating accumulated degree days, that is, the number of days above the threshold development temperature for an insect pest, makes the prediction of certain events, such as egg hatch, possible. University of California has an excellent Web site that uses weather station data from around the state to help California growers predict pest emergence.

Some growers gauge the emergence of insect pests by the flowering of certain non-crop plant species native to the farm. This method uses the "natural degree days" accumulated by plants. For example, a grower might time cabbage planting for three weeks after the *Amelanchier* species (also known as saskatoon, shadbush, or serviceberry) on their farm are in bloom. This will enable the grower to avoid peak egg-laying time of the cabbage maggot fly, as the egg hatch occurs about the time *Amelanchier* species are flowering. Using this information, cabbage maggot management efforts could be concentrated during a known time frame when the early instars (the most easily managed stage) are active. Optimum growing conditions are always important. Plants that grow quickly and are healthy can compete with and resist pests better than slow-growing, weak plants. Too often, plants grown outside their natural ecosystem range must rely on pesticides to overcome conditions and pests to which they are not adapted.

Mulches, living or non-living, are useful for suppression of weeds, insect pests, and some plant diseases. Hay and straw, for example, provide habitat for spiders. Research in Tennessee showed a 70 per cent reduction in damage to vegetables by insect pests when hay or straw was used as mulch. The difference was due to spiders, which find mulch more habitable than bare ground. Other researchers have found that living mulches of various clovers reduce insect pest damage to vegetables and orchard crops. Again, this reduction is due to natural predators and parasites provided habitat by the clovers. Vetch has been used as both a nitrogen source and as a weed suppressive mulch in tomatoes in Maryland. Growers must be aware that mulching may also provide a more friendly environment for slugs and snails, which can be particularly damaging at the seedling stage.

Mulching helps to minimize the spread of soil-borne plant pathogens by preventing their transmission through soil splash. Mulch, if heavy enough, prevents the germination of many annual weed seeds. Winged aphids are repelled by silver-or aluminum-coloured mulches. Recent springtime field tests at the Agricultural Research Service in Florence, South Carolina, have indicated that red plastic mulch suppresses root-knot nematode damage in tomatoes by diverting resources away from the roots (and nematodes) and into foliage and fruit. Biotech Crops. Gene transfer technology is being used by several companies to develop cultivars resistant to insects, diseases, and herbicides. An example is the incorporation of genetic material from *Bacillus thuringiensis* (Bt), a naturally occurring bacterium, into cotton, corn, and potatoes, to make the plant tissues toxic to bollworm, earworm, and potato beetle larvae, respectively.

Whether or not this technology should be adopted is the subject of much debate. Opponents are concerned that by introducing Bt genes into plants, selection pressure for resistance to the Bt toxin will intensify and a valuable biological control tool will be lost. There are also concerns about possible impacts of genetically-modified plant products (*i.e.*, root exudates) on non-target organisms as well as fears of altered genes being transferred to weed relatives of crop plants. Whether there is a market for gene-altered crops is also a consideration for farmers and processors. Proponents of this technology argue that use of such crops decreases the need to use toxic chemical pesticides.

ACHIEVING CROP HEALTH THROUGH ECOLOGICALLY BASED PEST MANAGEMENT

Nowhere are the consequences of biodiversity reduction more evident than in the realm of agricultural pest management. The instability of agroecosystems, which is manifested as the worsening of most insect pest problems, is increasingly linked to the expansion of crop monocultures at the expense of the natural vegetation, thereby decreasing local habitat diversity. Plant communities that are modified to meet the special needs of humans become subject to heavy pest damage and generally the more intensely such communities are modified, the more abundant and serious the pests. The inherent self-regulation characteristics of natural communities are lost when humans modify such communities through the shattering of the fragile thread of community interactions. Agroecologists maintain that this breakdown can be repaired by restoring the shattered elements of community homeostasis through the addition or enhancement of biodiversity.

The key is to identify the type of biodiversity that is desirable to maintain and or enhance in order to carry out ecological services, and then to determine the best practices that will encourage the desired biodiversity components. That there are many agricultural practices and designs that have the potential to enhance functional biodiversity, and others that negatively affect it. The idea is to apply the best management practices in order to enhance or regenerate the kind of biodiversity that subsidizes the health and sustainability of agroecosystems by providing ecological services such as biological pest control, nutrient cycling, water and soil conservation, etc. (Gliessman, 1998). Crop health can be achieved by regulating insect pests through two routes: a healthy soil and a rich natural enemy biodiversity harboured by a diversified agroecosystem.

HEALTHY SOILS-HEALTHY PLANTS

A key feature of modern cropping systems is the frequency of soil disturbance regimes, including periodic tillage and pesticide applications, which reduce soil biotic activity and species diversity in agroecosystems. Such soil biodiversity reductions are negative because the recycling of nutrients and proper balance between organic matter, soil organisms and plant diversity are necessary components of a productive and ecologically balanced soil environment.

The ability of a crop plant to resist or tolerate pests is tied to optimal physical, chemical and biological properties of soils. Adequate moisture, good soil tilth, moderate pH, right amounts of organic matter and nutrients and a diverse and active community of soil organisms all contribute to plant health. Organic rich soils generally exhibit good soil fertility as well as complex good webs and beneficial organisms that prevent infection by disease causing organisms as *Pythium* and *Rhizoctonia*.

On the other hand, farming practices that causes nutrition imbalances can lower crop resistance. High nitrogen fertilizer levels can enhance the incidence of diseases such as *Phytophtora* and *Fusarium* and stimulate outbreaks of

Homopteran insects such as aphids and leafhoppers (Campbell, 1989). In fact there is increasing evidence that crops grown in organic rich and biologically active soils are less susceptible to pest attacks. Many studies suggest that the physiological susceptibility of crops and pathogens may be affected by the form of fertilizer used (organic vs. chemical fertilizer).

Studies documenting lower diversity of several insect herbivores in low-input systems, have partly attributed such reduction to a low nitrogen content in organically farmed crops (Magdoff, 1992). In California, a series of comparative experiments conducted on various growing seasons between 1989-1996 where broccoli was subjected to varying fertilization regimes (conventional vs. organic) showed that agroecological techniques can reduce the abundance of key insect pests, cabbage aphid (*Brevicoryne brassicae*) and flea beetle (*Phyllotreta cruciferae*), while sustaining yields. Lower herbivore numbers in organically managed plots were attributed to low foliage nitrogen content in compost fed broccoli plants.

In Japan, density of immigrants of the planthopper *Sogatella furcifera* was significantly lower while the settling rate of female adults and survival rate of immature stages of ensuing generations were lower in organic rice fields. The number of eggs laid by a female of the invading and following generations was smaller, and the percentage of brachypterous females in the next generation was also lower. Consequently, the density of nymphs and adults in the ensuing generations decreased in organically farmed fields (Kajimura, 1995).

In England, conventional winter wheat fields developed a larger infestation of the aphid *Metopolophiunt dirhodum* than its organic counterpart. This crop also had higher levels of free protein amino acids in its leaves during June, which were believed to have resulted from a nitrogen top dressing of the crop in early April.

However, the difference in the aphid infestations between crops was attributed to the aphid's response to relative proportions of certain non-protein to protein amino acids in the leaves at the time of aphid settling in the crops. In greenhouse experiments when given a choice of maize grown on organic versus chemically fertilized soils, European corn borer females preferred to significantly lay more eggs in chemically fertilized plants.

Nutrition is also important in determining susceptibility or resistance of plants to pathogens. Mineral nutrients are essential metabolic regulators of plant growth, and most studies indicate that nutrition affects pathogens and diseases indirectly. Biological activity in the soil becomes very intense in response to organic amendments and increase fungistasis as well as populations of existing microbial antagonists. Composts of diverse organic materials have proven effective in controlling diseases. Many types of compost host beneficial organisms that feed directly on pathogens, compete with them for nutrients or produce antibiotics (Tjamos et al., 1992)

Balanced soil nutrition helps plants stay more vigorous, increase the growth rate, make better use of soil water, and improve anatomical or histological characteristics. These factors enable plants to produce greater numbers of roots, allowing more surface area for root absorption and nutrient uptake, and possibly, shortening susceptible stages of plant growth. This allows plants to function more efficiently even when some roots are infected. Histological changes strengthen plants and possibly create barriers more difficult for pathogens to breach. A healthier plant may also increase quality of exudates, which can stimulate increases in populations of antagonistic microorganisms. These, in turn, may compete with pathogens for nutrients or possibly produce toxins that directly affect pathogen development and survival (Palti, 1981).

Application of nutrients (especially N) may increase competitive suppression of crops by weeds, as most weeds, (especially C_4 species) are often more responsive to application of nitrogen than other crops. A study reported that application of chemical N fertilizer to wild oat-spring wheat mixtures increased wild oat growth and decreased wheat yields. On the other hand other studies have shown that N fertilizer can improve the competitive status of crops.

What seems critical in determining the outcome of competitive interactions, is the timing of nutrient availability relative to crop and weed demands upon nutrient supplies. Emerging research shows that the species composition and general ecology of weeds is radically different in organic versus conventionally fertilized systems. Yield reductions of wheat due to interference from Italian ryegrass (*Lolium multiflorum* Lam.) were greater under conventionally fertilized conditions than under organic fertilized conditions. In many cases weed suppression is related to delayed N release from the organic N source compared to nitrate fertilizer. In other cases soil incorporated organic residues increase phytotoxicity or pathogen activity, which suppress weed seed and/or seedlings.

DIVERSIFIED AGROECOSYSTEMS AND PEST MANAGEMENT

Diversified cropping systems, such as those based on intercropping and agroforestry or cover cropping of orchards, have been the target of much research recently. This interest is largely based on the new emerging evidence that these systems are more stable and more resource conserving (Vandermeer, 1995). Much of these attributes are connected to the higher levels of functional biodiversity associated with complex farming systems.

In fact, an increasing amount of data reported in the literature documents the effects that plant diversity have on the regulation of insect herbivore populations by favouring the abundance and efficacy of associated natural enemies. One of the key elements in diversified agroecosystems is the presence of flowering plants that provide pollen and nectar that serve as alternative food for natural enemies. A number of studies have shown that many parasitoids require nectar for normal fecundity and longevity, and thus spectacular parasitism increases have been observed in annual crops and orchards with rich floral undergrowth (Leius, 1967).

Flowering plants can also increase populations of non-pestiferous herbivores (neutral insects) in crop fields. Such insects serve as alternative hosts or prey to entomophagous insects thus improving the survival and reproduction of these beneficial insects in the agroecosystem. Many researchers have reported that the presence of neutral insects on flowering plants within or near crop fields increase predation and parasitism of specific crop pests.

Our research in northern California vineyards confirms the importance of maintaining full season floral diversity to enhance natural enemy abundance and diversity in the grape agroecosystem. Growing summer cover crops of buckwheat and sunflower had a substantial impact on the abundance of western grape leafhoppers and associated natural enemies. During two consecutive years (1996-1997) vineyard systems with flowering cover crops, were characterized by lower densities of leafhoppers. Such reduced pest numbers in diversified vineyards were due to the impacts of generalist predators. With increased plant diversity, insect pests remained at lower levels than in clean cultivated vineyards, partly because the summer cover crop vegetation harboured pollen, nectar and neutral insects that served as alternate food and hosts for important predators along with the parasitic wasp *Anagrus*.

Plant diversity around crop fields is also important in determining the diversity and abundance of natural enemies within agroecosystems. Emerging data demonstrate that there is enhancement of natural enemies and more effective biological control where wild vegetation remains at field edges in close association with crops (Altieri, 1994). These habitats are important as overwintering sites for predators, or they may provide increased resources, such as pollen and nectar for parasitoids from flowering plants. The presence of plant rich habitats enhances predator colonization and abundance of adjacent crop fields but this influence is limited to the distance to which natural enemies disperse into the vineyard. A corridor however could amplify this influence by allowing enhanced and timely circulation and dispersal movement of predators from edges into the centre of crop fields.

In northern California, taking advantage of an existing 600 m corridor connected to a riparian forest and that cut across a monoculture vineyard, we tested the idea whether such a corridor served as a biological highway for the movement and dispersal of natural enemies from the forest into the centre of the vineyard. The goal was to evaluate if the corridor acted as a consistent, abundant and well dispersed source of alternative food (pollen, nectar, and neutral insects) for a diverse community of generalist predators and parasitoids.

During 1996 and 1997, a 2.5 hectares organic vineyard dissected by a corridor composed of 65 flowering plants species which was connected to the surrounding riparian habitat, was monitored to assess the distributional and abundance patterns of the Western grape leafhopper and its parasitoid *Anargus* spp., and generalist predators. In both years, leafhopper adults and nymphs tended to be more numerous in middle rows of the vineyard and less abundant in border rows close to the forest and corridor where predators were more abundant. The complex of predators supported by the corridor moved to the adjacent vine rows and exerted a regulatory impact on herbivores present in such rows. Although it is suspected that the leafhopper parasitic wasp *Anagrus* depended on food resources of the corridor, it did not display a gradient from this rich flowering area into the middle of the field. Likewise no differences in rates of egg parasitism of leafhoppers could be detected in vines near the corridor or in the vineyard centre. The presence of riparian habitats enhanced predator colonization and abundance on adjacent vineyards, although this influence was limited by the distance to which natural enemies disperse into the vineyard. However, the corridor, amplified this influence by enhancing timely circulation and dispersal movement of predators into the centre of the field.

FARMER PARTICIPATION SCHEMES AND PLANT BREEDING

The overall aim of the USP taro breeding club is to produce high-yielding, good-quality taro cultivars that have high levels of horizontal resistance to TLB and other locally important taro pests, and that are adapted to a range of diverse environments. Most members are students but some are professionals, such as lecturers, crop researchers, technicians, and university administrators, while a small percentage are farmers.

A successful farmer participation scheme, the Taro Improvement Project (TIP) was initiated at USP in 1999. TIP aims to bring together taro growers and provide them with more options for improving production and managing TLB in Samoa. TIP supplies each participating farmer with planting material of several taro cultivars or seedlings. Information is provided on trial layout, labelling, and simple data collection. The trials are maintained and managed by farmers. To facilitate feedback and sharing of information regular monthly meetings held. These meetings help growers to learn about other growers' experiences.

NON-PESTICIDE METHODS OF CROP DISEASE MANAGEMENT

The continued promotion of non-pesticide methods for crop disease control is essential for the development of a more sustainable agriculture. However, there are important obstacles that at present minimize their impact. One of the most important factors is the power of the pesticide industry and its lobbyists often reinforced in countries that provide subsidies for chemicals. Even if subsidies are not provided directly they are through the costs required to treat those whose health is compromised by their use or adverse environmental impact. Governments have a significant role to play here by stronger regulation of the pesticide industry. This would include restriction on the use of very damaging pesticides and a price system that is a true reflection of their real impact.

At the same time governments need to actively support the use of non-pesticide methods through increased allocation of resources for research and staff and a policy environment that encourages the use of non-pesticide methods of control. More specific issues that need to be addressed at the individual method level are outlined below. Within the Asia-Pacific region there is a wide discrepancy in the capacity of national quarantine services. Most, if not all, quarantine services in the region are constrained by one or a few of the issues that follow. Many quarantine services in the region are under-resourced in terms of trained human capacity and modern technologies required to effectively police the movement of plant material.

This is more the case for Pacific countries but it applies also to many of the Asian countries that have much greater levels of trade and commerce to deal with. Many Asian countries share a common border and have multiple entry points that require more demands from quarantine services. This limited capacity makes it difficult for countries to adequately respond to a disease incursion and achieve eradication. Good quarantine requires effective regulations and laws. Although most countries in the region have adequate legislation there are still a few countries that do not. Pathogens pose a serious challenge to plant quarantine services compared to other pests such as insects because they may bc introduced as spores or eggs on unsuspected carriers or occur as latent infections. Plants or plant parts with virus infections may not show symptoms. Very often quarantine services do not have adequate personnel trained in disease recognition or the necessary serological/molecular diagnostic tools. In the case of the Pacific, few of the major pathogen problems have been adequately characterized such as viruses causing disease in taro, kava and yams. Sensitive diagnostic methods have not been developed and this seriously constrains the international movement of crop germ plasm. This has been a major constraint to crop improvement programmes in the Pacific. An important dilemma facing all countries of the region is the urgent need to find a viable alternative to methyl bromide which is to phase out by the Montreal Protocol in 2005. Certification schemes or programmes to produce pathogen-free planting material are costly and require a certain level of infrastructure and human capacity.

This is required to enforce regulations and ensure a degree of quality control. Schemes are also reliant on access to sensitive diagnostic methods. Not all countries in the region have this capacity. In some countries, *e.g.*, Thailand, there was a perception among most farmers that disease-free plants were resistant to pathogens when grown in orchards. This indicates the need for education and awareness to accompany programmes and the diseases being targeted. What do farmers need to know to ensure that pathogen-free planting material remains so for longer? Other constraints identified in such programmes include difficulty in getting farmers to remove healthy-looking trees and plants that are infected (again, education is very important).

There may be a requirement to use pesticides to control vector buildup to ensure the programme is effective. Molecular diagnostics are expensive and require expertise that may not be available in some countries. There is a need for simple and cheap diagnostics. The effectiveness of such certification strategies is dependent on the ability to accurately diagnose and detect viral pathogens. In certain countries, such as Thailand and Malaysia, there are problems related to accelerating the production of pathogen-free citrus seedlings to meet demand. Although cultural methods offer appropriate approaches to obtaining sustainable disease management and a sound alternative to pesticides it is important to remember that many of the methods are localityspecific. Some methods are also labour-intensive and generally have not become widely adopted by farmers due to economic constraints and the long-term nature of the methods. The benefits and advantages of biological control are well documented. However, there are relatively few successful examples of the widespread application in disease management. There have not been the same successes as have been achieved with insect management. Despite extensive research on the use of biological control of plant pathogens there are few practical applications, even fewer for the Asia-Pacific region.

Many of these antagonists are already present naturally in the soil and it may be best to encourage their activity through some of the other methods mentioned and which are common to traditional farming systems. Host plant resistance has been an important tool for achieving sustainable crop disease management. Despite this it is a relatively long-term activity. In addition, many countries in the Asia-Pacific region do not have the capacity to sustain breeding programmes. At present there are only two breeding programmes in the Pacific on taro and rice. The issues surrounding the use of biotechnology and its application to agricultural improvement are many and complex. There are many technical issues, such as the development of gene delivery and transformation systems, that are constraining developments for specific crops but these will be overcome with ongoing research.

Although biotechnology and its role in disease management appears to be progressing in Asia, it is worth discussing its possible role in the Pacific region. It has already been highlighted that Green Revolution technology had a minimal role in the Pacific region and unless things change considerably over the next few years it is possible that the technology gap between Asia and the Pacific will widen further. At present there is little capacity or resources to undertake biotechnology research in the Pacific. There is little that exists in the form of policy or a regulatory framework to facilitate biotechnology research or applications. Costs and resource requirements for biotechnology are high. Can Pacific countries afford the costs required to undertake such research? If they cannot, can they rely on biotechnology organizations, private or public, to address their relevant crop disease needs? If they can, will farmers be able to afford the end-product? New transgenic plants are the product of millions of dollars in investment. The price of the end-product is likely to be beyond what a typical farmer in the Pacific can afford to pay. Staple crops important in the Pacific have been neglected by past developments in crop protection technologies, will they continue to be? Who are these neglected crops going to interest? Some of these issues apply to countries in Asia with similar human capacity and resource constraints.

The other issues that concern the role of biotechnology are more general and have been summarized by Conway and Tripp as:

- *Environmental issues*: Gene transfer to wild relatives (genetic pollution), impact on natural fauna and flora, pathogen resistances.
- *Health issues:* Possibility of transferring antibiotic resistance to soil fauna and eventually to livestock and humans, allergens ('Frankenfoods'), toxins, long-term health effects.
- *Ethical issues*: Naturalness of genetic engineering, bypassing of public interest, domination by a few major private companies.
- *Benefits to the poor*: Are the poor going to benefit, will farmers and consumers in the Asia-Pacific benefit?
- *Sustainable agriculture issues:* What will be the impact on the pursuit of sustainable agriculture, will biotech lead to increased domination by large-scale agriculture? There is concern regarding the 'marriage' between the pesticide and biotechnology industries.

IMPACT OF BIOTECHNOLOGY IN ASIA

In Asia, countries like China, India, the Philippines, Malaysia and Thailand are committed to investing in the development of biotechnology for agriculture. Early impacts would seem to be positive. A study in 2002 found that farmers in China were benefiting substantially from the use of transgenic crops with pest and disease resistance. Farmers' production costs have decreased by 28 per cent and the average farmer income has increased by US$150 per year. In addition, there has been an 80-per cent decrease in the use of organophosphate pesticides and a significant reduction in the number of pesticide poisonings. Reduce pesticide use is reflected in a recorded increase in the diversity of natural enemies. The survey also found scientists in the country had introduced over 120 genes to more than 50 plant species, many of which have been largely ignored by Western companies. Also, most of the transgenic plants in China target insects and diseases.

Biotechnology offers the opportunity for decreased use of pesticides with the associated benefits to the environment and health. While there is reason to be optimistic about the potential of biotechnology to address many of the important disease constraints that we face in the Asia-Pacific region it has also raised widespread opposition that is not confined to developed countries but has also occurred in the Asia-Pacific region. Many of these concerns centre on a range of environmental, health, socioeconomic and ethical issues that will need to be addressed if we are to move ahead.

FUTURE OF CROP DISEASE MANAGEMENT

The rapid growth of biotechnology has generated considerable excitement and debate about its potential impact in the fight against plant disease and its contribution to the development of a more sustainable agriculture. There is no doubt that biotechnology has provided us with a very powerful new set of tools to manipulate novel traits for disease-resistance and to diagnose plant pathogen causes of disease.

It expands the germ plasm base from which plant breeders can draw on for disease-resistance genes.

Potentially, it represents a powerful precision tool that offers significant advantages over other disease-control strategies such as pesticides – better protection, reduced environmental impact, reduced health risks. Given the growing limitations imposed on agricultural productivity by land and water, novel biotechnologies will be a key factor in achieving the necessary productivity increases required over the next few decades to meet the food needs of a rapidly growing global population.

Some countries in the Asia-Pacific region are already realizing the benefits of this new technology. Transgenic plants have been reported for some of the crops important to the Asia-Pacific region.

Sugarcane plants with resistance to sugarcane mosaic virus have been reported. Potato, the most tested crop in field trials to evaluate transgenic plants, has been reported with genetically engineered resistance to PVX, PVY and PLRV (potato leaf roll virus).

There is still someway to go in terms of resistance to the two most important potato diseases in developing countries – late blight and bacterial wilt. In rice, transgenic plants with resistance to bacterial blight, blast, sheath blight and a number of viruses are available.

In Australia, efforts are underway to develop transgenic plants with resistance to banana viruses (although transgenic resistance to DNA viruses like BBTV has been difficult to achieve), taro viruses, papaya ring spot virus (PRSV), ZYMV and sugarcane Fiji disease virus (Rob Harding, personal communication). New transgenic varieties of banana with resistance to BBTV and papaya resistant to PRSV have been reported in the Philippines and Taiwan.

Bibliography

A.K. Dhawan, Balwinder Singh, Manmeet Brar Bhullar and Ramesh Arora: *Integrated Pest Management*, Scientific Publication, Delhi, 2013.

Fanindra Prasad Neupane: *Integrated Pest Management in Nepal : Proceedings of a National Seminar Kathmandu, Nepal 25-26 September 2002*, Himalayan Resources Institute, 2003.

G.D. Sharma: *Integrated Pest Management*, Swastik Publication, Delhi, 2010.

K R Nair: *Integrated Production and Pest Management*, Gene Tech Books, Delhi, 2007.

K.P. Srivastava and Dhamo K. Butani: *Pest Management in Vegetables (2 Vols-Set)*, Research Periodicals, 1998.

L.R. Verma, A.K. Verma and D.C. Gautam: *Pest Management in Horticulture Crops : Principles and Practices*, Asiatech Publication, Delhi, 2004.

M L Agarwal: *Pest Management : A Glossary*, Agrotech Publishing Academy, Delhi, 2007.

N. Venugopal Rao, T. Uma Maheswari, P. Rajendra Prasad, V. Govardhan Naidu and P. Savithri: *Integrated Insect Pest Management*, Agrobios Publication, Delhi, 2003.

P. Narayanasamy, S. Mohan and J.S. Awaknavar: *Pest Management in Store Grains*, Satish Serial Publication, Delhi, 2009.

P. Narayanasamy: *Organic Pest Management : Potentials and Applications*, Satish Serial Publication, Delhi, 2010.

Parvatha Reddy & Abraham Verghese: *Integrated Pest Management in Horticultural Ecosystems*, Capital Publication, Delhi, 2001.

Pravin Chandra Trivedi: *Plant Pest Management*, Aavishkar Publication, Delhi, 2002.

R. Arora, B. Singh and A.K. Dhawan: *Theory and Practice of Integrated Pest Management*, Scientific Publication, Delhi, 2012.

R.P. Srivastava: *Mango Insect Pest Management*, I.K. International, Delhi, 1997.

Rajesh Ravi: *Pest Management : Principles and Practices*, Anmol Publication, Delhi, 2007.

Rajinder Peshin and Rajinder Kalra: *Integrated Pest Management : Adoption and Its Impact on Agriculture*, Classical Publishing, Delhi, 2000.

Ram Prakash Srivastava: *Neem and Pest Management*, IBDCO, Delhi, 2001.

Robert L. Metcalf and William H. Luckmann: *Introduction to Insect Pest Management*, Wiley, 2011.

S Ignacimuthu & Alok Sen: *Strategies in Integrated Pest Management : Current Trends and Future Prospects*, Phoenix, 2002.

S. Ignacimuthu, S.j. and B.V. David: *Non-Chemical Insect Pest Management*, Elite Publishing House, Delhi, 2010.

S. Ignacimuthu, s.j. and S. Jayaraj: *Recent Trends in Insect Pest Management*, Elite Publication, Delhi, 2008.

S. Manisegaran and R.P. Soundararajan: *Pest Management in Field Crops : Principles and Practices*, Agrobios Publication, Delhi, 2008.

S.C. Dwivedi and Nalini Dwivedi: *Integrated Pest Management and Biocontrol*, Pointer Publication, Delhi, 2006.

S.N. Mundra and A.H. Shah: *Integrated Pest Management in Irrigated Agriculture*, Himanshu Publication, Delhi, 1998.

Shagufta: *Integrated Pest Management*, APH Publication, Delhi, 2012.

Shubhrata R. Mishra: *Plant Protection and Pest Management*, Discovery Publishing House, Delhi, 2005.

Sunday Ekesi and Nguya K. Maniania: *Use of Entomopathogenic Fungi in Biological Pest Management*, Research Signpost, 2007.

T.V. Sathe and A.D. Jadhav: *Sericulture and Pest Management*, Daya Publication, Delhi, 2001.

Uma Shankar; Satya Priya and Deepak Kumar: *Vegetable Pest Management Guide for Farmers*, International Book, Delhi, 2008.

Index